OXFORD LOGIC GUIDES

D0070852

General Editors

DOV GABBAY
ANGUS MACINTYRE
DANA SCOTT

OXFORD LOGIC GUIDES

---

1. Jane Bridge: *Beginning model theory: the completeness theorem and some consequences*
2. Michael Dummett: *Elements of intuitionism*
3. A. S. Troelstra: *Choice sequences: a chapter of intuitionistic mathematics*
4. J. L. Bell: *Boolean-valued models and independence proofs in set theory* (1st edition)
5. Krister Segerberg: *Classical propositional operators: an exercise in the foundations of logic*
6. G. C. Smith: *The Boole–De-Morgan correspondence 1842–1864*
7. Alec Fisher: *Formal number theory and computability: a work book*
8. Anand Pillay: *An introduction to stability theory*
9. H. E. Rose: *Subrecursion: functions and hierarchies*
10. Michael Hallett: *Cantorian set theory and limitation of size*
11. R. Mansfield and G. Weitkamp: *Recursive aspects of descriptive set theory*
12. J. L. Bell: *Boolean-valued models and independence proofs in set theory* (2nd edition)
13. Melvin Fitting: *Computability theory: semantics and logic programming*
14. J. L. Bell: *Toposes and local set theories: an introduction*
15. Richard Kaye: *Models of peano arithmetic*
16. J. Chapman and F. Rowbottom: *Relative category theory and geometric morphisms*
17. S. Shapiro: *Foundations without foundationalism*
18. J. P. Cleave: *A study of logics*
19. R. M. Smullyan: *Gödel's incompleteness theorems*
20. T. E. Forster: *Set theory with a universal set*
21. Colin McLarty: *Elementary categories, elementary toposes*
22. R. M. Smullyan: *Recursion theory for metamathematics*
23. Peter Clote and Jan Krajíček: *Arithmetic, proof theory, and computational complexity*
24. A. Tarski: *Introduction to logic and to the methodology of deductive sciences*
25. G. Malinowski: *Many-valued logics*
26. Alexandre Borovik and Ali Nesin: *Groups of finite Morley rank*
27. R. M. Smullyan: *Diagonalization and self-reference*
28. Dov M. Gabbay, Ian Hodkinson, and Mark Reynolds: *Temporal logic: mathematical foundations and computational aspects (Volume 1)*
29. Saharon Shelah: *Cardinal arithmetic*
30. Erik Sandewall: *Features and fluents: the representation of knowledge about dynamical systems (Volume 1)*

Elementary Categories, Elementary Toposes

Colin McLarty

Department of Philosophy
Case Western Reserve University

CLARENDON PRESS · OXFORD
1995

Oxford University Press, Walton Street, Oxford OX2 6DP

Oxford New York
Athens Auckland Bangkok Bombay
Calcutta Cape Town Dar es Salaam Delhi
Florence Hong Kong Istanbul Karachi
Kuala Lumpur Madras Madrid Melbourne
Mexico City Nairobi Paris Singapore
Taipei Tokyo Toronto
and associated companies in
Berlin Ibadan

Oxford is a trade mark of Oxford University Press

Published in the United States
by Oxford University Press Inc., New York

© *Colin McLarty, 1992*
First published in paperback 1995

All rights reserved. No part of this publication may be reproduced, stored in a retrieval system, or transmitted, in any form or by any means, without the prior permission in writing of Oxford University Press. Within the UK, exceptions are allowed in respect of any fair dealing for the purpose of research or private study, or criticism or review, as permitted under the Copyright, Designs and Patents Act, 1988, or in the case of reprographic reproduction in accordance with the terms of licences issued by the Copyright Licensing Agency. Enquiries concerning reproduction outside those terms and in other countries should be sent to the Rights Department, Oxford University Press, at the address above.

This book is sold subject to the condition that it shall not, by way of trade or otherwise, be lent, re-sold, hired out, or otherwise circulated without the publisher's prior consent in any form of binding or cover other than that in which it is published and without a similar condition including this condition being imposed on the subsequent purchaser.

A catalogue record for this book is available from the British Library

Library of Congress Cataloging in Publication Data
McLarty, Colin,
Elementary categories, elementary toposes/Colin McLarty.
Includes bibliographical references and index.
1. Categories (Mathematics) 2. Toposes. I. Title.
QA169.M45 1991 511.3—dc20 91-36574

ISBN 0 19 851473 5

Printed in Great Britain on acid-free paper by
Bookcraft (Bath) Ltd, Midsomer Norton, Avon.

To

Ray Nelson, for what he has taught me about things learned, things done, and the love of wisdom.

Preface

The introduction, which follows, introduces categories and toposes, while this preface contains remarks on exposition. The book is elementary in the sense that it stays close to first order category theory and in that it has few specific prerequisites. It does require skill at abstraction, say on the level of algebra through groups and rings, or axiomatic set theory, or abstract data structures. After Chapter 13, symbolic logic is indispensable. Some recursive function theory is assumed in Chapter 24.

In Part I the elementary focus merely means skipping the homset definitions of categories, limits, and colimits. In Part II some higher order constructions are given in naive category theory. For example, adjunctions are handled using comma categories, with an exercise on the set-valued functor definition. Chapter 12 offers both set-theoretic and category-theoretic formalizations.

In Part III the first order topos axioms are used to define the higher order internal language. The logic is called 'topos logic' here, although 'intuitionistic logic' is often referred to in the literature. I believe that 'intuitionism' is usually, and rightly, taken to mean Brouwer's epistemology of mathematics, which is unrelated to the origin or content of topos theory. Topos logic strikingly resembles formal intuitionist logic and the two have interacted, as in van der Hoeven and Moerdijk (1984) or many papers in Fourman *et al.* (1979). But topos logic coincides with no intuitionist logic studied before toposes, and this is due to real philosophical differences (see McLarty 1990*a*).

Our metatheory avoids excluded middle and choice so that it is sound in any topos, except when we are explicitly concerned with constructions in **Set**. Choice is omitted from the **Set** axioms in Chapter 23 as it adds nothing to the theorems there, but it is used in applications of **Set** as I consider it to be part of the idea of sets.

William Lawvere's influence shows throughout the book, and I relied on Charles Wells's constant advice. Michael Barr, Marta Bunge, and Gonzalo Reyes have taught me a great deal and supported my visits to the Centre Interuniversitaire en Etudes Catégoriques in Montréal. Peter Freyd, Martin Hyland, Anders Kock, Michael Makkai, John Powers, Dana Scott, and David Yetter contributed perspectives now found in and behind various chapters. André Joyal and Myles Tierney forcefully pressed the challenge to make toposes as intuitive from the beginning as they are to experts, especially as concerns topos logic. I could not do it, but the book was improved by their prodding.

I thank John Corcoran for debates on logic and for what I have learned at the Buffalo Logic Colloquium. Ti-Grace Atkinson, John Bell, Emily Grosholz, Gregory Moore, Charles Parsons, Michael Resnik, Stewart Shapiro, and Michael Wright helped shape the style and foci of the book. I especially thank Penelope Maddy, Jean-Pierre Marquis, and John Mayberry for talks on sets and categories in foundations of mathematics.

Other friends occasionally asked just what was the point of topos theory. Leaving too many out, I mention Peter Bing, Tom Bishop, Virginia Jones, Paul Kukuca, Marie-Pierre Le Hir, Jean Hrichus, John McNamara, Laura Mitchell, Sally Norman, Martha Woodmansee, and my parents, Colin McLarty Sr. and Sarah Suplee.

Cleveland C.M.
April 1991

Contents

Introduction

Eilenberg and Mac Lane created categories in the 1940s as a way of relating systems of algebraic structures and systems of topological spaces in algebraic topology. The spread of applications led to a general theory, and what had been a tool for handling structures became more and more a means of defining them. Grothendieck and his students solved classical problems in geometry and number theory using new structures—including toposes—constructed from sets by categorical methods. In the 1960s, Lawvere began to give purely categorical definitions of new and old structures, and developed several styles of categorical foundations for mathematics. This led to new applications, notably in logic and computer science.

This book focuses on purely categorical descriptions, on category theory 'from the ground up'. Few examples are given until they can be constructed categorically. Chapter 3 handles a little group theory in that way. In Part IV several examples are given: a topos **Set** of sets which can be handled more naively here in the introduction, a topos **Spaces** of differentiable spaces, and a topos **Eff** in which all functions from the natural numbers to themselves are recursive.

In this introduction some examples are considered less carefully. A reader unfamiliar with any one should glance through it but not worry unduly—they are illustrations for the chapters, not prerequisites. (All the mathematics of the examples, plus a chapter on categories and toposes is in Mac Lane (1986).) See the further advice on reading at the end of this introduction.

1 Individual categories

Consider sets and functions. Write $f\colon A \longrightarrow B$ to mean that f is a function defined on the set A and with all its values in B. Note that B need not be the set of values of f; it need only include them all. Call A the *domain* of f, and B the *codomain*. Every set A has an identity function $1_A\colon A \longrightarrow A$, with $1_A(x) = x$ for every x in A. And given functions $f\colon A \longrightarrow B$ and $g\colon B \longrightarrow C$ with common set B there is a *composite* $g \circ f\colon A \longrightarrow C$ with $(g \circ f)(x) = g(f(x))$ for every x in A. So there is a category **Set**, called the category of sets, with sets as *objects* and functions as *arrows*. Compare the axioms in Chapter 1.

Category theory treats functions differently from set theory in that every arrow in **Set** has a unique domain and codomain. For example, let Q be the set of rational numbers and R the set of reals. For most set theory texts there is a well-defined function on Q which takes each rational number x to its square x^2. In category theory a function, that is an arrow of **Set**, is not fully defined until a codomain is also specified. So there are distinct arrows

$$\_^2: Q \longrightarrow Q \qquad \text{and} \qquad \_^2: Q \longrightarrow R$$

and infinitely many others, all with domain Q and taking each x in Q to its square, but with different codomains.

For another category, consider the data types of a given computer programming language and the computable functions between them. For this example we write $f: A \longrightarrow B$ to say that f is a computable function from data type A to data type B. More precisely, this means that some program in our language accepts input of type A, and for any input x of that type returns the value $f(x)$ of data type B. For every data type A the identity function $1_A: A \longrightarrow A$ is computable. Given computable $f: A \longrightarrow B$ and $g: B \longrightarrow C$, compute the composite $g \circ f: A \longrightarrow C$ by the rule 'apply f to the input then apply g to the result'. We then have a category **Data** of the data types and functions in our programming language. For a survey of more sophisticated categorical semantics for programming languages, see Scedrov (1988).

Of course, a program may take several inputs. That can be modelled by a device, available in many actual languages, of forming *record types*: that is, for any data types A and B we suppose that there is a data type $A \times B$, and we think of a value of this type as being a pair $\langle x, y \rangle$, with x a value of type A and y of type B. Then a program which takes one input of type A and one of type B, and returns a value of type C, can be thought of as taking a single input of type $A \times B$, and so represented by an arrow $A \times B \longrightarrow C$ in the category **Data**. Neither we nor the language user cares how a computer actually stores records $\langle x, y \rangle$. We do care that we can recover the values in each field of a record. There must be a computable function $p_1: A \times B \longrightarrow A$ taking any record $\langle x, y \rangle$ to x, and another $p_2: A \times B \longrightarrow B$ taking $\langle x, y \rangle$ to y. Furthermore, given a data type T and computable $h: T \longrightarrow A$ and $k: T \longrightarrow B$ there must be a computable function from T to records of type $A \times B$ which agrees with h on the A field and with k on the B field. That is a function $\langle h, k \rangle: T \longrightarrow A \times B$, which takes any input z of type T to the record $\langle h(z), k(z) \rangle$. Compare the definition of a *product* for A and B in Chapter 2.

In **Set** the Cartesian product $A \times B$ of sets A and B has obvious projection functions p_1 and p_2 to A and B, such that for any set T and functions $h: T \longrightarrow A$ and $k: T \longrightarrow B$ there is a function $\langle h, k \rangle: T \longrightarrow A \times B$, as for record types. As with record types in **Data**, so with Cartesian products in **Set** these functions are what is used in practice, not some account of the make-up of $A \times B$.

In computer science rigorous attention to domains and codomains is called 'strong typing' and few languages use it consistently. Many languages, for example, treat addition '$x + y$' as one function, whether applied to a pair of integers, an integer and a real, or a pair of reals. But in the category **Data**, if I is the type of integers and R the type of reals, there are distinct arrows

$$+:I\times I \longrightarrow I \quad +:I\times R \longrightarrow R \quad +:R\times I \longrightarrow R \quad +:R\times R \longrightarrow R$$

One adds an integer to an integer for an integer sum, another adds an integer to a real for a real sum, and so on. They are distinct because they have different domains and codomains.

For a third example, consider (real, finite-dimensional) vector spaces and linear functions. Every identity function is linear and, given linear $f\colon V \longrightarrow W$ and $g\colon W \longrightarrow X$, the composite $g \circ f\colon V \longrightarrow X$ is linear. So there is a category **Vect** with vector spaces as objects and linear functions as arrows. This category also has products. Given spaces X and Y, the vectors in the product space $X \times Y$ are ordered pairs $\langle x, y\rangle$, with x a vector in X and y a vector in Y. We define vector addition and scalar multiplication on $X \times Y$ by the rules

$$\langle x, y\rangle + \langle x', y'\rangle = \langle x + x', y + y'\rangle, \qquad r.\langle x, y\rangle = \langle r.x, r.y\rangle$$

for x and x' vectors in X, and y and y' vectors in Y, and r any real number. The additions and multiplications inside the angle brackets on the right-hand side are those of X and Y. Note that these rules are just what they must be to make the projections $p_1\colon X \times Y \longrightarrow X$ and $p_2\colon X \times Y \longrightarrow Y$ linear. Again, this is a product for X and Y in **Vect**, as defined in Chapter 2.

There are categorical differences between **Set**, **Data**, and **Vect**. A *terminal object* in any category **C** is an object T of **C** such that every object of **C** has exactly one arrow to T. An *initial object* in **C** is an object S such that every object of **C** has exactly one arrow from S. So the terminal objects of **Set** are the singletons, and a set S is initial in **Set** iff it is empty. Every set A has exactly one function f to any singleton $\{*\}$, namely the one with $f(x) = *$ for every x in A. An empty set has one and only one function, namely the empty one, to each set A.

The category **Data** may have no terminal or initial objects, depending on just what programming language it is based on. Languages with limited means for constructing types generally have no terminal or initial ones. In such languages there are many computable functions between any two types.

In **Vect** there is the one-point space $\{0\}$, the vector space with only a zero vector. This is both terminal and initial in **Vect**. For any vector space V the constant zero function $V \longrightarrow \{0\}$ is linear, and so is the function $\{0\} \longrightarrow V$ inserting zero. These are clearly the only linear functions to or from $\{0\}$. An object that is both terminal and initial in its category is called a *zero object* because of this example.

In short, **Set** has terminal and initial objects, and they are distinct from one another: **Data** may well have neither; **Vect** has both, and they coincide.

For other examples, take topological spaces and continuous functions to obtain the category **Top**, and take differentiable manifolds and differentiable functions to obtain the category **Man**. These two resemble **Set** more than they do **Vect**. For example, in both these categories any single-point space is terminal and empty spaces are initial. Neither one has a zero object. This is the beginning of much further reaching differences, as **Vect** is an *Abelian* category while none of our other examples is (see Freyd 1964 or Mac Lane 1971, Chapter 8). There are also algebraic examples of categories, such as **Grp** with groups as its objects and group homomorphisms as arrows, or **Ring** with rings as objects and ring homomorphisms as arrows. Such examples are not in short supply, and are found in any discussion of category theory.

A topological space T is not studied by just looking at its points. Typically, the continuous curves and surfaces in it are examined. Let S^1 be the circle and S^2 the sphere (i.e. the outer surface of a ball). A closed curve in T is a continuous function $c: S^1 \longrightarrow T$. It is a picture of a circle in T, or a *map* of S^1 in T, in the usual terminology. A map of S^2 in T, i.e. a continuous function $s: S^2 \longrightarrow T$, is a closed surface. We can describe T in large part by describing the maps to it from S^1, S^2, and higher-dimensional spheres. Geometers often describe spaces by maps to and from other spaces, and they always give a map a specific domain and codomain. This is the origin of a considerable amount of category theory, and arrows in any category are sometimes called maps.

In any category **C** a *generalized element* of an object A is just an arrow to A. When we think of an arrow $h: C \longrightarrow A$ as a generalized element of A, we call C the *stage of definition*, or we may just call h a C-element of A. So a closed curve in a topological space T is an S^1-element of T, while a closed surface is an S^2-element. In the category **Data**, with I the type of integers, for any data type A an I-element of A is a computable function $I \longrightarrow A$ or, in other words, a computable sequence x_i of values in A, where i ranges over all integers.

A *global element* is a generalized element, the stage of definition of which is terminal, and these often have a special role. As before, let $\{*\}$ be a singleton set, with its single element $*$. Any set A has a function $x: \{*\} \longrightarrow A$ for each element x, namely the function taking $*$ to x. According to traditional set theory, functions from $\{*\}$ to A 'correspond to' elements of A but are not themselves the elements. On categorical foundations as in Chapter 22 we actually define an element of a set A to be a function from $\{*\}$ to A. (In that chapter we write 1 for a singleton rather than $\{*\}$.) We can do the same thing in a categorical treatment of **Top** or **Man**, and in fact we do for a variant of **Man** in Chapter 23, defining a point of a space to be an arrow to it from a singleton space. Of course, this does not work so well for vector spaces, since any vector space V has exactly one linear function from the terminal space $\{0\}$.

In any category **C** an arrow $A \longrightarrow B$ can be thought of as a kind of picture of A in B, but this is uselessly abstract without some initial picture of A. For most categories it is useful to see a terminal object as a point. (That is, a terminal object is a point when looked at internally to its category. From outside it may be much more complex. See Exercise 16.15 and examples in Chapter 22). When we define natural number objects in Chapter 19 it will be useful to see a given one N as a sequence of points, so that an N-element $s: N \longrightarrow A$ is a sequence of points in A. In specific categories such as the category of sets or the category of spaces in Chapter 23, the subject matter conveys its own images.

It is often convenient to view a monic arrow $i: A \rightarrowtail C$, defined in Chapter 1, as showing that A is a copy of a part of C, and that i maps the copy on to that part. For example, in **Set**, the monic arrows are the one-to-one functions and, clearly, if i is one-to-one then A is a copy of a subset of C, namely of the image of i. This works well in all toposes and in the category of categories, but not in all categories. (For example, in **Top** the monics are also the one-to-one functions, but a continuous function $i: A \longrightarrow C$ may be one-to-one, while A has a finer topology than the image of i as a subspace of C.) In any category we call a monic arrow to an object C a *sub-object* of C. Note that one object A may copy many different parts of C, and so it is the monic i that identifies the sub-object, and not just its domain A. In other words, we have to know 'where' in C the copy A is being mapped. Sub-objects $i: A \rightarrowtail C$ and $j: B \rightarrowtail C$ of a single object C are *equivalent*, as defined in Chapter 3, if i and j map on to the same part of C.

2 The category of categories

In Part II *functors*, which are maps of categories, are introduced. A functor **F** from a category **A** to another category **B** is a structure-preserving function from **A** to **B** (see Chapter 8). Intuitively, if **A** is seen as a network of arrows between objects and then **F** maps that network onto the network of arrows of **B**, this gives a picture of **A** in **B**.

Every category **A** has an identity functor $1_{\mathbf{A}}: \mathbf{A} \longrightarrow \mathbf{A}$, which leaves the objects and arrows of **A** unchanged, and given functors $\mathbf{F}: \mathbf{A} \longrightarrow \mathbf{B}$ and $\mathbf{G}: \mathbf{B} \longrightarrow \mathbf{C}$ there is a composite $\mathbf{G} \circ \mathbf{F}: \mathbf{A} \longrightarrow \mathbf{C}$. So it is natural to speak of a category of all categories, which we call **CAT**, the objects of which are all the categories, and the arrows of which are all the functors. This raises genuine problems. Is **CAT** a category in itself? Our answer here is to treat **CAT** as a regulative idea; that is, an inevitable way of thinking about categories and functors, but not a strictly legitimate entity. (Compare the self, the universe, and God in Kant 1781.) Of course, general category theory applies to **CAT**, and this category that we do not quite believe in is the single one that we investigate the most. In Chapter 12 several alternative rigorous foundations are given for our results, but no actual 'category of all categories'.

The central notion in Part II is that of an *adjoint* to a functor, defined in Chapter 10. Examples figure heavily in Chapters 11 and 17. In the exercises of Chapter 12 a series of adjoint functors are developed between a category of sets and a category of categories. Further mathematical examples require more background than we assume, but the literature on categories is replete with them (see, for example, Mac Lane 1971 and Barr and Wells 1985).

3 Toposes

Toposes are sometimes explained as categories 'with **Set**-like properties', but we can be more specific. Toposes are categories which allow the constructions used in ordinary mathematics. A topos is a category **E** such that it has a terminal object, and any pair of objects A and B of **E** have a product $A \times B$. For any parallel pair of arrows, that is any arrows $f: A \longrightarrow B$ and $g: A \longrightarrow B$ with the same domain and the same codomain, there is a sub-object $e: E \rightarrowtail A$ of A representing the solutions to the equation $f = g$. (This is an equalizer, defined in Chapter 2.) Any pair of objects A and B have an *exponential* B^A, as defined in Chapter 6. In short, B^A is an object of **E** representing all arrows from A to B in **E**. And there is a *sub-object classifier*, also called a *truth value object*, Ω, as defined in Chapter 13.

In the example of **Set** we have already seen the terminal objects and products $A \times B$. Given parallel arrows f and g, the equalizer E is the subset of A containing all x with $f(x) = g(x)$. The exponential B^A is the set of functions from A to B. The sub-object classifier Ω is a two-element set. Let it be $\{t, fa\}$, where 't' is pronounced 'true' and 'fa' is pronounced 'false'. The idea is that, given any set A and a subset of it S, there is a function χ_S from A to $\{t, fa\}$ which *classifies* S; namely, the function defined on A with

$$\chi_S(x) = t \qquad \text{iff } x \text{ is in } S.$$

Conversely, any function $\chi: A \longrightarrow \{t, fa\}$ classifies a subset of A, namely the subset of all x such that $\chi(x) = t$.

Singleton sets and $\{t, fa\}$ are finite, and if A and B are both finite then so are $A \times B$, all subsets of A, and B^A. So the finite sets already form a topos, which we call $\mathbf{Set}_{\mathrm{fin}}$. An axiom of infinity, the existence of a *natural number object* or set of natural numbers, which we require **Set** to satisfy while $\mathbf{Set}_{\mathrm{fin}}$ does not, is described in Chapter 19.

Any object A of a topos **E** has an exponential Ω^A. It represents arrows from A to Ω, which classify sub-objects of A, so it is an **E** object representing all sub-objects of A. We call it the power object of A. In the case of **Set** or $\mathbf{Set}_{\text{fin}}$ the exponential $\{t, fa\}^A$ is the power set of A. Set-theory texts say that $\{t, fa\}^A$ is merely isomorphic to the power set, but category theory knows no such invidious distinction. Instead, we think of a sub-object of an object A in a topos indifferently as a part of A, an arrow from A to Ω, and an element of Ω^A.

In Chapter 16 we prove that every topos **E** has an initial object $\emptyset$ with the typical properties of an empty set; and it has unions, and quotients of equivalence relations. If **E** also has a natural number object we can do all of classical mathematics, but not always with classical results, since **E** may be quite different from **Set**, as we see with the examples in Part IV.

To handle the rich structure of toposes, we follow Johnstone (1977) in combining elementary category theory and the *internal language* of a topos. For each object A of the topos we introduce quantifiers $(\forall x . A)$ and $(\exists x . A)$ over A. We regard each arrow $f: A \longrightarrow B$ as a function. We give interpretations for negation ($\sim$), conjunction (&), disjunction ($\vee$), and conditional ($\rightarrow$) with familiar logical rules, but not quite classical ones. Notably, the law of excluded middle fails. In some toposes there are formulas φ such that '$\varphi \vee \sim \varphi$' is not true, and neither is '$\sim \sim \varphi \longrightarrow \varphi$'. In any topos, if φ implies a contradiction then we can conclude $\sim \varphi$. But, in many toposes, if $\sim \varphi$ implies a contradiction we can only conclude $\sim \sim \varphi$. In other toposes, such as **Set** or $\mathbf{Set}_{\text{fin}}$, the law of excluded middle is sound. The axiom of choice fails in many toposes, including all those in which excluded middle fails, since it implies excluded middle (see Section 17.5).

I consider the internal language of each topos an actual language, describing structures in the topos. Formal rules are provided for the languages, but the reader should go on to use them colloquially, as Joyal and Tierney (1984) or Moerdijk (1985) do.

The point is that toposes describe objective structures. The world around us has a geometric structure that can be idealized in the notion of smooth spaces and maps, as indeed it was in classical analysis in the service of Newton's and later Einstein's physics. The smooth topos **Spaces** formalizes that structure. Another abstraction moves away from geometry to view the world in terms of pure cardinality. This is Cantor's set theory, and is formalized in the topos **Set**. Yet another treats functions as procedures, and so requires them to have algorithms. This is formalized in part of the effective topos **Eff**. These are not competing theories, much less contradictory: they are alternatives suited to different purposes.

There is no meaningful question of whether all functions from the line to itself are 'really' differentiable, or all functions from the natural numbers to themselves are recursive. Rather, we need to study both of these and other idealizations, and the relations between them, whether we model them in sets or toposes or whatever. The most important relations between toposes are geometric morphisms, defined in Chapter 17. We do not study them in detail in this book, but we give the most important means of constructing them—namely Grothendieck toposes.

Let **A** be a small category in a topos **E**. A functor from **A** to **E** is called a *diagram* on **A**. Picture the functor as a 'diagram' of objects and arrows of **E**, shaped like the network of objects and arrows of **A**. The category of all diagrams on **A** is itself a topos which we call $\mathbf{E}^{\mathbf{A}}$ (see Chapter 20). The diagram category has certain *topologies* on it. For each topology j on $\mathbf{E}^{\mathbf{A}}$ there is category $(\mathbf{E}^{\mathbf{A}})_j$ of *sheaves* on j, and this too is a topos (see Chapter 21). A *Grothendieck topos* over **E** is any category of sheaves for some topology on some category of diagrams in **E**. The two-chapter proof that these are toposes is the central result of Part III.

If we simply posit two toposes, say a topos **Set** satisfying the axioms in Chapter 22 and one **Spaces** satisfying those of Chapter 23, it does not follow that there are any geometric morphisms between them. But Chapter 23 sketches a way of beginning with **Set** and constructing a Grothendieck topos over it that satisfies that chapter's axioms for a category of spaces. Geometric morphisms between **Set** and this Grothendieck topos fruitfully formalize relations between sets and spaces. Conversely, McLarty (1987) shows how to take **Spaces** and construct a Grothendieck topos over it that satisfies the axioms for a topos of sets. Either method gives the same geometric morphisms between **Set** and **Spaces**. For geometric morphisms toposes constructed one from the other, or both from a common base, are needed. Grothendieck toposes are not the only means of doing this either, as shown in Chapter 24 with **Eff**.

4 Advice on reading

The reader can often profit from glancing ahead in this book. Part IV should make some sense after the Introduction and Chapters 1 and 2, more sense after Chapter 6, and a great deal more after the definition of toposes in Chapter 13. The reader might look at Chapter 8 on functors immediately after Chapter 1, and could read Chapter 19 on natural number objects in almost full detail after reading Chapter 6. The definition of a topos in Chapter 13 should make some sense after Chapter 4, and more after Chapter 6. There may be other suitable short cuts.

The exercises contain definitions and results that are used later on, and so must be studied, except that details of formalization should be studied only so far as they are of interest.

Through Part I we usually work with one category at a time. To speak of 'an object' or 'an arrow' means an object or arrow of that category. Some chapters only assume that the category is a category, while others assume that it has further structure (such as selected products for pairs of objects). Each chapter's assumptions are given at its beginning and not repeated in individual theorems. We call the category we work with the 'base category'. Part II concerns relations between categories, but in Part III most chapters again start with one base category, in fact one topos. These chapters often construct new toposes from the base.

Parentheses are punctuation in this book, used for clarity and not a requisite part of function notation. Thus $f(x)$ or fx mean the same thing, and the choice in any occurrence is a matter of readability.

Part I

CATEGORIES

1

Rudimentary structures in a category

In this chapter we study axioms for a category, and possible structures within it. We introduce the 'generalized element' notation, which allows us to restate some of our axioms and definitions in a more familiar form.

1.1 Axioms

A category has *objects* $A, B, C, \ldots$ and *arrows* $f, g, h \ldots$. Each arrow goes from an object to an object. To say that g goes from A to B we write $g: A \longrightarrow B$, or say that A is the *domain* of g, and B the *codomain*. We may write $\mathrm{Dom}(g) = A$ and $\mathrm{Cod}(g) = B$.

Two arrows f and g with $\mathrm{Dom}(f) = \mathrm{Cod}(g)$ are called *composable*. If f and g are composable, then they must have a *composite*, an arrow called $f \circ g$. Every object A has an *identity arrow*, 1_A.

The axioms read as follows. For every composable pair f and g the composite $f \circ g$ goes from the domain of g to the codomain of f. For each object A the identity arrow 1_A goes from A to A. Composing any arrow with an identity arrow (supposing that the two are composable) gives the original arrow. And composition is associative.

The axioms can be displayed in diagrams or in equations. In the equations we assume that any arrows we compose are composable. In the diagrams this is explicit.

Domain and codomain:

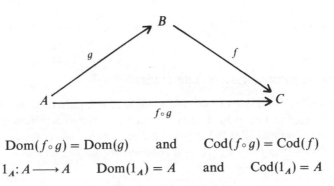

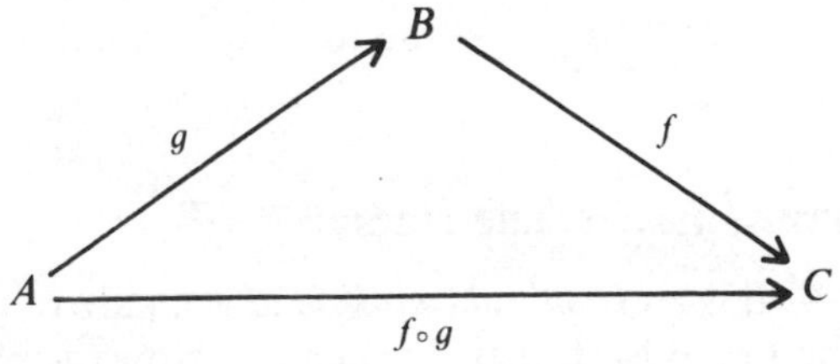

$$\mathrm{Dom}(f \circ g) = \mathrm{Dom}(g) \quad \text{and} \quad \mathrm{Cod}(f \circ g) = \mathrm{Cod}(f)$$

$$1_A: A \longrightarrow A \quad \mathrm{Dom}(1_A) = A \quad \text{and} \quad \mathrm{Cod}(1_A) = A$$

Identity

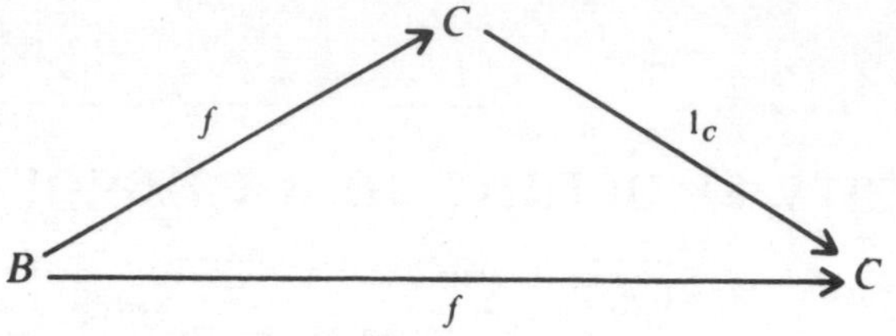

$1_C \circ f = f$

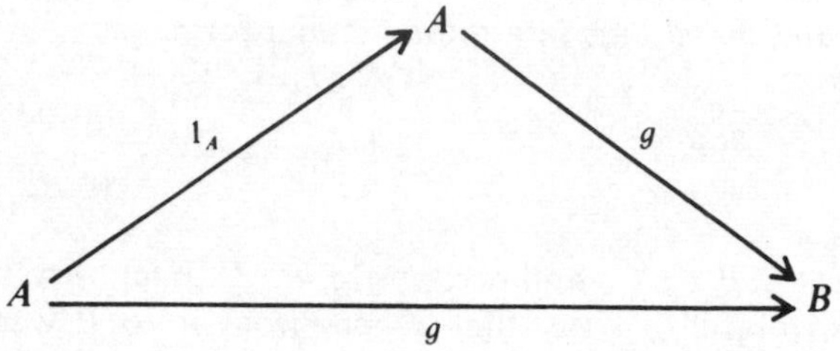

$g \circ 1_A = g$

Associativity:

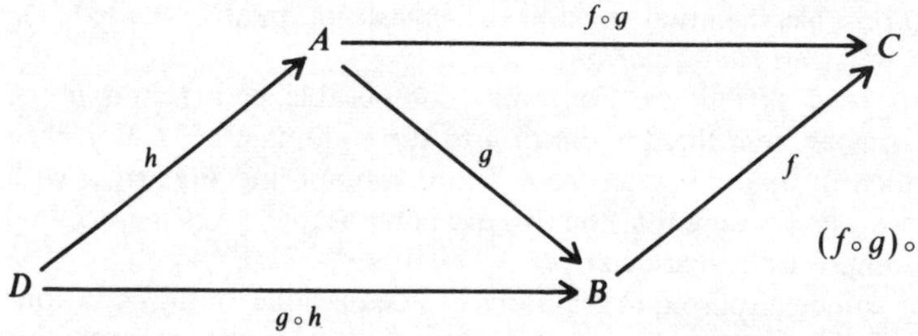

$(f \circ g) \circ h = f \circ (g \circ h)$

Because of associativity we can omit parentheses in composites of several arrows without ambiguity. We can write $f \circ g \circ h$ for the triple composite above.

We will often use just 'A' to label the identity arrow 1_A.

1.2 Isomorphisms, monics, and epics

An arrow $f: A \longrightarrow B$ is an *isomorphism* if there is an arrow $g: B \longrightarrow A$ such that $g \circ f = 1_A$ and $f \circ g = 1_B$. If there is such an arrow there is only one (see Exercise 1.1) and we call it the *inverse* of f, or $f^{-1}: B \longrightarrow A$. We may say that f is *iso*, for short, or write $f: A \xrightarrow{\sim} B$ to show that f is an isomorphism. In diagrams:

We say that A is *isomorphic* to B, and write $A \cong B$, to say that there is some isomorphism from A to B. There may be many different isomorphisms between two objects.

This is a generalization of the various notions of isomorphism that mathematicians have used since the 19th century. It has long been traditional to say that 'isomorphic groups have all the same group properties' or that 'isomorphic topological spaces (often called homeomorphic) have all the same topological properties'. The latter was often taken to define 'topological properties', in the spirit of Klein's Erlanger program. These slogans are abstracted and unified in category theory, where isomorphic objects in a category have all the same categorical properties. What this means will become clear as we go along.

An arrow $f: A \longrightarrow B$ is *monic* if it meets the following condition. For every object T and any arrows $h: T \longrightarrow A$ and $k: T \longrightarrow A$, if $f \circ h = f \circ k$ then $h = k$. This is sometimes expressed by saying that f is 'cancellable on the left'. We will often prove that two arrows, say p and q, are equal by finding a known monic f and proving that $f \circ p = f \circ q$. We may write $f: A \rightarrowtail B$ to show that f is monic. Notice that every iso is monic, since $f \circ h = f \circ k$ implies $f^{-1} \circ f \circ h = f^{-1} \circ f \circ k$.

$$T \underset{k}{\overset{h}{\rightrightarrows}} A \overset{f}{\rightarrowtail} B$$

A monic f can be thought of as a 'one-to-one' arrow. It will not identify h with k unless $h = k$. We make this precise in the section on generalized elements. A monic f can also be thought of as making A a 'part of' B. We make this precise when discussing sub-objects in Chapter 4.

An arrow $f: A \longrightarrow B$ is *epic* if it meets the following condition. For every object T and any arrows $h: B \longrightarrow T$ and $k: B \longrightarrow T$, if $h \circ f = k \circ f$ then $h = k$. We may say that f is 'cancellable on the right', and we may write $f: A \twoheadrightarrow B$ to show that f is epic. Every iso is epic. It is not accurate to think of an epic arrow as 'onto', as we shall see when discussing generalized elements. It is better to think of an epic as 'covering enough of B' that any two different arrows out of B must disagree somewhere within the part covered by f.

In general, an arrow may be monic and epic without being iso (see Exercise 1.5). In some categories though, every monic–epic is iso. Such a category is called *balanced*. Every topos is balanced, but not the category of categories.

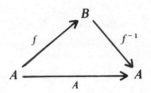

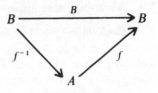

1.3 Terminal and initial objects

A *terminal object*, or *terminator*, is an object P such that every object has exactly one arrow to P. A category may have any number of terminators, including none. However many it has, they are *determined up to a unique isomorphism*, meaning that the following two theorems hold. The simple proofs are paradigms of categorical methods.

THEOREM 1.1 If P is a terminator and there is an iso $f: Q \xrightarrow{\sim} P$ then Q is also a terminator.

PROOF Any object T has a unique $u: T \longrightarrow P$, so there is at least $f^{-1} \circ u: T \longrightarrow Q$. Given any $v: T \longrightarrow Q$ we have $f \circ v: T \longrightarrow P$. By uniqueness, $u = f \circ v$, and so $v = f^{-1} \circ u$. Thus $f^{-1} \circ u$ is the unique arrow from T to Q. □

THEOREM 1.2 If P and Q are both terminators, then the unique arrow from P to Q is an isomorphism (with the unique arrow from Q to P as inverse).

PROOF Suppose that P and Q are both terminators, with the unique arrows $u: P \longrightarrow Q$ and $v: Q \longrightarrow P$. Then $v \circ u$ goes from P to P, but 1_p is the only arrow from P to P. So $v \circ u = 1_P$. Similarly, $u \circ v = 1_Q$. □

Therefore the definition of a terminator does not pick out a unique object (if there is a terminator at all). But when we are talking about a category that does have at least one terminator it is handy to suppose that we have selected one which we will call '1' for the duration of the discussion. Then for each A we write $!_A: A \longrightarrow 1$ for the unique arrow from A to 1. We may even omit the name of the arrow, since it is unique.

An *initial object* is an object I such that every object has exactly one arrow from I. From the symmetry of this definition with the definition of a terminal object one can see that initial objects are determined up to isomorphism. Any object iso to an initial object is also initial, and any two initial objects have a unique isomorphism between them. One can simply repeat the proofs of Theorems 1.1 and 1.2, reading 'arrow from I' each time they say 'arrow to P', and so on. We often use $\emptyset$ to name a selected initial object, and $!_A: \emptyset \longrightarrow A$ for the unique arrow.

An object which is both terminal and initial is called a *zero object*. Again, a category may have no zero objects, or any number. Since all the terms are defined up to isomorphism, if a category has one zero object then all initial or terminal objects in the category are zero objects.

1.4 Generalized elements

We can think of an arrow to any object B, say $x: A \longrightarrow B$, as a kind of 'element' of B. Specifically, we call x a *generalized element* of B defined over A, or an A-element of B. When thinking in these terms we write $x \in_A B$, but this is just another notation for $x: A \longrightarrow B$.

We also say that A is the *stage of definition* of x. It is often useful to think of all the generalized elements of B at one stage of definition A at one time.

For example, we can restate the definition of a terminal object: P is terminal iff, at each stage of definition A, P has exactly one A-element. At each stage of definition, P is a singleton. (There is no similar definition of an initial object in terms of its generalized elements: see Exercise 1.6.)

Thinking of $x \in_A B$ as a generalized element, for any $f: B \longrightarrow C$, we may write $f(x)$ for the composite $f \circ x$. In this notation the first domain–codomain axiom reads as follows: for any $x \in_A B$ and $f: B \longrightarrow C$ there is a well-defined $f(x) \in_A C$; that is, at each stage A, f takes A-elements of B to A-elements of C. The first identity axiom reads as follows: for any $x \in_B C$, $1_C(x) = x$. The associativity axiom takes a familiar form: for any $x \in_D A$, and $g: A \longrightarrow B$, and $f: B \longrightarrow C$, we have $(f \circ g)(x) = f(g(x))$.

On the other hand, this notation does not work so well with the second identity axiom, which would appear as $g(1_A) = g$. The apparent asymmetry whereby g appears once as a function symbol and once as a value is an artefact of the notation. The original notation of arrows and composites remains fundamental.

An arrow is fully determined by its effect on generalized elements.

THEOREM 1.3 Take any $f: A \longrightarrow B$ and $g: A \longrightarrow B$. Then $f = g$ iff, for every stage of definition T and every $x \in_T A$, $f(x) = g(x)$.

PROOF The 'only if' is obvious, since $f = g$ implies $f(x) = g(x)$. Conversely, assume that f and g agree on all generalized elements of A. Then take A itself as stage of definition and let x be 1_A. By assumption, $f(1_A) = g(1_A)$ and this gives $f = g$. □

This is typical of proofs using generalized elements, in that one half makes pivotal use of 1_A as an A-element of A. For any object B we call 1_B the *generic element* of B.

If a category has terminal object 1, then 1-elements may be of special interest, so 1-elements are often called *global elements.* We call an arrow $f: A \longrightarrow B$ *constant* if it factors through a global element; that is, $f = x \circ !_A$ for some $x: 1 \longrightarrow B$. The idea is that f then takes the single value x.

1.5 Monics, isos, and generalized elements

The definition of a monic takes a familiar form in generalized elements. An arrow $f: A \longrightarrow B$ is monic iff, at each stage T and for any $x \in_T A$ and $y \in_T A$, if $f(x) = f(y)$ then $x = y$. Monics are one-to-one on generalized elements.

Epics have no such definition in terms of their action on generalized elements. Let an arrow $f: A \longrightarrow B$ be called *onto on generalized elements* if, for each stage T and each $y \in_T B$ there is some $x \in_T A$ with $f(x) = y$. Then we have the following theorem.

THEOREM 1.4 An arrow $f: A \longrightarrow B$ is onto on generalized elements iff there is some $g: B \longrightarrow A$ with $f \circ g = 1_B$.

PROOF For 'if' suppose that there is such a g. Then for any $y \in_T B$ the T-element $g(y)$ of A has $f(g(y)) = y$. For 'only if' suppose that f is onto on generalized elements. Taking the generic element 1_B of B, there is some $g \in_B A$ with $f(g) = 1_B$. □

When $f \circ g = 1_B$ we call g a *right inverse* to f, and f a *left inverse* to g. An arrow may have many different right inverses. An arrow with a right inverse is epic, since $h \circ f = k \circ f$ implies $h \circ f \circ g = k \circ f \circ g$. But having a right inverse is much stronger than being epic (compare with Exercise 1.5). An arrow that has a right inverse is called a *split epic*.

Obviously every iso is split epic. Conversely, if f is monic and split epic then it is iso. For proof consider a right inverse g. Then $f \circ g \circ f = f$, and since f is monic we can cancel it on the left to obtain $g \circ f = 1_A$. So we have the following theorem.

THEOREM 1.5 An arrow is iso if it is one-to-one and onto on generalized elements.

Exercises

1.1 (An arrow with a left inverse and a right inverse is iso.) Given $f: A \longrightarrow B$, suppose that there is some $h: B \longrightarrow A$ with $h \circ f = 1_A$ and some $g: B \longrightarrow A$ with $f \circ g = 1_B$. Show that $g = h$. Conclude that $g = f^{-1}$ is the unique inverse of f.

1.2 Show, if $f: A \longrightarrow B$ and $g: B \longrightarrow C$ are both iso, that $g \circ f$ is then also iso. What is its inverse?

1.3 Show that $\cong$ is an equivalence relation: every object is iso to itself: if A is iso to B then B is to A; and if $A \cong B$ and$B \cong C$ then $A \cong C$.

1.4 Suppose that there is a terminator 1. Show that every arrow $f: 1 \longrightarrow B$ is monic. Explain in terms of generalized elements why every arrow with domain 1 is one-to-one on generalized elements.

1.5 Describe a category **2** with two objects, 0 and 1, their identity arrows, and one other arrow $\alpha: 0 \longrightarrow 1$. Composition can be defined in just one way verifying the axioms. Think of this category as an isolated arrow:

$$0 \xrightarrow{\alpha} 1$$

We do not draw identity arrows because we know there is one at each object. Show that the arrow α is monic and epic, but not iso (and not split epic).

(For those who know the relevant mathematics.) Some examples of monic–epics that are not iso are instructive: in the category of rings and ring homomorphisms, the inclusion of the ring of integers into the ring of rationals; in the category of topological spaces and continuous maps, any one-to-one onto function the domain of which has finer topology than its codomain; and in the category of Hausdorff spaces and continuous maps, any one-to-one map with dense image but not onto.

1.6 Show that, if there is an initial object $\emptyset$, a terminal object 1, and an arrow $1 \longrightarrow \emptyset$, then 1 and $\emptyset$ are zero objects.

In general, an initial object may have generalized elements at some or all stages of definition. However, in any Cartesian closed category (see Chapter 4) an initial object has a T-element iff T is also initial. In such categories, which include toposes and the category of categories, it is useful to think of an initial object as 'empty'. The only objects with arrows to an empty object are themselves empty.

(For those who know the relevant mathematics.) In the category of real vector spaces R^n and linear maps, R^0 is a zero object. The category of groups and group homomorphisms, and categories of modules and homomorphisms, have similar zero objects. This plus Theorem 6.4 proves that none of these is Cartesian closed. In the category of rings and ring homomorphisms, the ring of integers is initial, while some rings have many homomorphisms to it and others have none.

2

Products, equalizers, and their duals

2.1 Commutative diagrams

Definitions more intricate than that of a monic or a terminator are easier to give using diagrams and commutativity. We will not be formal about these. A diagram is an arrangement of objects and arrows, such as:

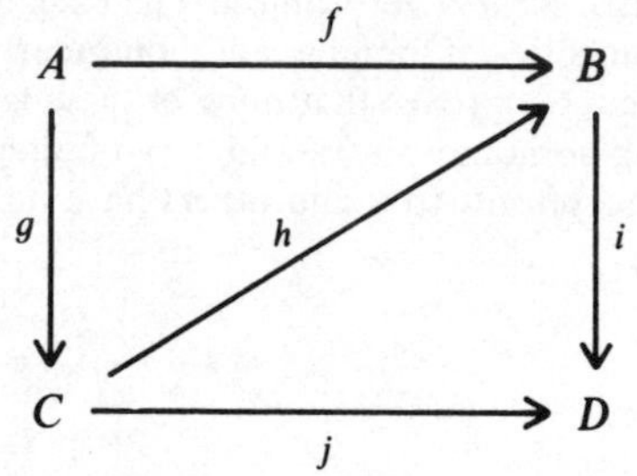

We say that the objects occupy *vertices* of the diagram, and the arrows occupy its *edges*. One object may occupy several different vertices. In the example, we might have $A=B=C=D$, so that the same object occupies all four vertices. That would still be different from the following one-vertex diagram:

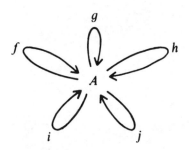

Similarly, one arrow may occupy several edges.

A diagram *commutes* if the following holds. Taking any two vertices, the composite of the arrows along any path from the first vertex to the second is equal to the composite along any other path from the first to the second. We count the empty path at any vertex, and say that the composite along it is the identity arrow for the object at that vertex. Thus to say that the four-vertex example above commutes means that $f = h \circ g$ and $i \circ f = i \circ h \circ g = j \circ g$ and $i \circ h = j$. The one-vertex example commutes iff all the arrows in it are 1_A.

Notice that to prove that the four-vertex diagram commutes it suffices to prove that both triangles contained in it do; that is, $f = h \circ g$ and $i \circ h = j$ imply all the other equations. One can prove that any diagram commutes by proving that each of its parts does, in a sense that we will not bother making precise. A diagram such as

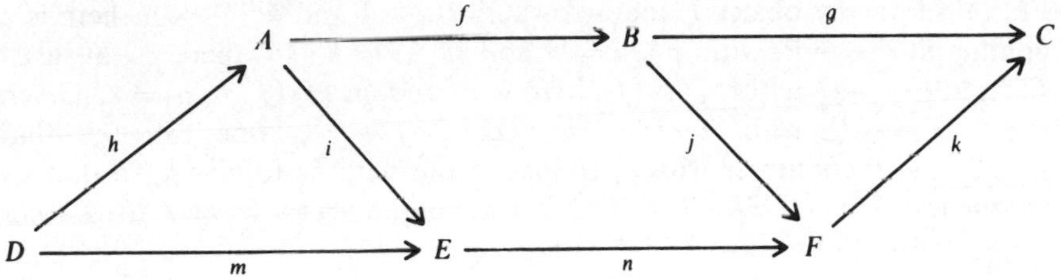

commutes if each triangle and square in it does; that is, $m = i \circ h$ and $j \circ f = n \circ i$ and $g = k \circ j$ suffice to imply all the equations required by commutativity of the diagram.

2.2 Products

Given objects A and B, a *product diagram* for A and B consists of an object P and two arrows p_1 and p_2 as shown:

$$A \xleftarrow{p_1} P \xrightarrow{p_2} B$$

with the following property. For any object T and arrows as shown:

$$A \xleftarrow{h} T \xrightarrow{k} B$$

there is a unique arrow $u: T \longrightarrow P$ that makes the following diagram commute:

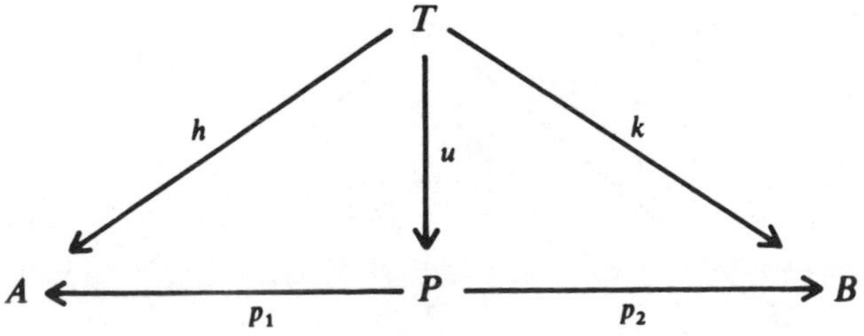

In equations, there is a unique u with $p_1 \circ u = h$ and $p_2 \circ u = k$.

The arrows p_1 and p_2 are the *projection arrows* of the product diagram. They are essential, since the object P alone is not enough to determine the diagram (see Exercise 2.1).

In a given category a given pair of objects might have no product diagrams, or any number. Product diagrams are determined up to isomorphism as follows:

THEOREM 2.1 If P, p_1, p_2 is a product diagram for A and B, and there is an iso $f: Q \xrightarrow{\sim} P$, then Q, $p_1 \circ f$, $p_2 \circ f$ is also a product diagram for A and B.

PROOF For any object T and arrows $h: T \longrightarrow A$ and $k: T \longrightarrow B$ there is a unique $u: T \longrightarrow P$ with $p_1 \circ u = h$ and $p_2 \circ u = k$, so there is at least $f^{-1} \circ u: T \longrightarrow Q$ with $p_1 \circ f \circ (f^{-1} \circ u) = h$, and $p_2 \circ f \circ (f^{-1} \circ u) = k$. Given any $v: T \longrightarrow Q$ with $p_1 \circ f \circ v = h$ and $p_2 \circ f \circ v = k$, one can see that $f \circ v: T \longrightarrow P$ composes with p_1 to give h and with p_2 to give k, so that by uniqueness $f \circ v = u$. So $v = f^{-1} \circ u$ is the unique arrow from T to Q composing with $p_1 \circ f$ to give h, and with $p_2 \circ f$ to give k. □

THEOREM 2.2 If P, p_1, p_2 is a product diagram for A and B, and Q, q_1, q_2 is another, then the unique arrow $u: Q \longrightarrow P$ with $p_1 \circ u = q_1$ and $p_2 \circ u = q_2$ is an isomorphism.

PROOF Take the two product diagrams, $u: Q \longrightarrow P$ and the obvious $v: P \longrightarrow Q$. Then $p_1 \circ u \circ v = q_1 \circ v = p_1$. Similarly, $p_2 \circ u \circ v = q_2 \circ v = p_2$. But 1_P is the unique arrow from P to P with those composites with p_1 and p_2, so $u \circ v = 1_P$. By the same reasoning, $v \circ u = 1_Q$. □

The reader should draw diagrams keeping track of the arrows in these proofs, and should compare the proofs with those for Theorems 1.1 and 1.2.

As with terminal objects, if we know that objects A and B have at least one product diagram it is convenient to select one to talk about. We will use the notation $A \times B$, p_1, p_2 to name the one we are talking about, or draw it:

$$A \xleftarrow{p_1} A \times B \xrightarrow{p_2} B$$

We use $\langle f, g \rangle$ to name the unique arrow induced by $f: T \longrightarrow A$ and $g: T \longrightarrow B$, so the following diagram commutes by definition:

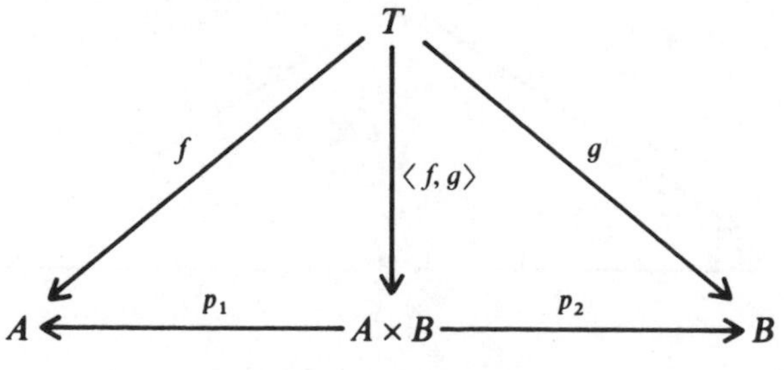

Given that A' and B' also have a product diagram, and given any arrows $h: A \longrightarrow A'$ and $k: B \longrightarrow B'$, we use $h \times k$ to name the arrow that makes the following diagram commute:

$$\begin{array}{ccccc} A & \xleftarrow{p_1} & A \times B & \xrightarrow{p_2} & B \\ \Big\downarrow h & & \Big\downarrow h \times k & & \Big\downarrow k \\ A' & \xleftarrow{p_1} & A' \times B' & \xrightarrow{p_2} & B' \end{array}$$

That is, $h \times k$ is the arrow to $A' \times B'$ induced by $h \circ p_1: A \times B \longrightarrow A'$ and $k \circ p_2: A \times B \longrightarrow B'$. Here we use '$p_1$' to name two different projection arrows, one in the diagram for A and B, and one in the diagram for A' and B': we do the same with 'p_2'. We will do this whenever the context makes it clear which projection arrow is which.

When there is more than one product diagram for each pair of objects then the choice of one for each pair is arbitrary; but once those choices are made the arrows $\langle f, g \rangle$ and $h \times k$ are uniquely determined.

Products have a familiar description in generalized elements, as follows. Given a product diagram $A \times B$, p_1, p_2, at each stage of definition T, any pair of T-elements $x \in_T A$ and $y \in_T B$ determines a unique T-element of $A \times B$, namely $\langle x, y \rangle \in_T A \times B$. And any T-element of $A \times B$ gives such a pair by applying p_1 and p_2. In short, a T-element of $A \times B$ corresponds with an ordered pair, one T-element of A and one of B (the fact that it is an ordered pair is important for products $A \times A$). A product arrow $h \times k$: $A \times B \longrightarrow A' \times B'$ acts on an ordered pair $\langle x, y \rangle \in_T A \times B$ componentwise: $h \times k(\langle x, y \rangle) = \langle h(x), k(y) \rangle$ (see Exercises 2.2 and 2.3).

The projection arrows are essential here. It is composition with them that defines the correspondence between T-elements of $A \times B$ and pairs of T-elements of A and B. But we will often refer to 'the product $A \times B$' and so on, leaving the projection arrows implicit.

2.3 Some natural isomorphisms

If a category has a terminator, 1, then every object is its own product with 1. To be precise:

THEOREM 2.3 Given a terminator 1, for every object A this is a product diagram for 1 and A:

$$1 \xleftarrow{!_A} A \xrightarrow{1_A} A$$

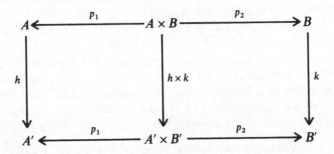

PROOF The proof proceeds by direct application of the definitions of terminators and product diagrams. □

Thus A is isomorphic to the object $1 \times A$ in whatever product diagram for 1 and A we might discuss. But that in itself does not mean that A *is* the object $1 \times A$ in the selected diagram. We might want to make a convention that whenever we select a product diagram for 1 and any object A we will select the one given in Theorem 2.3, so that A is always $1 \times A$. But this could conflict with other conventions we might want to make for other purposes as we go along, and in any case it is an unnatural, uncategorical solution. We do not need to define a specific product $1 \times A$—we can just define the product up to isomorphism and use the isomorphism!

COROLLARY Whatever the selected product diagram for 1 and A may be, the arrow u making this diagram commute is an isomorphism:

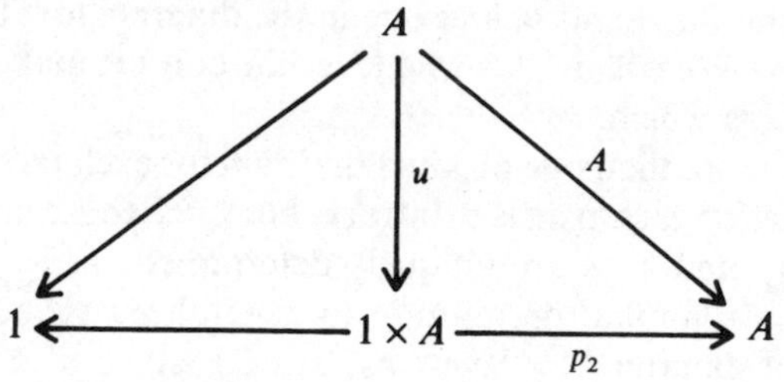

PROOF Use Theorems 2.2 and 2.3. □

We call u the *unitary isomorphism* for A. Note that it depends on the choice of product diagram for 1 and A. It is the inverse to the selected projection $p_2: 1 \times A \xrightarrow{\sim} A$. These are the isos that we mean whenever we write $A \xrightarrow{\sim} 1 \times A$ or $1 \times A \xrightarrow{\sim} A$.

The projections and unitary isomorphisms are called *natural*, because they get along with arrows. Specifically, in this case, for any $f: A \longrightarrow B$ the following square commutes:

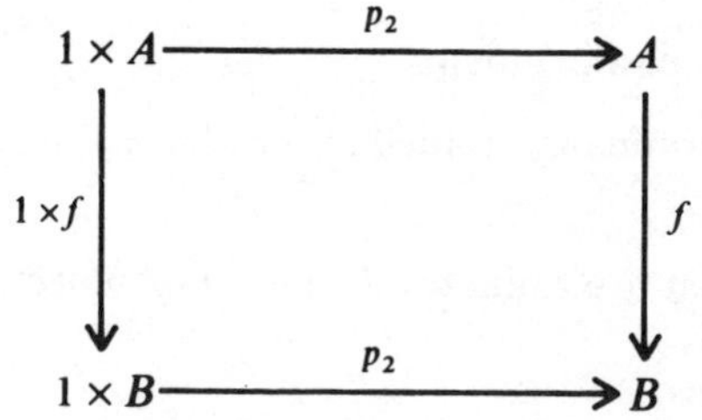

by definition of $1 \times f$. Then by Exercise 2.6 the following square commutes:

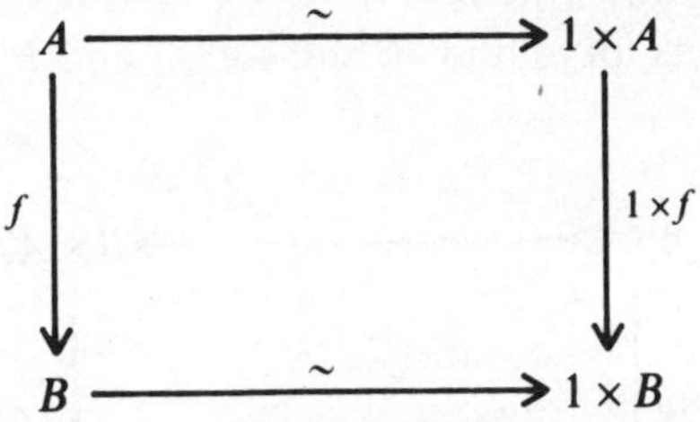

Products are symmetrical up to isomorphism:

THEOREM 2.4 If $B \times A$, q_1, q_2 is a product diagram for B and A:

$$B \xleftarrow{q_1} B \times A \xrightarrow{q_2} A$$

then $B \times A$, q_2, q_1 is a product diagram for A and B:

$$A \xleftarrow{q_2} B \times A \xrightarrow{q_1} B$$

PROOF The proof is a direct application of the definition of products. □

Again, we cannot conclude that the selected product diagram for A and B *is* exactly the one for B and A with the projection arrows switched. But we do have the following:

COROLLARY If $A \times B$, p_1, p_2 is a product diagram for A and B, while $B \times A$, q_1, q_2 is a product diagram for B and A, then the unique arrow tw that makes the following diagram commute is iso:

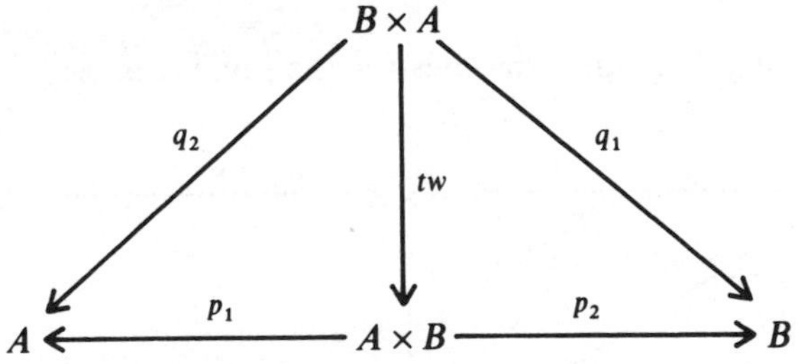

PROOF Use Theorems 2.4 and 2.2. □

We will call $tw: B \times A \longrightarrow A \times B$ the *twist arrow*. As with the unitary isomorphism, it depends on the selection of product diagrams. It follows directly from the definition that the twist arrow from $A \times B$ to $B \times A$ is the inverse to that from $B \times A$ to $A \times B$.

Twist arrows are natural. To be precise:

THEOREM 2.5 Take any arrows $f: A \longrightarrow A'$ and $g: B \longrightarrow B'$ and suppose that there are products for A and B, and for A' and B'. Then the following square commutes:

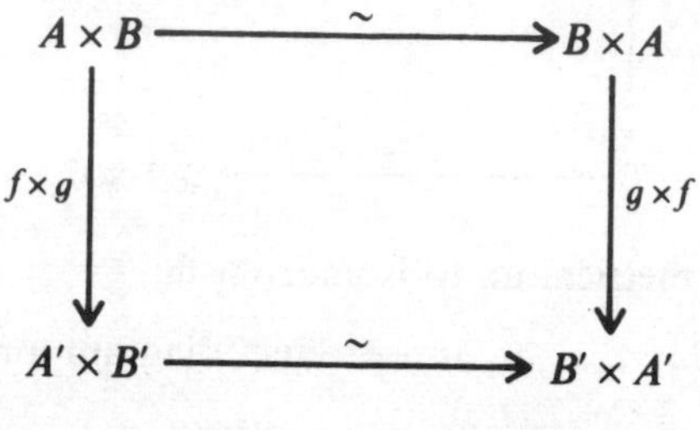

where the top and bottom arrows are the twist arrows.

PROOF The two composites go to a product, so we prove them equal by showing that they have the same composite with q'_1 and with q'_2. For the first, the triangle below commutes by definition of the twist arrow, and the square by definition of $f \times g$:

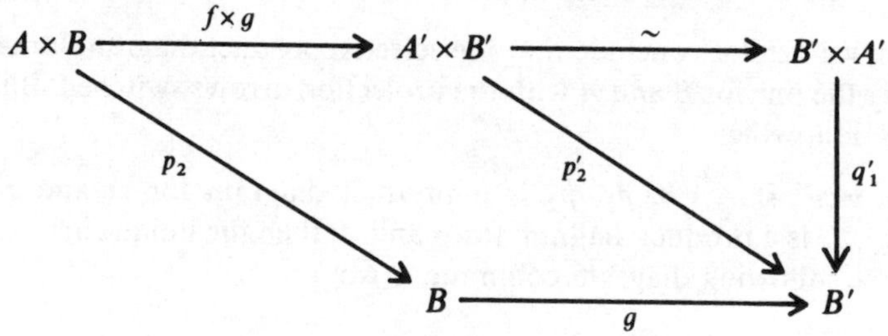

and the following diagram commutes for the same reasons:

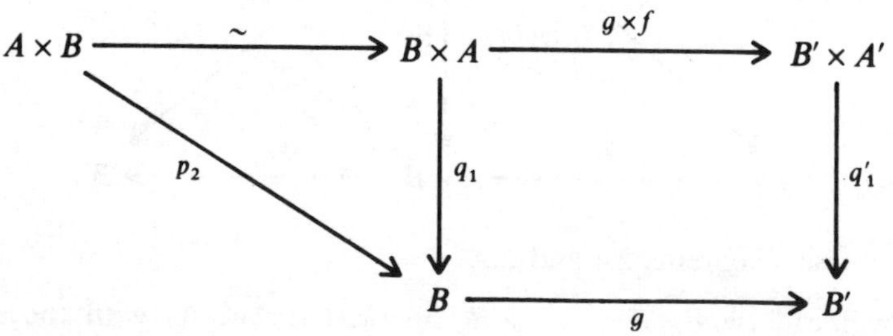

Therefore both composites compose with q'_1 to give $g \circ p_2$. Similar reasoning shows that both compose with q_2 to give $f \circ p_1$. □

2.4 Finite products

A product diagram for three objects A, B, and C is a diagram such that for any object T and arrows $f: T \longrightarrow A$, $g: T \longrightarrow B$, and $h: T \longrightarrow C$ there is a unique $u: T \longrightarrow P$ with $p_1 \circ u = f$, and so on. The generalization to any number of objects is obvious. Such products are determined up to isomorphism, by easy analogues to Theorems 2.1 and 2.2.

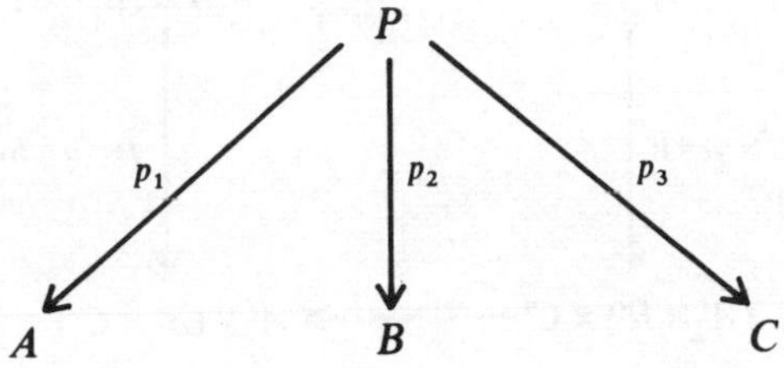

A terminator is a product diagram for no objects: it is an object 1 with no specified arrows, such that given any object T and no specified arrows there is a unique arrow $u: T \longrightarrow 1$. For any object A, A itself and the identity arrow 1_A is a product diagram for the single object A. Up to now this chapter has treated products of pairs of objects, and if a category has products for all pairs then we can use those to build product diagrams for any larger finite number of objects; that is, a category with a terminator and with all binary products has all finite products. But we have to look closer at how products of three or more objects are built.

Suppose that every pair of objects has a product, and consider three objects A, B, and C. The product diagram for A and B can be combined with that for $A \times B$ and C in this way:

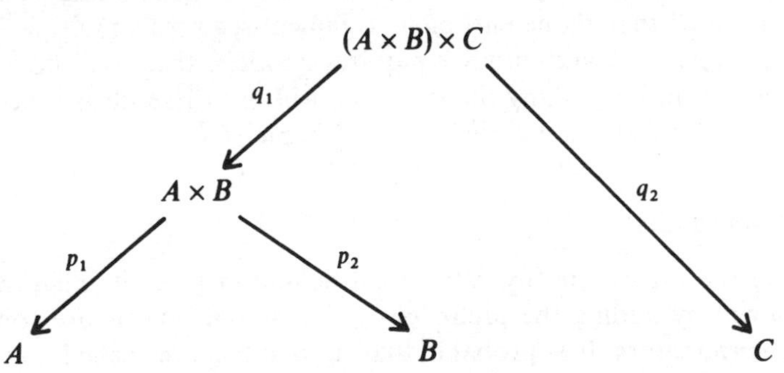

with the obvious projections. That this is a product diagram for A, B, and C can be verified directly. On the other hand, we could just as well have used $A \times (B \times C)$. By uniqueness up to isomorphism of products, the unique $a: (A \times B) \times C \xrightarrow{\sim} A \times (B \times C)$ which gets along with the projections in the obvious way is iso. Its inverse is the unique arrow from $A\times(B\times C)$ to $(A\times B)\times C$ that gets along with the projections. We will call either of these isos *associativity arrows.*

Associativity arrows are natural. That is, for any $f\colon A\longrightarrow A'$, $g\colon B\longrightarrow B'$ and $h\colon C\longrightarrow C'$ the following square commutes, where the top and bottom are the associativity arrows:

$$\begin{array}{ccc}
(A\times B)\times C & \xrightarrow{\ \sim\ } & A\times(B\times C)\\
{\scriptstyle (f\times g)\times h}\Big\downarrow & & \Big\downarrow{\scriptstyle f\times(g\times h)}\\
(A'\times B')\times C' & \xrightarrow{\ \sim\ } & A'\times(B'\times C')
\end{array}$$

The proof is very much like that of Theorem 2.5. We take $A'\times(B'\times C')$ as a product of three objects and show that both sides have the same composite with the projection to A', and also to B' and to C'. However, Theorem 2.5 dealt with four projections from binary products, while this deals with six from binary products plus four composites of them. Therefore the proof is left to the assiduous reader, who can also generalize to products for any finite number of objects.

In practice, multiple products are often built up from binary ones, so there is no particular advantage to introducing 'selected ternary products' and so on. Nor would introducing them spare us the trouble of keeping track of associativities, since we would then have to keep track of how each ternary product $A\times B\times C$ related to the iterated binary $A\times(B\times C)$ and so on. So we will suppose that all multiple products are built from binary ones. The result just given shows that, if one part of an argument requires a multiple product to be built one way, and another requires another, there will be a unique natural isomorphism linking the two. (We will use these ideas especially in connection with Cartesian closedness, in Chapter 6.)

2.5 Co-products

Reversing the arrows in any categorical definition gives its *dual,* which is often named by adding the prefix 'co-'. Thus initial objects are sometimes called coterminators. It is probable that no one has ever called epics 'comonic', but an obsessive systematizer could. Notice that reversing the arrows in the category axioms gives exactly those axioms back again; so reversing them in any theorem gives a dual theorem (see Exercises 2.7, 2.8, and 2.9).

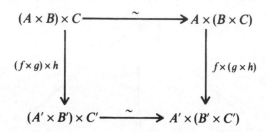

Reversing the arrows in the definition of a product gives the following. A *coproduct diagram* for A and B consists of a third object P and arrows as shown:

$$A \xrightarrow{i_1} P \xleftarrow{i_2} B$$

with the following property. For any object T and arrows as shown:

$$A \xrightarrow{h} T \xleftarrow{k} B$$

There is a unique $u: P \longrightarrow T$ that makes the following diagram commute:

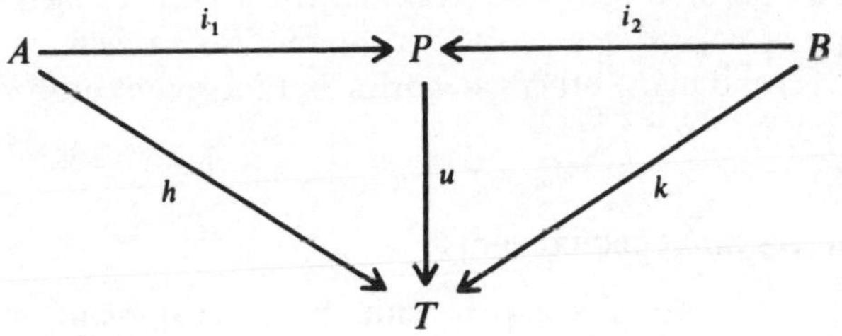

Reverse the arrows in the proofs of Theorems 2.1 and 2.2 to show that coproducts are unique up to isomorphism. If objects A and B have at least one coproduct diagram we write $A + B, i_1, i_2$ to name one we have selected to discuss. We write $\binom{h}{k}$ to name the arrow that makes the following diagram commute:

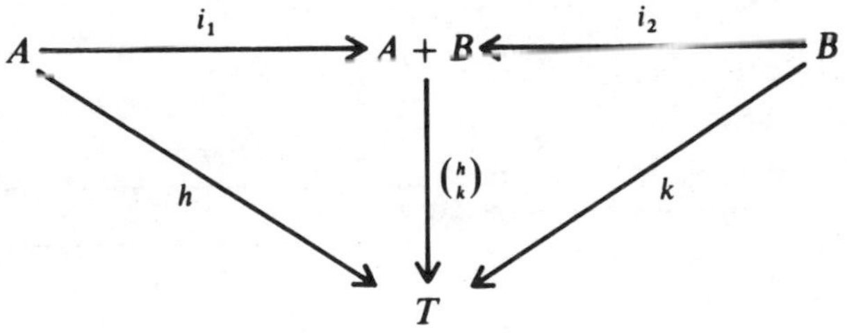

and $f + g: A + B \longrightarrow A' + B'$ for the obvious analogue to $f \times g$. The duals to Theorem 2.3 and its corollary show that if there is an initial object (dual to a terminal object) $\emptyset$ then for every object A there is a natural isomorphism $A \cong \emptyset + A$. If A and B have a coproduct there is a natural isomorphism $A + B \cong B + A$; and similarly for associativity.

Of course, the dual to 'generalized element of B' is *not* 'generalized element of B'. The dual to a T-element of $A \times B$ is an arrow *from* $A + B$ to T. Therefore, knowledge of generalized elements of products is no help with coproducts. In fact, coproducts have no general description in terms of their generalized elements.

In many categories (including all toposes and the category of categories) one can think of $A + B$ as a disjoint union of A and B—a copy of A plus a separate copy of B. This picture fits with the definition of a coproduct: an arrow $\binom{h}{k}$ from $A + B$ to T acts like h on the copy of A and like k on the copy of B; an arrow $f + g: A + B \longrightarrow A' + B'$ takes the copy of A to the copy of A' by f, and the copy of B to that of B' by g.

It is important to realize that the picture of $A + B$ as a disjoint union does not mean that every T-element of $A + B$ either comes from a T-element of A or else from a T-element of B. A T-element of $A + B$ might factor partly through A and partly through B—consider the generic element!

This image of coproducts as disjoint unions works well for toposes (cf. Theorem 16.15) and many other categories, but there are categories in which it does not work.

2.6 Equalizers and coequalizers

A *parallel pair* of arrows is a pair with the same domain and the same codomain. An *equalizer* for a parallel pair of arrows $f: A \longrightarrow B$ and $g: A \longrightarrow B$ is an arrow $e: E \longrightarrow A$ with these properties: $f \circ e = g \circ e$, and given any arrow $h: T \longrightarrow A$ with $f \circ h = g \circ h$ there is a unique arrow $u: T \longrightarrow E$ such that $e \circ u = h$. This is easier to keep track of in a diagram:

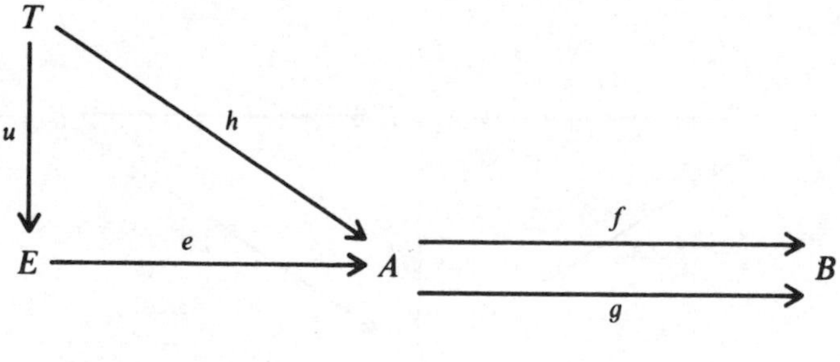

In other words, say that an arrow h *equalizes* the parallel pair f and g if $f \circ h = g \circ h$. Then e is an equalizer for f and g if it equalizes them and any h that equalizes them factors uniquely through e.

A parallel pair of arrows may have no equalizer or any number, but equalizers are determined up to isomorphism. Precisely: if $e: E \longrightarrow A$ is an equalizer for f and g, and there is an iso $i: F \longrightarrow E$, then $e \circ i: F \longrightarrow A$ is also an equalizer for f and g—and if $e: E \longrightarrow A$ and $e': F \longrightarrow A$ are both equalizers for f and g, then the unique arrow $u: F \longrightarrow E$ with $e \circ u = e'$ is iso.

In terms of generalized elements we can restate the definition of an equalizer in this way: $f \circ e = g \circ e$, and for any T and any $x \in_T A$ with $f(x) = g(x)$ there is a unique $y \in_T E$ with $e(y) = x$. The equalizer can be thought of as the 'part of A on which $f = g$'. We will make this more precise when we discuss sub-objects. Here we can establish the key point:

THEOREM 2.6 Every equalizer is monic.

PROOF Suppose that $e: E \longrightarrow A$ is an equalizer for $f, g: A \longrightarrow B$, and that h and k have $e \circ h = e \circ k$. Then $e \circ h$ equalizes f and g, so it factors uniquely through e. But since it also factors as $e \circ k$, $h = k$. □

Coequalizers are dual to equalizers. A *coequalizer* for $f, g: B \longrightarrow A$ is an arrow $q: A \longrightarrow Q$ such that for any $h: A \longrightarrow T$ with $h \circ f = h \circ g$ there is a unique $u: Q \longrightarrow T$ with $u \circ q = h$. That is, q coequalizes f and g, and any h that coequalizes them factors uniquely through q.

Coequalizers cannot be defined in terms of their generalized elements, but some remarks might be helpful. Consider any object T. Suppose that there are some $x \in_T A$ and $y \in_T A$ such that there is some $z \in_T B$ with $f(z) = x$ and $g(z) = y$. Then, from $q \circ f = q \circ g$, it follows that $q(x) = q(y)$. Intuitively, the T-elements of A fall into classes, where all the T-elements in any one class are related to each other either directly as x and y in the last sentence or indirectly through a chain of such connections. The arrow q will take all the T-elements of A in such a class to the same T-element of Q. Therefore Q can be thought of as containing 'equivalence classes of generalized elements of A', with q taking any T-element of A to its equivalence class in Q. We return to this picture in Exercise 5.6.

Exercises

2.1 Suppose that P, p_1, p_2 is a product diagram for A and B, and suppose that $i: P \longrightarrow P$ is an isomorphism other than 1_P. Then show that P is the object in another product diagram for A and B with different projections.

2.2 Suppose that A and B have a product and so do A' and B'. Consider arrows $i: S \longrightarrow T$, $h: T \longrightarrow A$, $k: T \longrightarrow B$, $f: A \longrightarrow A'$, and $g: B \longrightarrow B'$. Show that $\langle h, k \rangle \circ i = \langle h \circ i, k \circ i \rangle$ and $(f \times g) \circ \langle h, k \rangle = \langle f \circ h, g \circ k \rangle$.

2.3 Keeping the arrows and products from Exercise 2.2, take another two arrows, $m: A' \longrightarrow A''$ and $n: B' \longrightarrow B''$, and suppose that A'' and B'' have a product. Show that $(m \times n) \circ (f \times g) = (m \circ f) \times (n \circ g)$. Show that $(f \times 1_{B'}) \circ (1_A \times g)$ equals $(1_{A'} \times g) \circ (f \times 1_B)$. [Draw diagrams.]

2.4 Take arrows $r: B \longrightarrow C$, $t: C \longrightarrow E$, $s: B \longrightarrow D$, and $v: D \longrightarrow E$ and suppose that A has a product with each of B, C, D, and E. Show that if $t \circ r = v \circ s$, then $(t \times A) \circ (r \times A) = (v \times A) \circ (s \times A)$. Draw the diagrams for this, the commutative squares. Also interpret this as following from Exercise 2.3.

2.5 Suppose that A and B have a product, and take any arrow $f: A \longrightarrow B$. Define the *graph* of f to be the arrow Γ_f that makes the following diagram commute:

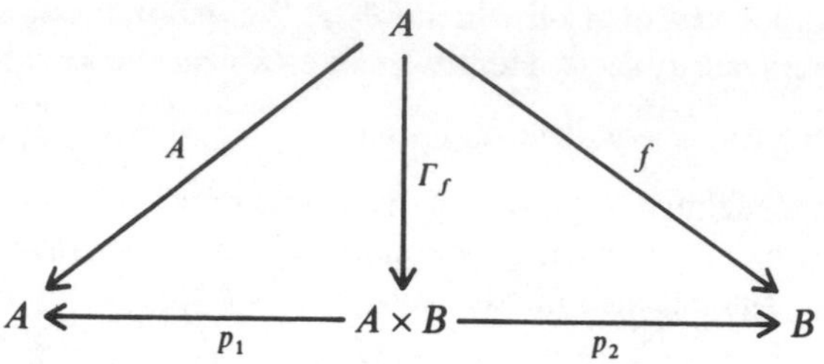

Show that Γ_f is an equalizer for $f \circ p_1$ and p_2. For any A with a product with itself, the *diagonal* $\Delta_A : A \rightarrowtail A \times A$ is the graph of 1_A. For any $h: T \longrightarrow A$ and $k: T \longrightarrow B$ show that $\langle h, k \rangle = (h \times k) \circ \Delta_T$, assuming that the products exist.

2.6 Suppose that $h: A \longrightarrow B$ and $k: C \longrightarrow D$ are both iso. Show, for any $f: A \longrightarrow C$ and $g: B \longrightarrow D$, that $k \circ f = g \circ h$ implies $f \circ h^{-1} = k^{-1} \circ g$.

2.7 Notice that when the arrows are reversed in a statement the order of each composite must also be reversed. Given $f: A \longrightarrow B$, $g: B \longrightarrow C$, and $g \circ f: A \longrightarrow B$, reversing the arrows gives $f: B \longrightarrow A$ and $g: C \longrightarrow B$, and thus $f \circ g: C \longrightarrow A$.

We say that the operator Dom(_) is *dual* to Cod(_) and vice versa, and that $f \circ g$ is dual to $g \circ f$; while $1\_$ and $f = g$ are dual to themselves. Any statement using those expressions has a dual, formed by replacing each expression in the statement by its dual; that is, replace 'Dom' by 'Cod', and vice versa, and switch the arrows in any composite.

Show that each category axiom in Section 1.1 is its own dual.

2.8 Show that the definition of isomorphism is its own dual. Show that the dual to the definition of monic is the definition of epic. Give the definition of split monic, dual to split epic (Section 1.5).

2.9 A categorical statement with defined expressions also has a dual, formed by replacing all expressions with their duals. Demonstrate that, since the dual to any axiom is an axiom, the dual to any theorem of category theory is a theorem. [Warning: this will not be true in all extensions of the category axioms. Neither the Cartesian closed category axioms nor the topos axioms include all of their own duals.] Give the proof that coproducts are determined up to isomorphism, by taking the dual to each statement in the proof that products are. Conclude, from the fact that we have proved the dual statements, that any arrow both epic and split monic is iso, and that any coequalizer is epic.

2.10 Let e be an equalizer for f and g. Prove that the following are equivalent: (a) e is epic; (b) $f = g$; (c) e is iso. Note that e is iso iff the identity arrow on the domain of f is also an equalizer. Formulate the dual result on coequalizers.

2.11 Show that any split monic $e: E \rightarrowtail A$ is an equalizer for 1_A and $e \circ h$, where h is any left inverse for e. This and Exercise 2.10 give another proof that any monic split epic is iso.

2.12 Define a *subterminator* to be an object P such that for every object A there is at most one arrow from A to P. Note how the argument that any two terminators are isomorphic fails to show that any two subterminators are. Show the two objects in **2**

(Exercise 1.5) are both subterminators. Assuming that there is a terminator 1, show that P is a subterminator iff the unique arrow $P \longrightarrow 1$ is monic.

Show that P is a subterminator iff P, 1_P, 1_P is a product diagram for P and P. Show that P is a subterminator iff there is some product diagram for P and P with $p_1 = p_2$, and that in that case $p_1 = p_2$ in every product diagram for P and P.

The dual notion of a *subinitial object* is little used but it exists, and every theorem on subterminators has a dual theorem.

3

Groups

This chapter treats groups as an example of structures in a category. It presumes some familiarity with groups, and is not a prerequisite to later chapters. We begin with a category **C** with all finite products, called the *base* category. For convenience we suppose that **C** has a selected product $A \times B$, p_1, p_2 for each pair of objects A and B. If **C** is in fact **Set** the definition gives the usual groups. A group in **Data** is a data type together with computable multiplication and inverse rules, a group in **Top** is a continuous group, and in **Man** it is a Lie group.

3.1 Definition

A group is a structure with a unit, e, a multiplication, $x.y$, and an inverse operation, x^{-1}, satisfying these equations:

unit: $x.e = x$ and $e.x = x$

inverse: $x.x^{-1} = e$ and $x^{-1}.x = e$

associativity: $(x.y).z = x.(y.z)$

More precisely, define a *group* in **C** to be an object G of **C** together with arrows

$$e: 1 \longrightarrow G \qquad m: G \times G \longrightarrow G \qquad \_^{-1}: G \longrightarrow G$$

such that these diagrams commute:

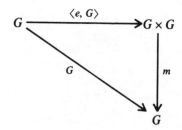

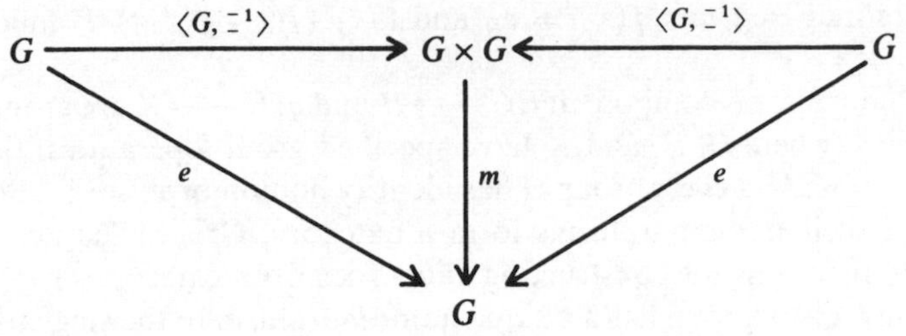

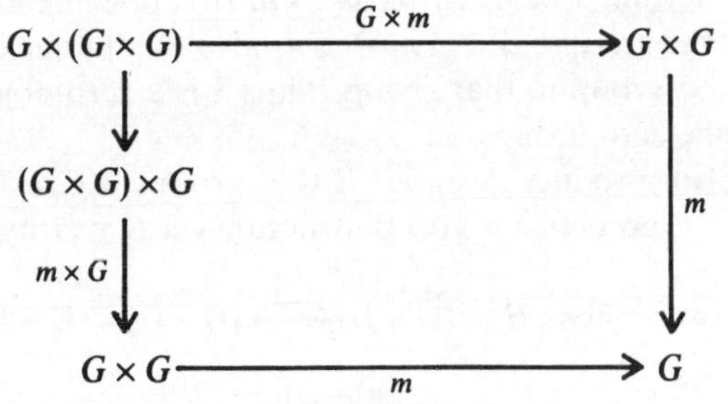

where we have written e as a shorthand for $e \circ !_G : G \longrightarrow G$.

In terms of generalized elements, the first diagram says, for every object T and T-element x of G, $m(\langle e, x \rangle) = x$. If we abbreviate $m(\langle h, k \rangle)$ by $h.k$ this becomes $e.x = x$. The second says, again for every T and T-element x of G, $x.x^{-1} = e$ and $x^{-1}.x = e$. The third gives associativity.

I gave no diagram for $x.e = x$, but this follows from the others (see Exercise 3.1).

There is a *singleton group* the object of which is 1. Its arrows e, m, and $\_^{-1}$ are the only ones they could be, since their co-domain is 1. This is the only group that we can prove exists if we only assume that **C** has all finite products. With stronger assumptions on **C** we obtain more groups.

3.2 Homomorphisms

A *group homomorphism* from one group, $\langle G, e_G, m_G, \_{G}^{-1} \rangle$, to another, $\langle H, e_H, m_H, \_{H}^{-1} \rangle$, is an arrow $f: G \longrightarrow H$ that makes the following diagrams commute:

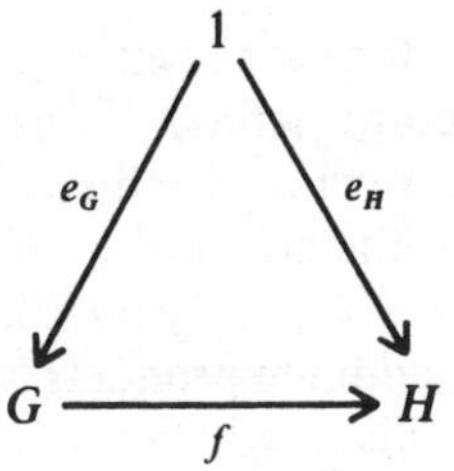

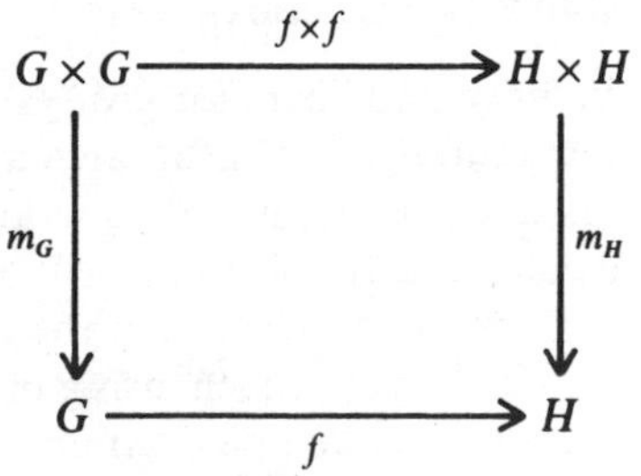

In generalized elements, $f(e_G) = e_H$ and $(fx).(fy) = f(x.y)$. It follows that $(fx)^{-1} = f(x^{-1})$.

Homomorphisms compose. If $f: G \longrightarrow H$ and $g: H \longrightarrow K$ are group homomorphisms (where G, H, and K have specified group operations) then so is $g \circ f: G \longrightarrow K$. And every group G has identity homomorphism 1_G. So groups in **C** and their homomorphisms form a category, $\mathbf{Grp_C}$. (Chapter 12 gives formal foundations for constructing categories from categories.)

Trivially, every group has a unique homomorphism to the singleton group 1: that is, 1 is a terminal object in $\mathbf{Grp_C}$. On the other hand, for any group $\langle G, e, m, \_^{-1}\rangle$, the group unit $e: 1 \longrightarrow G$ is also the unique homomorphism from the singleton group to that group. Thus 1 is a zero object in $\mathbf{Grp_C}$ and we may call it the zero group.

There are also product groups. Take groups $\langle G, e_G, m_G, \__G^{-1}\rangle$ and $\langle H, e_H, m_H, \__H^{-1}\rangle$ and define a group structure on $G \times H$ by these arrows:

$$1 \xrightarrow{\langle e_G, e_H\rangle} G \times H \qquad (G \times H) \times (G \times H) \xrightarrow{\sim} (G \times G) \times (H \times H) \xrightarrow{m_G \times m_H} G \times H$$

$$G \times H \xrightarrow{\__G^{-1} \times \__H^{-1}} G \times H$$

The isomorphism from $(G \times H) \times (G \times H)$ to $(G \times G) \times (H \times H)$ is the arrow that, for any T and any x, y T-elements of G and w, z T-elements of H, takes $\langle\langle x, w\rangle, \langle y, z\rangle\rangle$ to $\langle\langle x, y\rangle, \langle w, z\rangle\rangle$ (see Exercise 3.2). It is easy to verify that this is a group by using generalized elements. The same method shows that $p_1: G \times H \longrightarrow G$ is a homomorphism, as is p_2. Given a group and a homomorphism from it to $\langle G, e_G, m_G, \__G^{-1}\rangle$ and one to $\langle H, e_H, m_H, \__H^{-1}\rangle$, there is a unique homomorphism to the product group, with the obvious property. So $\mathbf{Grp_C}$ has all finite products.

If **C** has equalizers for all parallel pairs then so does $\mathbf{Grp_C}$ (Exercise 3.3).

Even if **C** has coproducts or coequalizers the category $\mathbf{Grp_C}$ may not. It does if the base **C** is the category of sets or any other topos with a natural number object, but those are much stronger assumptions than we are making now.

3.3 Algebraic structures

What we have done here for groups could also be done for abelian groups, rings, commutative rings, or any algebraic structure defined entirely by operators and equations. The operators become arrows and the equations become commutative diagrams. All of the above results hold for all of those as well, except for the result that the terminal object of **Grp** is also initial (see Exercise 3.6). This approach, using only diagrams and finite products, would not work for more complex axioms, such as for fields in which the axioms require not an inverse for every element but an inverse for every element except 0.

Exercises

3.1 Consider the usual algebraic proof that $x.e = x$ follows from the other group equations:

$$x.e = x.(x^{-1}.x) = e.x = x$$

Draw the diagram for $x.e = x$ and conclude that it commutes, by showing that those equations hold for all generalized elements. Show that $f(x^{-1}) = (fx)^{-1}$ follows from the definition of a group homomorphism f.

3.2 Show that the isomorphism used in constructing a product group is $\langle\langle p_1 \circ p_1, p_1 \circ p_2\rangle, \langle p_2 \circ p_1, p_2 \circ p_2\rangle\rangle$. Show that it is iso by using generalized elements, or more conceptually using *product diagrams for four objects*. Show that multiplication in $G \times H$ is defined co-ordinate-wise: in generalized elements, $\langle x, w\rangle . \langle y, z\rangle = \langle x.y, w.z\rangle$, where each multiplication is in the appropriate group.

3.3 Consider two groups $\langle G, e_G, m_G, -_G^{-1}\rangle$ and $\langle H, e_H, m_H, -_H^{-1}\rangle$ and homomorphisms h and k from the first to the second. Suppose that there is an equalizer i, as in the following diagram:

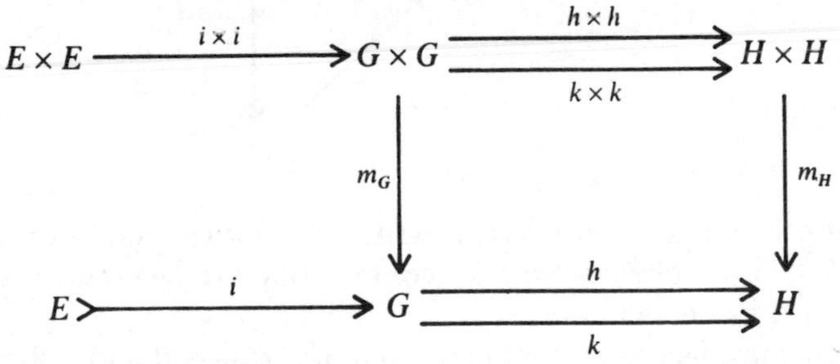

Show that $i \times i$ is also the equalizer of $h \times h$ and $k \times k$. Define m_E by showing that $m_G \circ (i \times i)$ factors through e. Show that there is a unique way to define e_E and $-_E^{-1}$ so that they commute with i; and that these definitions make E a group and i a group homomorphism. Show that i is an equalizer for h and k in **Grp**.

3.4 Prove that any two groups $\langle G, e_G, m_G, -_G^{-1}\rangle$ $\langle H, e_H, m_H, -_H^{-1}\rangle$ have a trivial homomorphism $G \longrightarrow 1 \xrightarrow{e} H$.

3.5 Show that the category **2** has all finite products, and that the only group in **2** is the singleton group.

3.6 Show how the proof that the singleton group is also initial involves the fact that the group axioms use only one constant, one global element $e: 1 \longrightarrow G$, and the operations carry e to itself; that is, $e.e = e$ and $e^{-1} = e$. The ring axioms use two constants, $0: 1 \longrightarrow R$ and $1: 1 \longrightarrow R$, and the terminal ring is not also initial (unless all rings in the base category are singletons).

4

Sub-objects, pullbacks, and limits

4.1 Sub-objects

To work with the idea that a monic, say $i: S \rightarrowtail A$, can be thought of as giving a 'part of A', we define a *sub-object* of A to be a monic with codomain A. We say that a T-element $x \in_T A$ is a *member* of i, and write $x \in i$, if x factors through i; that is, if there is some $h: T \longrightarrow S$ that makes the following triangle commute:

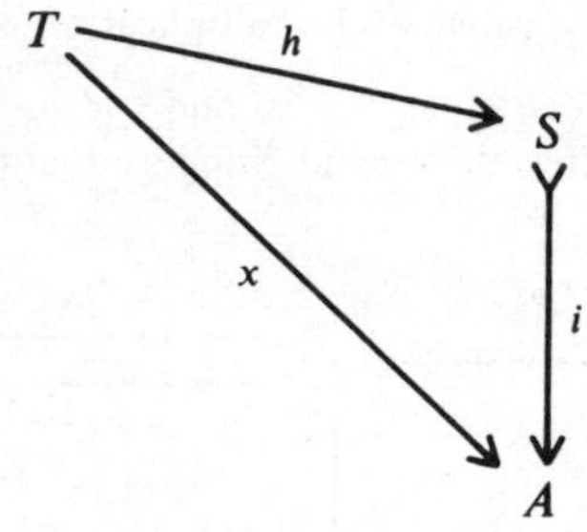

A common abuse of notation is to write $x \in S$ rather than $x \in i$, when the monic $i: S \rightarrowtail A$ is obvious from the context. But it is the monic that matters here, not just its domain.

We need some facts about monics. First, if $i: A \rightarrowtail B$ and $j: B \rightarrowtail C$ are monic then so is their composite $j \circ i: A \rightarrowtail C$; since for any $h, k: T \longrightarrow A$, $j \circ i \circ h = j \circ i \circ k$ implies that $i \circ h = i \circ k$ and so $h = k$. There is a kind of converse:

THEOREM 4.1 Given any arrows $s: A \longrightarrow B$ and $g: B \longrightarrow C$, if $g \circ s$ is monic then s is also.

PROOF Suppose that $g \circ s$ is monic, and $h, k: T \longrightarrow A$ have $s \circ h = s \circ k$. Then $g \circ s \circ h = g \circ s \circ k$, and so $h = k$. □

It is worthwhile to restate Theorem 4.1 in generalized elements: if $g \circ s$ is one-to-one on generalized elements, then s must be.

THEOREM 4.2 Given monics $i: A \rightarrowtail C$ and $j: B \rightarrowtail C$, if there is an arrow s with $i = j \circ s$ and an arrow t with $j = i \circ t$, then t and s are inverse to one another.

PROOF The assumptions imply that $i = i \circ t \circ s$, and since i is monic this implies $1_A = t \circ s$. Similarly, $s \circ t = 1_B$. □

Theorem 4.2 says that if i and j are monic and both triangles below commute, then the whole diagram does. Specifically, $t = s^{-1}$ and so $A \cong B$:

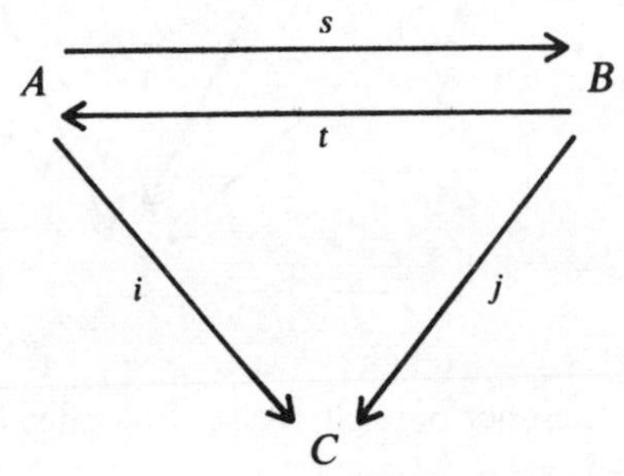

We can define an *inclusion* relation on monics to any one object. Given monics $i\colon A \rightarrowtail C$ and $j\colon B \rightarrowtail C$ we write $i \subseteq j$, or say i is *included in* j, if i factors through j; that is, if there is an arrow s that makes the following diagram commute:

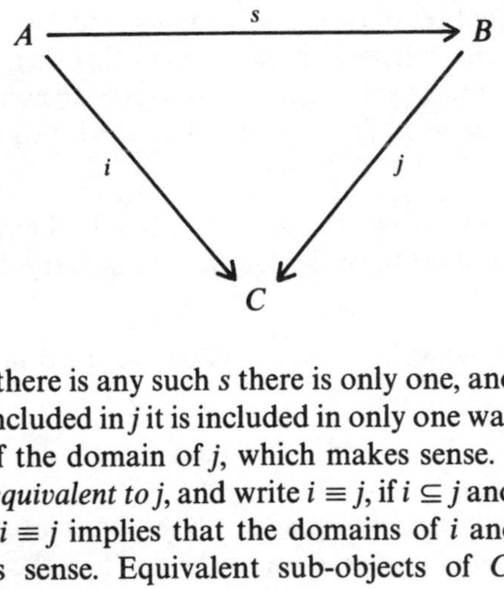

Since j is monic, if there is any such s there is only one, and by Theorem 4.1 it is monic. So if i is included in j it is included in only one way, and the inclusion s is a sub-object of the domain of j, which makes sense.

We say that i is *equivalent to* j, and write $i \equiv j$, if $i \subseteq j$ and $j \subseteq i$. By Theorem 4.2 we know that $i \equiv j$ implies that the domains of i and j are isomorphic, which also makes sense. Equivalent sub-objects of C have isomorphic domains.

THEOREM 4.3 Given sub-objects $i\colon A \rightarrowtail C$ and $j\colon B \rightarrowtail C$, we have $i \subseteq j$ iff every member of i is a member of j.

PROOF For 'only if' suppose that $i \subseteq j$. Then for any T and any $x \in_T C$, $x \in i$ implies $x \in j$ as shown:

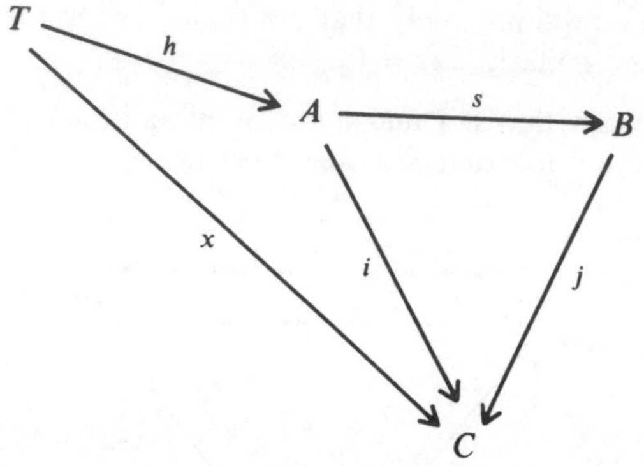

For 'if' note that i is a member of itself, by 1_A. But since i is monic, $i \in j$ means the same thing as $i \subseteq j$. □

COROLLARY For any sub-objects i and j of any object C, $i \equiv j$ iff i and j have exactly the same members. □

Again, the notation is often abused by writing $A \subseteq B$ or $A \equiv B$ when the context shows which monics are meant.

Categorical constructions of sub-objects are all defined up to equivalence; for example, any equalizer $e: E \rightarrowtail A$ is monic. The fact that equalizers are determined up to isomorphism can be stated in this way: if $e: E \rightarrowtail A$ is an equalizer for f and g, then a sub-object $i: S \rightarrowtail A$ is an equalizer for f and g iff $e \equiv i$ as sub-objects of A. In fact, we can characterize equalizers in terms of their members:

THEOREM 4.4 Given arrows $f, g: A \longrightarrow B$, a sub-object $e: E \rightarrowtail A$ is an equalizer for f and g iff we have the following: for every T and every $x \in_T A$, $x \in e$ iff $f(x) = g(x)$.

PROOF This is the definition of an equalizer restated in the new notation. Remember that $e \in e$. □

Note that inclusion is defined up to equivalence: if $i \equiv i'$ and $i \subseteq j$ then $i' \subseteq j$; and if $j \equiv j'$ then $i' \subseteq j'$.

In fact, category theorists often define a 'sub-object' of A to be not a single monic to A but an equivalence class of them. Then the 'sub-object' determined by a monic $i: S \rightarrowtail A$ is the class of all monics j to A with $i \equiv j$. Mac Lane discusses the two possible definitions of 'sub-object' and, with the no-nonsense attitude of a working mathematician, picks both (Mac Lane 1971, p. 122). In practice, one almost always works with monics anyway, and not directly with equivalence classes, and the conclusions almost always apply to whole equivalence classes.

4.2 Pullbacks

A *pullback* of two arrows with common codomain, say $f: A \longrightarrow C$ and $g: B \longrightarrow C$, consists of an object P and two arrows as shown:

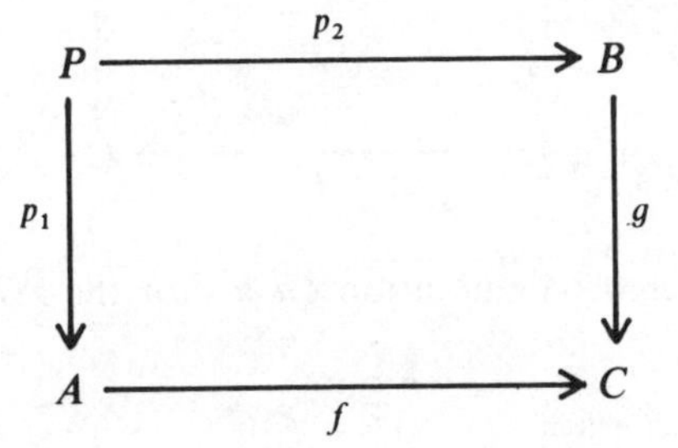

with the following property. The square commutes, and for any object T and arrows h and k that make the outer square below commute, there is a unique $u: T \longrightarrow P$ that makes the whole diagram commute:

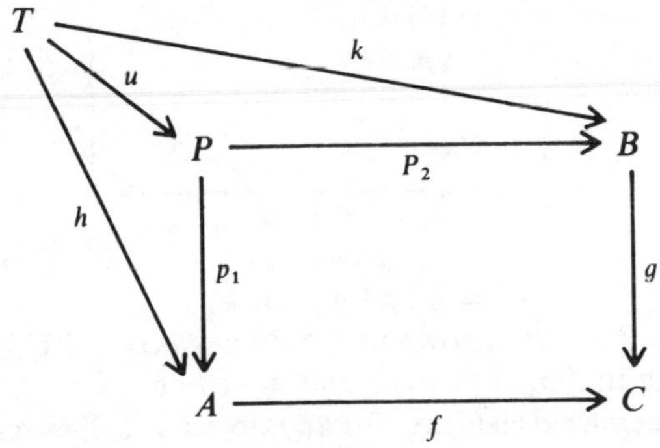

That is, $f \circ p_1 = g \circ p_2$ and, for any T and $h: T \longrightarrow A$ and $k: T \longrightarrow B$, if $f \circ h = g \circ k$ there is a unique u with $p_1 \circ u = h$ and $p_2 \circ u = k$. We call the arrows p_1 and p_2 *projections*.

We refer to a pair of arrows with the same codomain as a *corner* of arrows. As one would expect, a given corner of arrows $f: A \longrightarrow C$ and $g: B \longrightarrow C$ may have no pullback or any number, and pullbacks are defined up to unique isomorphisms that get along with the projections. Given two pullbacks for f and g, say P, p_1, p_2 and Q, q_1, q_2, the unique $u: Q \longrightarrow P$ with $q_1 = p_1 \circ u$ and $q_2 = p_2 \circ u$ is iso. The reader should state and prove precise analogues to Theorems 2.1 and 2.2.

If f and g have at least one pullback, we will use this notation to indicate one:

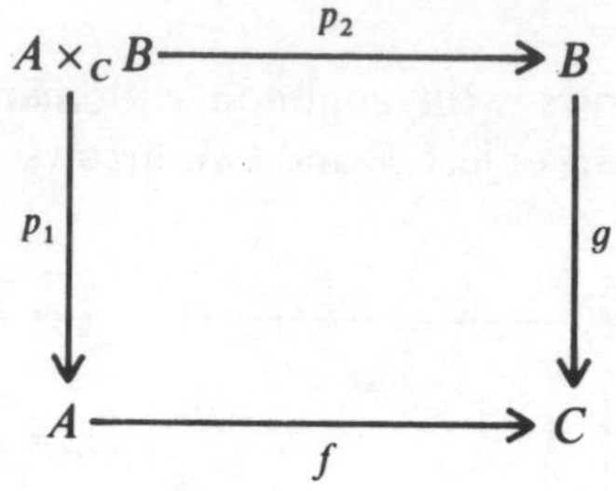

and we use the ordered pair notation $\langle h, k \rangle$ for the arrow that makes this diagram commute:

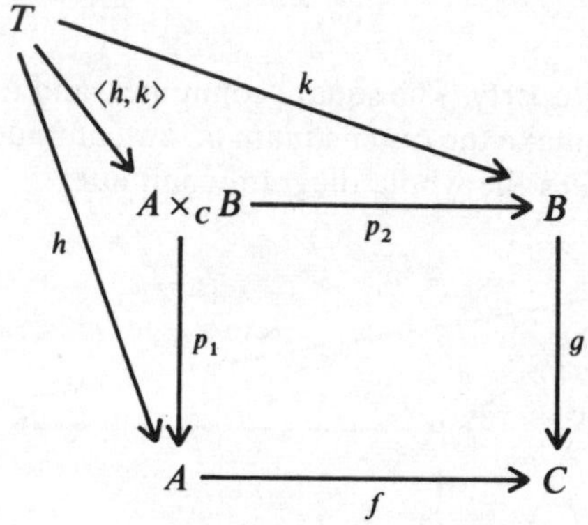

So, by definition $p_1 \circ \langle h, k \rangle = h$ and $p_2 \circ \langle h, k \rangle = k$.

We can prove that two arrows to a pullback, say $r, s : Q \longrightarrow A \times_C B$, are equal by showing that $p_1 \circ r = p_1 \circ s$ and $p_2 \circ r = p_2 \circ s$.

In terms of generalized elements, for any object T, a T-element of $A \times_C B$ is an ordered pair $\langle x, y \rangle$ with $x \in_T A$ and $y \in_T B$ and $f(x) = g(y)$. This makes $A \times_C B$ look like a sub-object of the product $A \times B$, in fact an equalizer, and it is so if the product $A \times B$ exists:

THEOREM 4.5   Take any corner of arrows $f : A \longrightarrow C$ and $g : B \longrightarrow C$ and suppose that there is a product $A \times B$. If there is an equalizer for the arrows $f \circ p_1$ and $g \circ p_2$,

$$E \rightarrowtail \xrightarrow{\quad e \quad} A \times B \; \begin{array}{c} \xrightarrow{\quad p_1 \quad} A \xrightarrow{\quad f \quad} \\ \\ \xrightarrow{\quad p_2 \quad} B \xrightarrow{\quad g \quad} \end{array} \; C$$

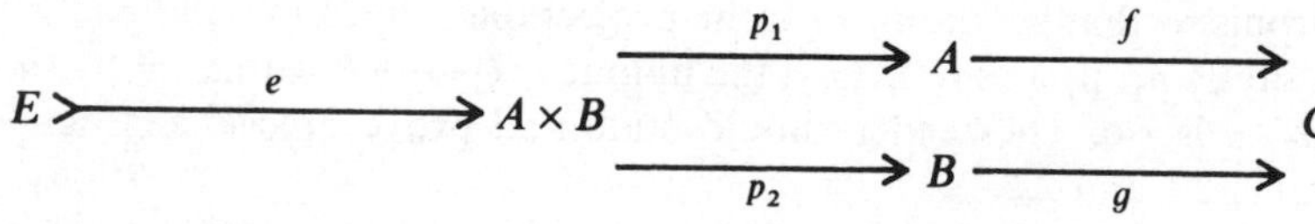

then this is a pullback (with p_1 and p_2 as in the equalizer):

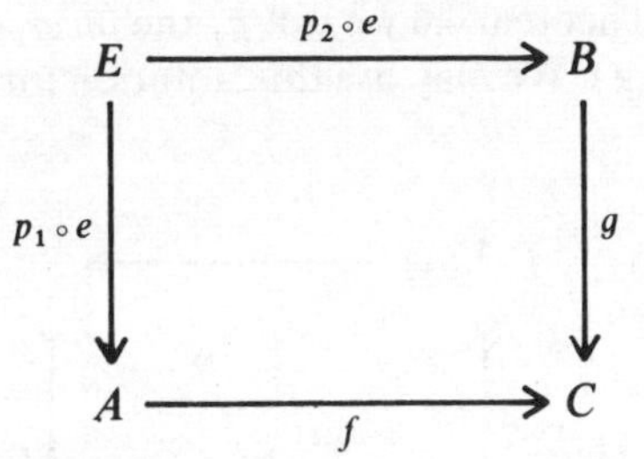

Conversely, if there is a pullback for f and g, then it has an arrow to $A \times B$, and that arrow is an equalizer as shown above.

PROOF Take any T and arrows $h: T \longrightarrow A$ and $k: T \longrightarrow B$, with $f \circ h = g \circ k$. Then $\langle h, k \rangle: T \longrightarrow A \times B$ factors through the equalizer e by a unique $u: T \longrightarrow E$. This u is also the unique arrow showing that the square is a pullback. The converse is similar. □

Thus any category with products for all pairs of objects and equalizers for all parallel pairs of arrows also has pullbacks for all corners of arrows.

4.3 Guises of pullbacks

Pullback is a powerful flexible construction which appears throughout mathematics. We will use two special cases in particular, both of which depend on this theorem:

THEOREM 4.6 If $g: B \rightarrowtail C$ is monic and the following is a pullback square:

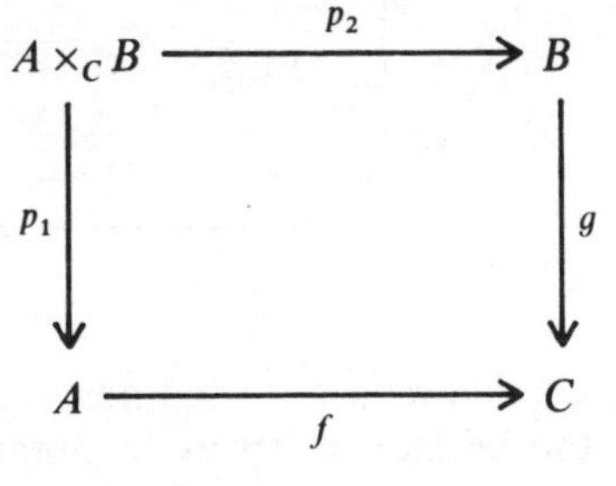

then p_1 is also monic.

PROOF Suppose that $h, k: T \longrightarrow A \times_C B$ have $p_1 \circ h = p_1 \circ k$. Since $f \circ p_1 = g \circ p_2$, we have $g \circ p_2 \circ h = g \circ p_2 \circ k$. Since g is monic, $p_2 \circ h = p_2 \circ k$. Thus h and k have the same composite with both projections, and so $h = k$. □

In the situation of Theorem 4.6 we call p_1 the *inverse image* of g along f, or the *preimage* of g along f. We may use this notation for a pullback of a monic $i: S \rightarrowtail C$ along f:

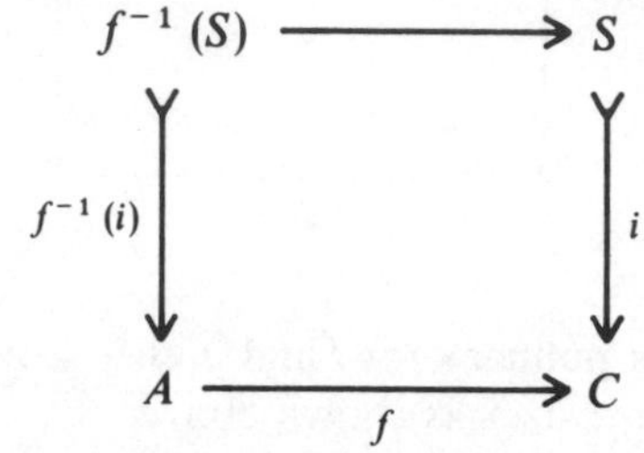

where $f^{-1}(i)$ does not refer to any inverse arrow to f, but to the inverse image of i along f. The top arrow has no special name.

The term 'inverse image' is justified by the following:

THEOREM 4.7 For any object T and any $x \in_T A$, we have $x \in f^{-1}(i)$ iff $f(x) \in i$.

PROOF Suppose that all the arrows in this diagram exist as shown, except possibly h and k:

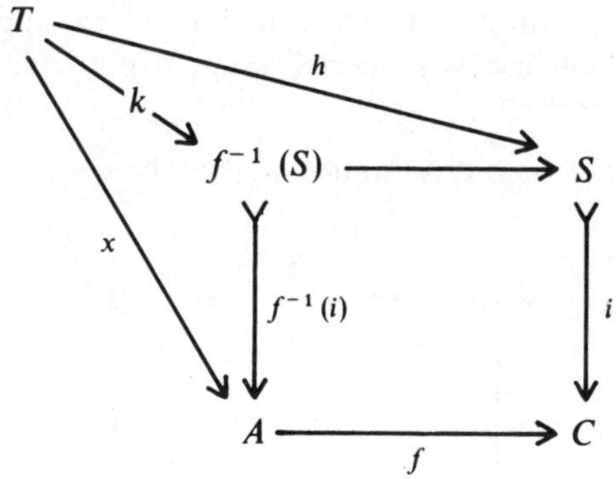

Then if there is an arrow k that makes the leftmost triangle commute, it can be composed with the top horizontal arrow to obtain h, thus making the outer square commute; while by definition of a pullback if there is an h that makes the outer square commute then there is a unique k that makes the whole diagram commute. □

When both arrows $i: S \rightarrowtail A$ and $j: T \rightarrowtail A$ are monic and have at least one pullback, we may refer to their pullback as their *intersection*, and use this notation:

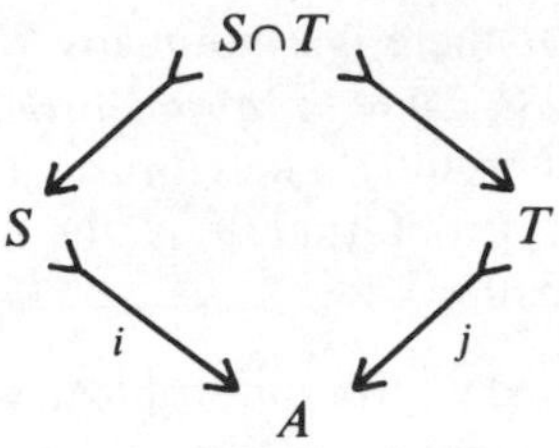

The composites along either side from $S \cap T$ to A are equal, and monic, and we call that monic $i \cap j$. This is an intersection because, for any object T, any $x \in_T A$ has $x \in i \cap j$ iff $x \in i$ and $x \in j$. This is just the definition of a pullback rewritten in this notation.

4.4 Theorems on pullbacks

Three theorems on pullbacks will be especially important.

THEOREM 4.8 Suppose that the diagram below commutes, and that the right-hand square is a pullback (for j and g). Then the left-hand square is a pullback (for i and f) iff the outer rectangle is a pullback (for j and $g \circ f$):

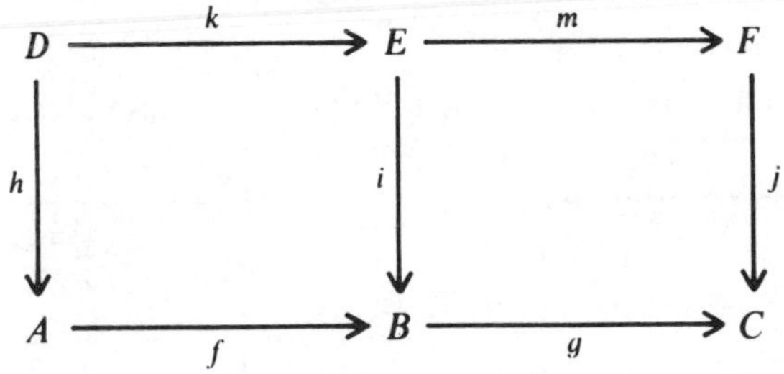

PROOF The reader should draw diagrams following the steps of this proof.

First, suppose that the left-hand square is a pullback and consider any $r: T \longrightarrow F$ and $s: T \longrightarrow A$, with $j \circ r = g \circ f \circ s$. Since the right-hand square is a pullback there is a unique $u: T \longrightarrow E$ with $r = m \circ u$ and $f \circ s = i \circ u$. Then, by the left-hand pullback, there is a unique $v: T \longrightarrow D$ with $u = k \circ v$ and $s = h \circ v$. Thus $r = (m \circ k) \circ v$ and $s = h \circ v$. To show that v is the unique arrow with these composites with $m \circ k$ and h, suppose that $w: T \longrightarrow D$ has $r = (m \circ k) \circ w$ and $s = h \circ w$. Then $k \circ w$ has the same composites with m and i as u does, and so by uniqueness $u = k \circ w$. So w has the same composites with h and k as v does, and so $w = v$. Thus the outer square is a pullback for j and $g \circ f$.

Conversely, suppose that the outer square is a pullback and consider any $q: T \longrightarrow E$ and $t: T \longrightarrow A$ with $i \circ q = f \circ t$. Then $m \circ q$ and t have composites

$j \circ (m \circ q) = (g \circ f) \circ t$, so that there is a unique $u\colon T \longrightarrow D$ with $t = h \circ u$ and $m \circ q = (m \circ k) \circ u$. Thus $m \circ (k \circ u) = m \circ q$ and $i \circ (k \circ u) = f \circ t$, but then q has these same composites with m and i. Therefore $k \circ u = q$; and we have $t = h \circ u$ and $q = k \circ u$, as desired. Proof that u is the unique arrow with these composites is left to the reader. □

Since pullbacks are defined up to isomorphism, we can restate Theorem 4.8 as saying that pulling back along $g \circ f$ gives the same result, up to a unique isomorphism, as pulling back along g and then pulling the result back along f. If we have selected a pullback for each of the corners that we are discussing, it might happen that i is the selected pullback of j along g, and h is the selected pullback of i along f, but that h is not the selected pullback of j along $g \circ f$. But that is a detail. The results are the same up to isomorphism.

THEOREM 4.9 Suppose that there is a commutative triangle

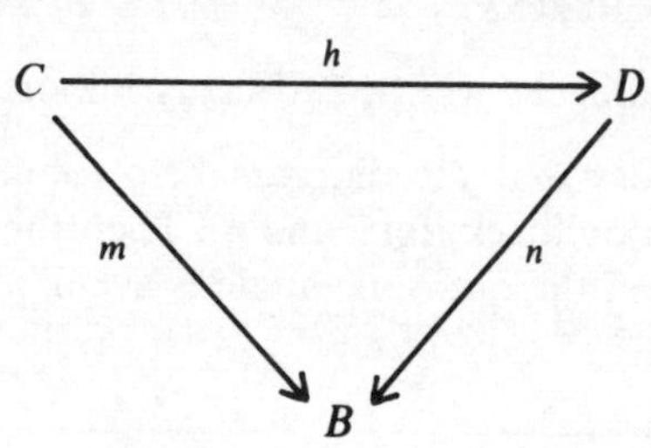

and suppose that m and n have the following pullbacks along f:

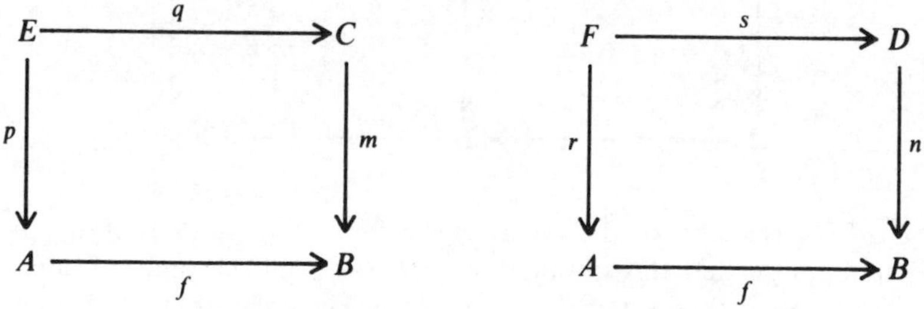

Then there is a unique $u\colon E \longrightarrow F$ that makes the following diagram commute:

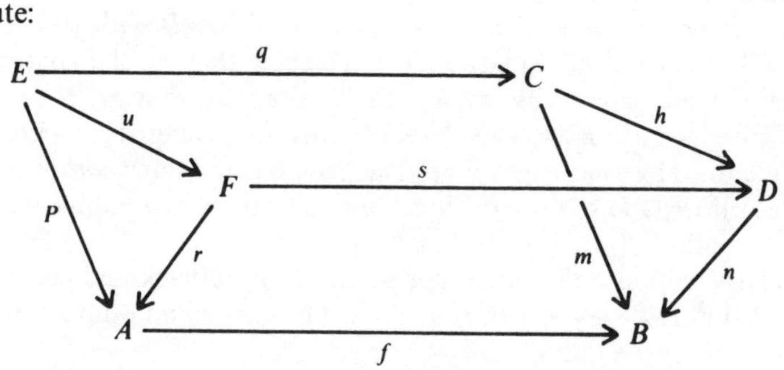

and then E, u, q is a pullback for h and s.

PROOF Consider the arrows $h \circ q$ and p. Since $n \circ (h \circ q) = m \circ q = f \circ p$ and F is a pullback, there is a unique $u: E \longrightarrow F$ with $p = r \circ u$ and $h \circ q = s \circ u$. That is enough to make the whole diagram commute. Then Theorem 4.8 shows that E, u, q is a pullback for h and s, as follows. Take r, s, n, f as the right-hand square, and q, u, h, s as the left-hand square. Since $p = r \circ u$ and $m = n \circ h$, the outer square is q, p, m, f, which is a pullback by assumption. □

In words, a commutative triangle over B pulls back along f to a unique commutative triangle over A, assuming that the two arrows to B themselves have pullbacks along f.

THEOREM 4.10 Suppose that each of the following diagrams is a pullback:

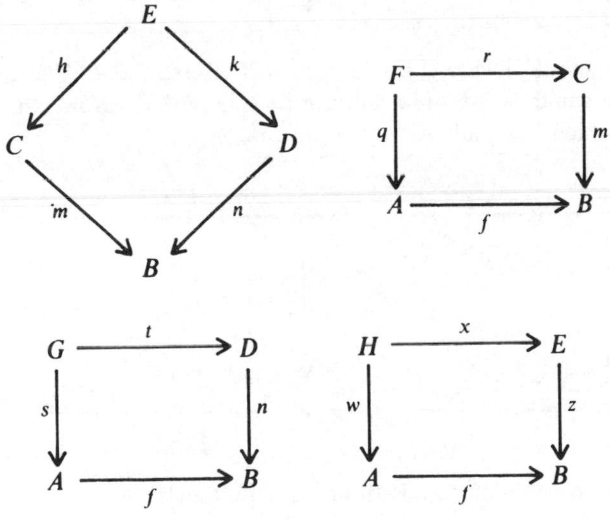

where $z = m \circ h$ (and, since the first square commutes, $z = n \circ k$). Then there are unique arrows u and v that make the following diagram commute:

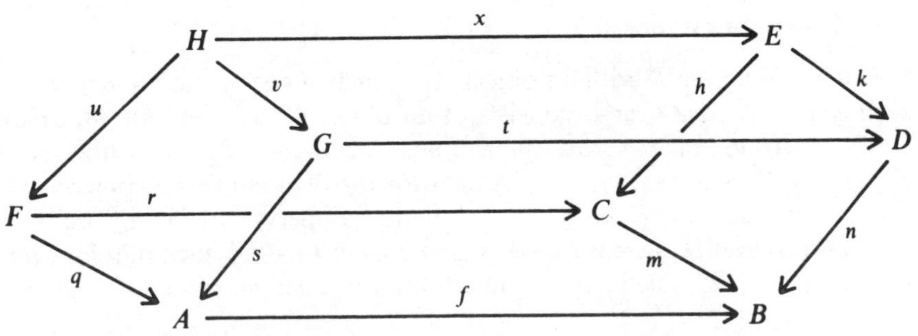

and every square in the diagram is a pullback. In particular, H, u, v is a pullback of s and q.

PROOF To obtain u apply Theorem 4.9 to m, h, and z. To obtain v apply Theorem 4.9 to n, k, and z. That theorem also shows that the squares x, u, h, r and x, v, k, t are pullbacks. So by Theorem 4.8 the outer square here is a pullback:

$$\begin{array}{ccccc} H & \xrightarrow{\;x\;} & E & \xrightarrow{\;k\;} & D \\ {\scriptstyle u}\downarrow & & \downarrow{\scriptstyle h} & & \downarrow{\scriptstyle n} \\ F & \xrightarrow[\;r\;]{} & C & \xrightarrow[\;m\;]{} & B \end{array}$$

The reader should locate this diagram in the larger one. The outer square there is the same as the outer square in this next diagram, the right-hand square of which is a pullback by assumption:

$$\begin{array}{ccccc} H & \xrightarrow{\;v\;} & G & \xrightarrow{\;t\;} & D \\ {\scriptstyle u}\downarrow & & \downarrow{\scriptstyle s} & & \downarrow{\scriptstyle n} \\ F & \xrightarrow[\;q\;]{} & A & \xrightarrow[\;f\;]{} & B \end{array}$$

By Theorem 4.8 the left-hand square is a pullback. □

In words, a pullback square over B pulls back along f to give a pullback square over A, if each of the arrows to B has a pullback along f.

4.5 Cones and limits

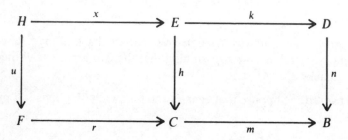

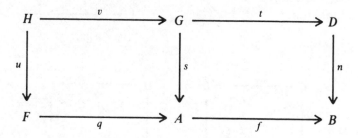

Consider a diagram D with an object A_i at each vertex i, and an arrow f_e at each edge e. A *cone* over D consists of an object C together with an arrow $p_i: C \longrightarrow A_i$ to each vertex such that, for every edge e with arrow $f_e: A_i \longrightarrow A_j$ we have $f_e \circ p_i = p_j$. A *limit* for the diagram D is a cone over D, say C with arrows p_i, with the following property: given any cone T, $q_i: T \longrightarrow A_i$ over D, there is exactly one arrow $u: T \longrightarrow C$ such that for every vertex i, $p_i \circ u = q_i$. The arrows p_i of a limit are called *projection arrows*. The diagram D may have any number of limits, but limits are defined up to a unique isomorphism that gets along with the projection arrows.

For example, let the diagram have two vertices and no edges:

$$A \qquad\qquad B$$

A cone over it is an object and two arrows:

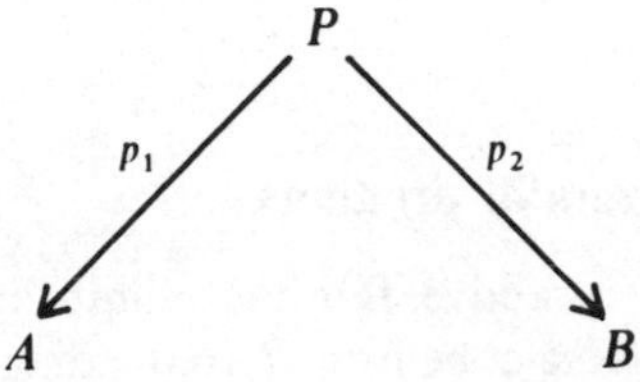

and this cone is a limit iff it is a product diagram for A and B. This generalizes to products for any number of objects, including terminal objects as limits of empty diagrams.

Let the diagram be:

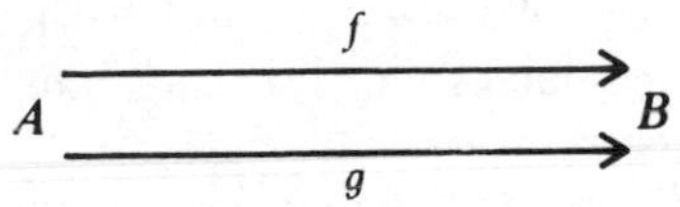

A cone over it is

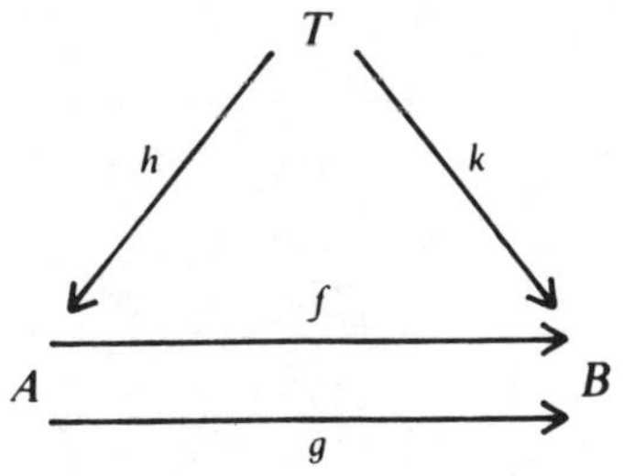

with $f \circ h = k$ and $g \circ h = k$; that is, any $h\colon T \longrightarrow A$ with $f \circ h = g \circ h$ determines a unique cone. The cone is a limit iff h is an equalizer for f and g. Similarly, a pullback is a limit for a diagram:

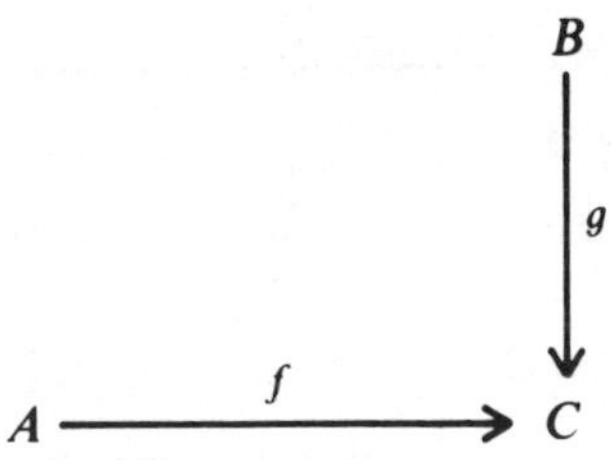

A *cocone* on the diagram D is an object C together with arrows $q_i: A_i \longrightarrow C$ for each vertex i, such that for every edge e with arrow $f_e: A_i \longrightarrow A_j$ we have $q_i \circ f_e = p_j$. A *colimit* for the diagram D is a cone on D, say C with arrows p_i, with the following property: given any cone T, $r_i: A_i \longrightarrow T$ on D, there is exactly one arrow $u: C \longrightarrow T$ such that for every vertex i, $u \circ q_i = r_i$. These are dual to the definitions of cones and limits, so initial objects, coproducts, coequalizers, and pushouts (see Exercise 4.10) are all colimits.

4.6 Limits as equalizers of products

Consider the diagram D as above. If it has a limit then for any T a T-element of the limit corresponds to a cone from T to the diagram; that is, a T-element $x_i \in_T A_i$ for each vertex i, such that for every $f_e: A_j \longrightarrow A_k$ we have $f_e(x_j) = x_k$. Thus it seems that the limit ought to be a kind of equalizer of arrows from the product of all the A_i, and it is so if the requisite products exist.

Suppose that there is a product $\Pi_i A_i$ of all the A_i, with projection $p_j: \Pi_i A_i \longrightarrow A_j$ for each vertex j. Use the notation A_e for the codomain of the arrow at edge e, so that if $f_e: A_j \longrightarrow A_k$ then $A_e = A_k$. Suppose that there is a product $\Pi_e A_e$ with one factor A_e for each edge e, and a projection $p_e: \Pi_e A_e \longrightarrow A_k$ for each $f_e: A_j \longrightarrow A_k$.

Let $r: \Pi_i A_i \longrightarrow \Pi_e A_e$ be the arrow such that, for each edge e with $f_e: A_j \longrightarrow A_k$, the following triangle commutes:

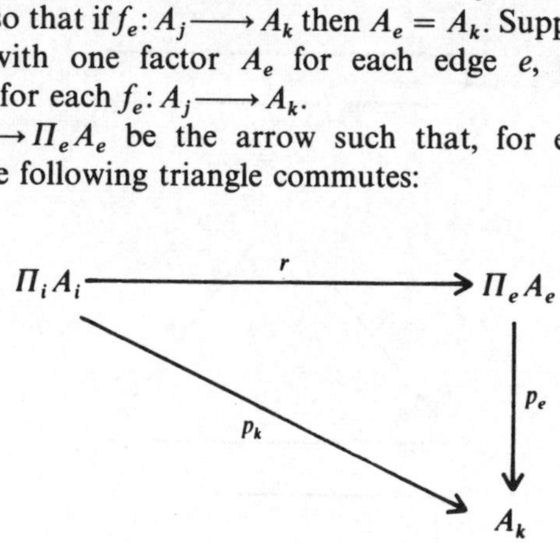

Let $s: \Pi_i A_i \longrightarrow \Pi_e A_e$ be the arrow such that for each edge e the following square commutes:

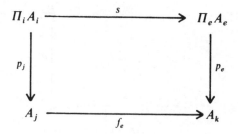

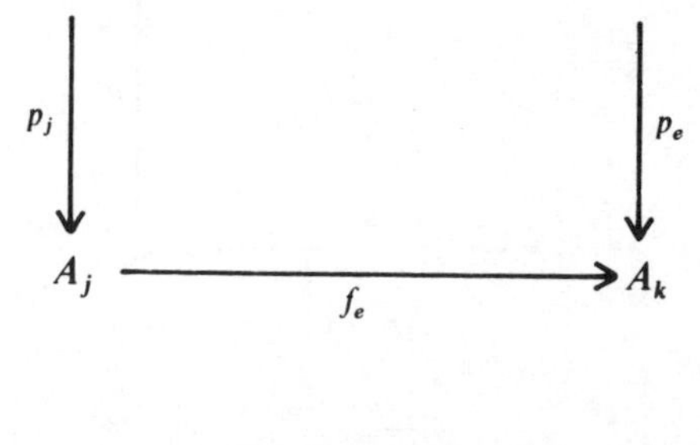

Then D has a limit iff there is an equalizer for r and s. More specifically, $l: L \longrightarrow \Pi_i A_i$ is an equalizer for r and s iff L together with all $p_i \circ l: L \longrightarrow A_i$ forms a limit cone for D.

For proof, consider any cone T, $h_i: T \longrightarrow A_i$ over D. There is a unique $h: T \longrightarrow \Pi_i A_i$ such that for each vertex k we have $p_k \circ h = h_k$. Then $r \circ h = s \circ h$ iff for every edge e, $p_e \circ r \circ h = p_e \circ s \circ h$. But, by the definitions of r, s, and h, that last equation means that $h_k = f_e \circ h_j$, where $f_e: A_j \longrightarrow A_k$. Thus T, h factors uniquely through L, l iff the cone T, h_i factors uniquely through the cone L, $p_i \circ l$.

The dual result for colimits is obvious.

This result applies to any size of diagram. In particular:

THEOREM 4.11 If our category has a terminal object, binary products, and an equalizer for every parallel pair of arrows, then it has a limit for every finite diagram. □

When the conditions for Theorem 4.11 are met we say that our category has *all finite limits.* By Exercises 4.5 and 4.6, our category has all finite limits iff it has a terminal object and pullbacks for all corners of arrows.

Exercises

4.1 Pullback preserves inclusion of sub-objects. Suppose that $i: S \rightarrowtail C$ and $j: T \rightarrowtail C$ both have pullbacks along $f: A \longrightarrow C$:

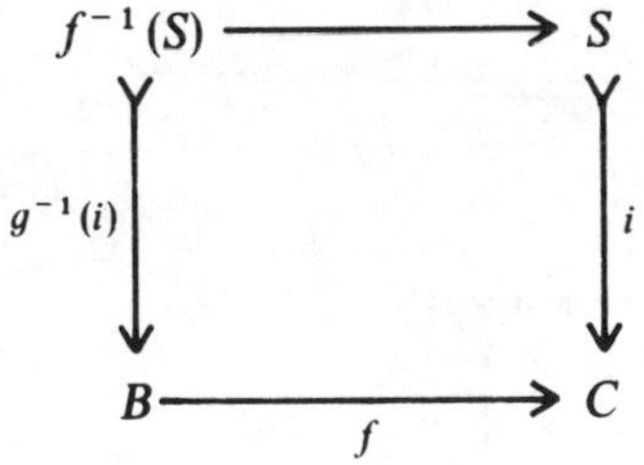

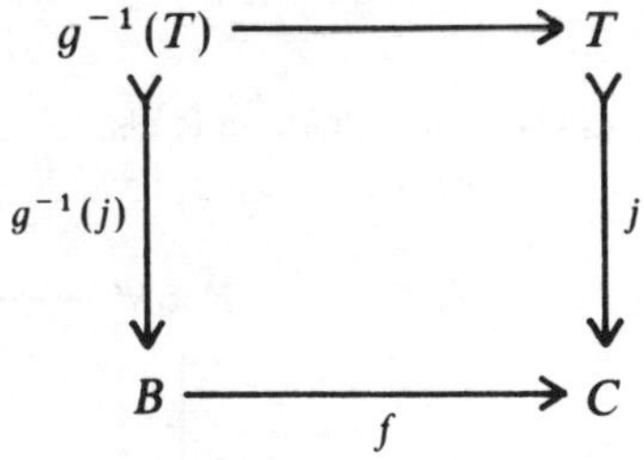

Use the theorems on pullbacks to show that if $i \subseteq j$ then $f^{-1}(i) \subseteq f^{-1}(j)$ and if $i \equiv j$ then $f^{-1}(i) \equiv f^{-1}(j)$.

Show that if i has a pullback along f then every sub-object h of C with $i \equiv h$ also has a pullback along f.

4.2 Prove for any $f: A \longrightarrow B$ in any category that this is a pullback:

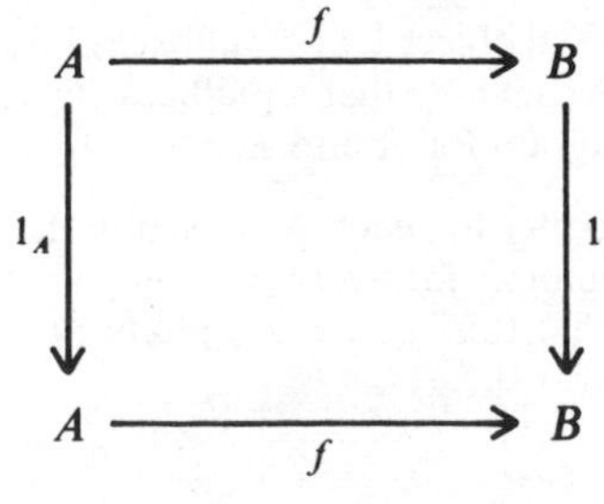

4.3 Show that for any arrow $f: A \longrightarrow B$ and object C, if A and B each have a product with C then this square is a pullback, where the arrows 'p_1' are both projection arrows for the products:

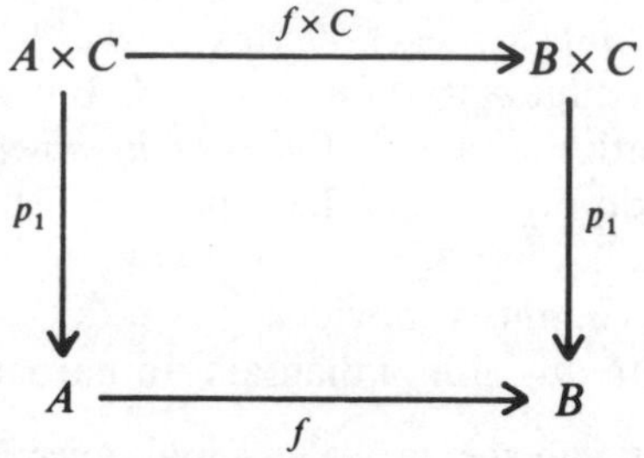

4.4 Assuming that every two objects have a product, suppose that both of these squares are pullbacks:

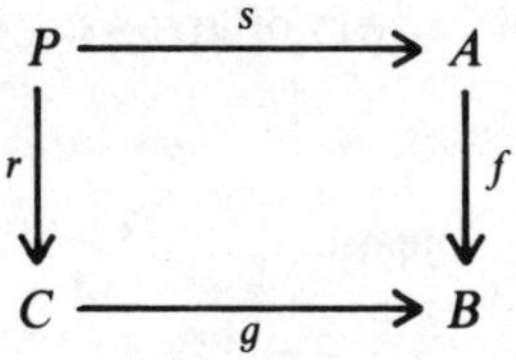

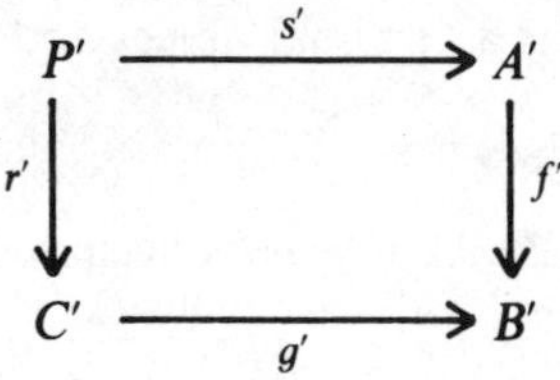

and show that this one is also:

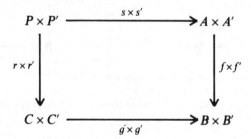

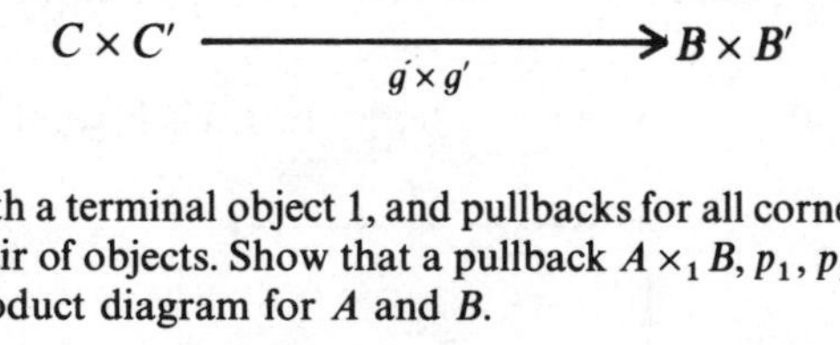

4.5 A category with a terminal object 1, and pullbacks for all corners of arrows, has a product for each pair of objects. Show that a pullback $A \times_1 B, p_1, p_2$ for arrows $!_A$ and $!_B$ to 1 is also a product diagram for A and B.

4.6 A category with a product for each pair of objects, and a pullback for every corner of arrows, has an equalizer for every parallel pair. Consider any parallel pair $f, g: A \longrightarrow B$ and suppose that this square is a pullback:

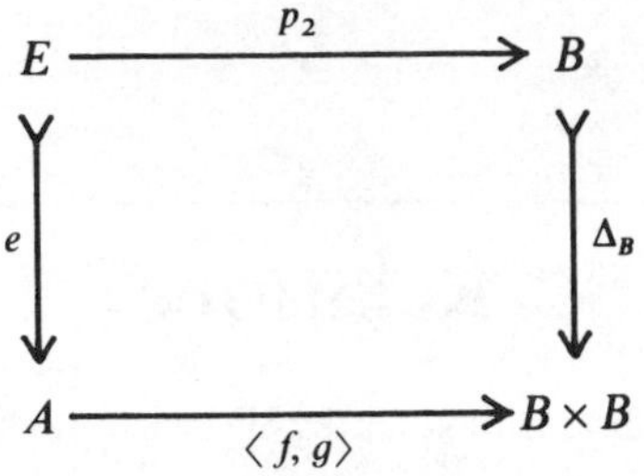

where Δ_B is the diagonal arrow defined in Exercise 2.5. Show that e is an equalizer for f and g.

4.7 Assume that A and B have a product and consider an arrow $f: A \longrightarrow B$ and its graph Γ_f as defined in Exercise 2.5. Show, for any T and $x \in_T A$ and $y \in_T B$, $\langle x, y\rangle \in \Gamma_f$ iff $y = f(x)$.

4.8 Suppose that $h: C \rightarrowtail D$ is monic and take any $f: A \longrightarrow C$ and $g: B \longrightarrow C$. Show that any P, p_1, p_2 form a pullback diagram for f and g iff they form a pullback diagram for $h \circ f$ and $h \circ g$.

4.9 Show that an arrow $h: A \longrightarrow B$ is monic iff A, 1_A, 1_A is a pullback of h along itself. (A sub-object is its intersection with itself.) Conclude that h is monic iff it has some pullback with itself with $p_1 = p_2$, and in that case $p_1 = p_2$ in every pullback diagram for f and itself.

4.10 The dual to a pullback is a *pushout*. Give the precise definition by reversing the direction of all the arrows (including h, k, and u) in the diagram defining pullbacks.

Dualize the result on pullbacks to show how a pushout is a coequalizer of a coproduct. Consider the duals to the other facts that have been proved about pullbacks. We adapt the coproduct notations to pushouts just as we did the product notations for pullbacks.

4.11 Compare the construction of a pullback as an equalizer of a product in Theorem 4.5 with the one given as a case of Theorem 4.11.

4.12 (Limits preserve products or, products preserve limits.) Suppose that L, $l_i: L \longrightarrow A_i$ is a limit for a diagram A_i, f_e. Suppose that L', $l'_i: L' \longrightarrow A_i$ is a limit for another diagram A'_i, f'_e with the same vertices i and edges e (but not necessarily the same object at each vertex or arrow at each edge). Show that $L \times L'$ and $p_i \times p'_i$ form a limit for the diagram with $A_i \times A'_i$ at each vertex i, and $f_e \times f'_e$ at each edge e. Compare this with Exercise 4.4.

5

Relations

In this chapter we assume a base category with all finite limits.

5.1 Definition

By a *relation* from A to B we mean a sub-object of $A \times B$, say $r: R \rightarrowtail A \times B$. The idea is that, for any object T and $x \in_T A$ and $y \in_T A$, x is in the relation r to y iff $\langle x, y \rangle \in r$. For example, the diagonal arrow for any object, $\Delta_B: B \rightarrowtail B \times B$, now appears as the equality relation on B. For any object T and $x \in_T B$ and $y \in_T B$, $\langle x, y \rangle \in \Delta_B$ iff $x = y$. Any relation $r: R \rightarrowtail A \times B$ has a *converse* $r^\circ: R \rightarrowtail B \times A$ defined by composing with the twist arrow, $r^\circ = tw \circ r$. So, for any $y \in_T B$ and $x \in_T A$, $\langle y, x \rangle \in r^\circ$ iff $\langle x, y \rangle \in r$.

There is another way to look at relations. A pair of arrows with common domain, $v: S \longrightarrow A$ and $w: S \longrightarrow B$, is *jointly monic* if: for any T and parallel pair of arrows $h, k: T \longrightarrow S$, if $v \circ h = v \circ k$ and also $w \circ h = w \circ k$ then $h = k$:

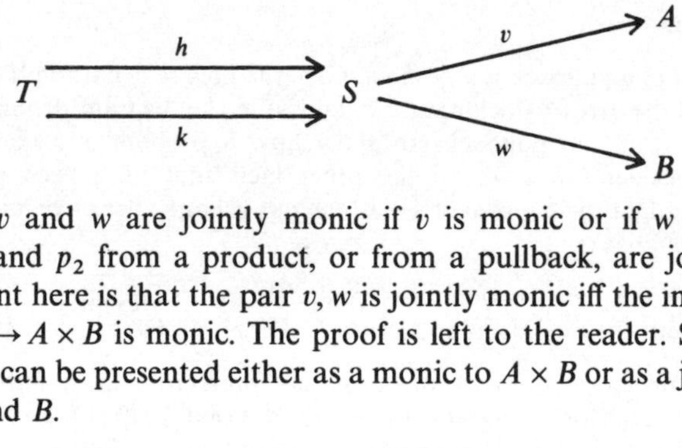

Obviously, v and w are jointly monic if v is monic or if w is. The projections p_1 and p_2 from a product, or from a pullback, are jointly monic. The key point here is that the pair v, w is jointly monic iff the induced arrow $\langle v, w \rangle: S \longrightarrow A \times B$ is monic. The proof is left to the reader. So a relation from A to B can be presented either as a monic to $A \times B$ or as a jointly monic pair to A and B.

5.2 Equivalence relations

A relation $r: R \rightarrowtail A \times A$ is called *reflexive* if $\Delta_A \subseteq r$. In a diagram:

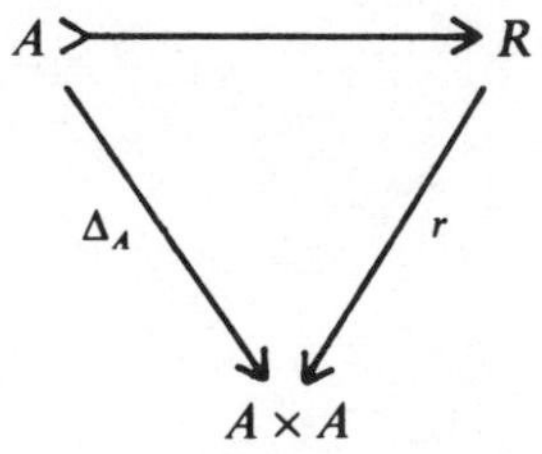

In terms of members, r is reflexive iff, for every T and $x \in_T A$, $\langle x, x\rangle \in r$. (For a quick proof consider the generic element 1_A.) Every T-element of A bears r to itself.

A relation $r: R \rightarrowtail A \times A$ is *symmetric* if $r^{\circ} \subseteq r$. The diagram that expresses this should be obvious. In terms of members then, r is symmetric iff $\langle x, y\rangle \in r$ implies $\langle y, x\rangle \in r$.

We call r *transitive* if, for every T and x, y, and z all T-elements of A, $\langle x, y\rangle \in r$ and $\langle y, z\rangle \in r$ implies that $\langle x, z\rangle \in r$. (See Exercise 5.1 for a diagram.)

For any arrow $f: A \longrightarrow B$ the projection arrows in a pullback of f along itself are called a *kernel pair* for f:

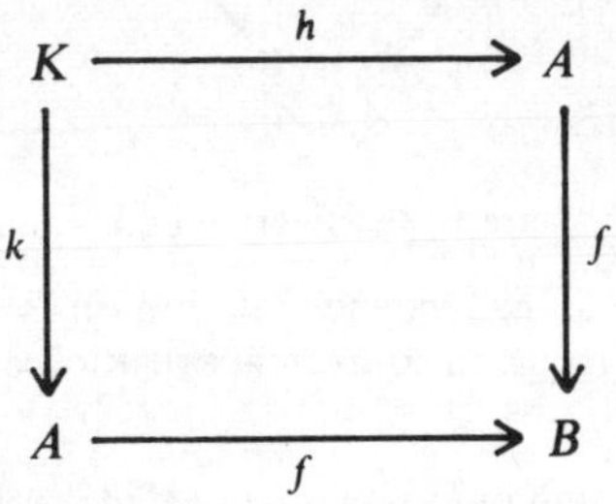

Kernel pairs are determined up to isomorphism. Perhaps the most germane way to express this precisely is to notice that the kernel pair h, k is jointly monic, so that $\langle h, k\rangle$ is a sub-object of $A \times A$. Then any pair $h': K' \longrightarrow A$ and $k': K' \longrightarrow A$ is a kernel pair for f iff $\langle h, k\rangle \equiv \langle h', k'\rangle$ as sub-objects of $A \times A$.

In fact, we can describe a kernel pair in terms of its members. It is just a restatement of the definition to say that $\langle h, k\rangle$ is a kernel pair for $f: A \longrightarrow B$ iff the following holds: for every T and every x and y T-elements of A, $\langle x, y\rangle \in \langle h, k\rangle$ iff $f(x) = f(y)$. It follows that every kernel pair is an *equivalence relation*. It is reflexive, symmetric, and transitive.

On the other hand, an equivalence relation is called *effective* if it is the kernel pair of some arrow.

Exercises

5.1 Verify that the pullback T below is the sub-object of $A \times A \times A$, which at any stage of definition contains all triples $\langle x, y, z\rangle$ such that x has r to y, and y has r to z.

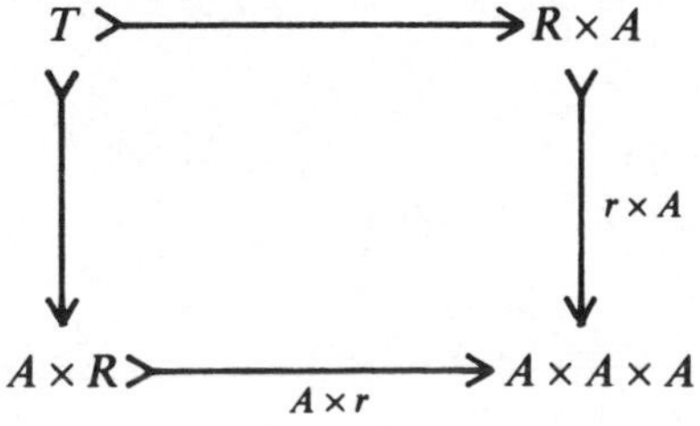

(The scrupulous reader will want to write the lower right corner as either $A \times (A \times A)$ or $(A \times A) \times A$ and include an associativity arrow in either the vertical or horizontal arrow to it accordingly.)

Show that r is transitive iff there is an arrow h that makes the following diagram commute:

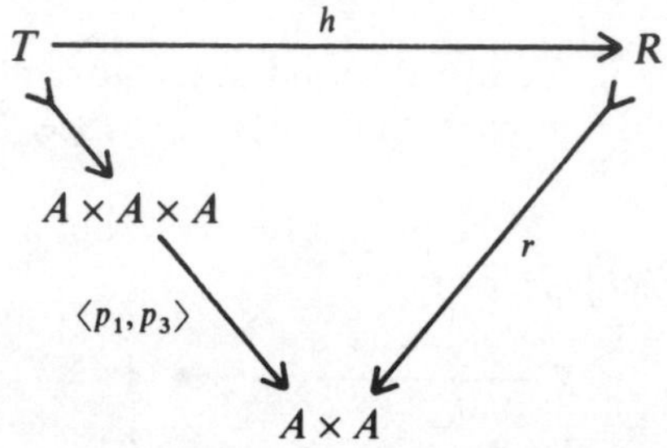

5.2 Show that Δ_A and $1_{(A \times A)}$ are both effective equivalence relations on A.

5.3 Let $r: R \rightarrowtail B \times B$ be an equivalence relation on B. For any $f: A \longrightarrow B$, show that the pullback of r along $f \times f$ is an equivalence relation on A. Describe it in terms of generalized elements of A.

5.4 Show that the intersection of any two equivalence relations on A is an equivalence relation.

5.5 Take any arrow $f: A \longrightarrow B$ and monic $i: B \rightarrowtail C$. Show that h, k is a kernel pair for f iff it is a kernel pair for $i \circ f$.

5.6 Let $\langle h, k \rangle$ be a kernel pair for $f: A \longrightarrow B$. For any object T and $x \in_T A$ and $y \in_T A$ we know that $f(x) = f(y)$ iff there is some $z \in_T K$ with $h(z) = x$ and $k(z) = y$, and in that case z is unique. So, in a sense, B can be thought of as 'including' a certain quotient of A. Exercise 5.5 shows that there is no reason to think that B is a close fit to that quotient.

Precisely, we define a *quotient* for an equivalence relation $\langle h, k \rangle$ to be a coequalizer for h and k which has them as its kernel pair. Show that an arrow $q: A \longrightarrow B$ is a coequalizer for some parallel pair into A iff it is a quotient for its own kernel pair.

6

Cartesian closed categories

This chapter assumes a base category with finite limits, and after defining exponentials we assume that every two objects have one. An intuitive discussion and examples follow in Section 6.5.

6.1 Exponentials

Given objects A and B, an *exponential of B by A* consists of an object I and an arrow $e: I \times A \longrightarrow B$ with the following property. For any object C and arrow $g: C \times A \longrightarrow B$ there is a unique arrow $\bar{g}: C \longrightarrow I$ that makes this triangle commute:

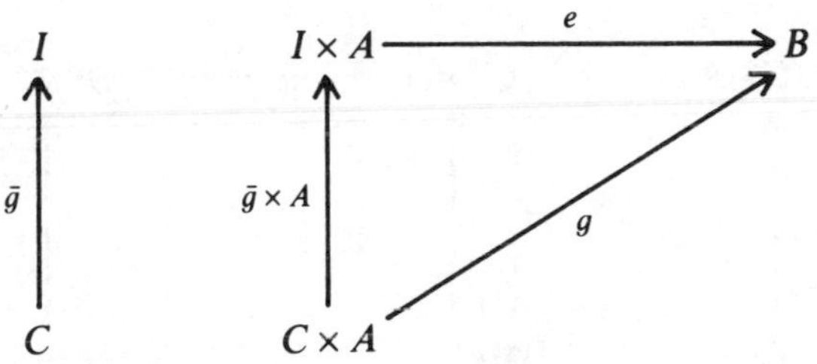

Exponentials are unique up to isomorphism (see Exercise 6.1). A category is called *Cartesian closed* if it has all finite products, and each two objects in it have an exponential. In this chapter our base category is Cartesian closed.

For any objects A and B we write B^A, and $ev_{A,B}: B^A \times A \longrightarrow B$ to indicate an exponential of B by A. We usually omit the subscripts on $ev_{A,B}$. Exponentials are also called function sets, function spaces, or internal homs.

The idea is that B^A represents arrows from A to B. An arrow $A \longrightarrow B$ is (up to the natural isomorphism $A \cong 1 \times A$) an arrow $1 \times A \longrightarrow B$, and that gives a unique arrow $1 \longrightarrow B^A$. But we have to make this correspondence between arrows $A \longrightarrow B$ and global elements of B^A precise, and extend it to generalized elements of B^A.

For each $g: C \times A \longrightarrow B$, call the corresponding $\bar{g}: C \longrightarrow B^A$ the *transpose* of g. Each $g: C \times A \longrightarrow B$ uniquely determines its transpose, by definition, but it is also determined by its transpose since the definition says that $g = ev \circ (\bar{g} \times A)$. Furthermore, every $f: C \longrightarrow B^A$ is the transpose of an arrow (and thus of a unique arrow) from $C \times A$ to B, namely $ev \circ (f \times A)$. We will also call $ev \circ (f \times A)$ the *transpose* of f, and write it as $f: C \times A \longrightarrow B$. So arrows from $C \times A$ to B correspond exactly to arrows from C to B^A. We pass in either direction by forming the 'transpose', and the transpose of the transpose is the original arrow.

For any arrow $f: A \longrightarrow B$ the *name* of f, 'f': $1 \longrightarrow B^A$, is defined to be the transpose of

$$1 \times A \xrightarrow{\sim} A \xrightarrow{f} B$$

and it follows that every global element $x: 1 \longrightarrow B^A$ is the name of an arrow; specifically,

$$A \xrightarrow{\sim} 1 \times A \xrightarrow{\bar{x}} B$$

6.2 Internalizing composition

In addition to exponentiating objects, we can exponentiate an arrow by an object. For any arrow $f: B \longrightarrow C$ define $f^A: B^A \longrightarrow C^A$ to be the transpose of $f \circ ev: B^A \times A \longrightarrow C$. In other words, f^A is the unique arrow that makes the following square commute:

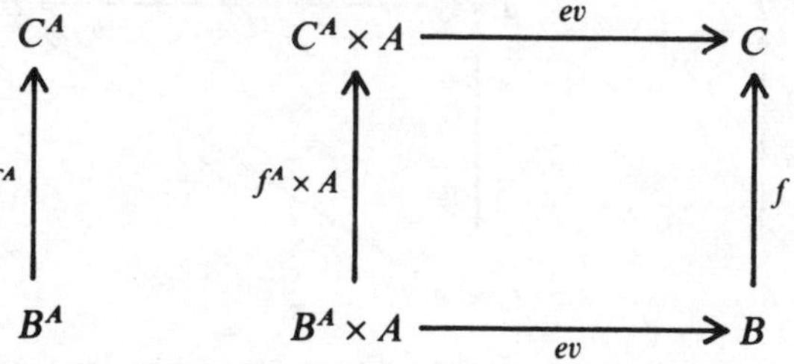

Intuitively, f^A takes any arrow from A to B and composes it with f, giving an arrow from A to C. We can now show this for global elements of B^A:

THEOREM 6.1 Take any $h: A \longrightarrow B$. Applying f^A to the name of h gives the name of $f \circ h$; that is, f^A('h') = '$f \circ h$'.

PROOF Consider the following diagram:

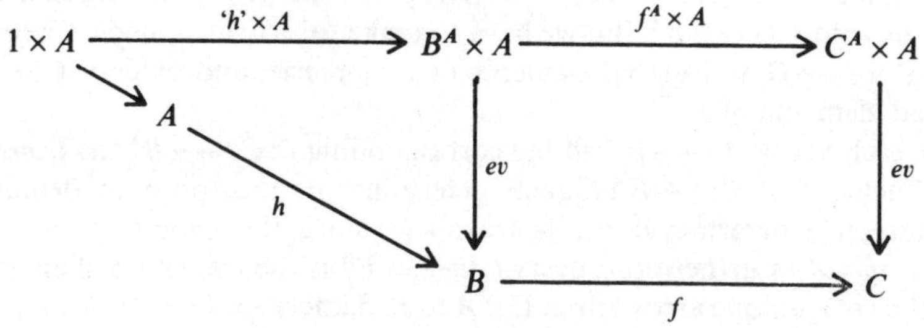

Composing along the top edge and down to C gives $ev \circ ((f^A \circ `h') \times A)$ which, by definition, is the transpose of $f^A \circ `h'$. But the triangle and the square commute by definition of 'h' and f^A respectively, and the composite along the bottom is the transpose of '$f \circ h$'. Since $f^A \circ `h'$ and '$f \circ h$' have the same transpose, they are the same arrow. □

We can say that f^A *internalizes* composition of arrows from A to B with f. Notice that, as usual, the choice of B^A, $ev_{A,B}$ and C^A, $ev_{A,C}$ among exponentials of B and C by A is arbitrary, but once it is made the arrow f^A is uniquely determined.

It is easy to see that, for any objects A and B, the exponential by A of the identity arrow on B, $(1_B)^A$, equals the identity arrow on B^A, $1_{(B^A)}$, since the identity arrow makes the relevant square commute. We say that exponentiation by any A *preserves identity arrows.*

It is not much harder to see, for any A and any $f: B \longrightarrow C$ and $g: C \longrightarrow D$, that $(g \circ f)^A = g^A \circ f^A$. Consider the following diagram:

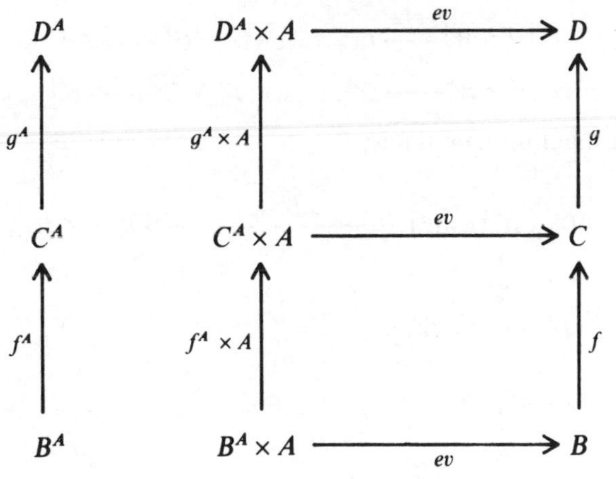

The upper and lower squares commute by definition, so the outer square commutes. But $(g^A \times A) \circ (f^A \times A) = (g^A \circ f^A) \times A$ trivially. And by definition $(g \circ f)^A$ is the only arrow the product of which with A makes the outer square commute. So $(g \circ f)^A = (g^A \circ f^A)$. We say that exponentiation by A *preserves composition.*

It is worth noticing that Theorem 6.1 has already shown that $(g \circ f)^A$ acts the same way on global elements 'h' of B^A as does $(g^A \circ f^A)$. Either arrow takes 'h' to '$g \circ f \circ h$'. But that is not enough to prove that the arrows are equal. We would have to know that they have the same effect on all generalized elements of B^A. We return to this point in Sections 6.5 and 6.6.

6.3 Further internalizing composition

Another arrow internalizes composition of arrows from A to B with arbitrary arrows from B to C. Define $c: C^B \times B^A \longrightarrow C^A$ as the transpose of:

$$(C^B \times B^A) \times A \xrightarrow{\sim} C^B \times (B^A \times A) \xrightarrow{C^B \times ev} C^B \times B \xrightarrow{ev} C$$

For any arrows $f: A \longrightarrow B$ and $g: B \longrightarrow C$, applying c to the names of f and g gives the name of $g \circ f$; that is, $c(\text{‘}g\text{’}, \text{‘}f\text{’}) = \text{‘}g \circ f\text{’}$. Here, $c(\text{‘}g\text{’}, \text{‘}f\text{’})$ abbreviates the official generalized element notation $c(\langle \text{‘}g\text{’}, \text{‘}f\text{’} \rangle)$. We will freely use this multivariable function notation to eliminate useless angle brackets. We could prove this directly by showing that both arrows have the same transpose. The transpose of ‘$g \circ f$’ is obvious, and the transpose of $c(\text{‘}g\text{’}, \text{‘}f\text{’})$ is found by taking its product with A and composing with ev. It helps to factor $\langle \text{‘}g\text{’}, \text{‘}f\text{’} \rangle$ as the composite $(\text{‘}g\text{’} \times B^A) \circ u \circ \text{‘}f\text{’}$, with u the unitary isomorphism for B^A. But we will derive the equation from a more general result in Section 6.6. There we will also show that internal composition is associative. Given composition arrows

$$c: C^B \times B^A \longrightarrow C^A \qquad c': D^C \times C^B \longrightarrow C^B$$

$$c'': D^B \times B^A \longrightarrow D^A \qquad c''': D^C \times C^A \longrightarrow C^A$$

the following diagram commutes:

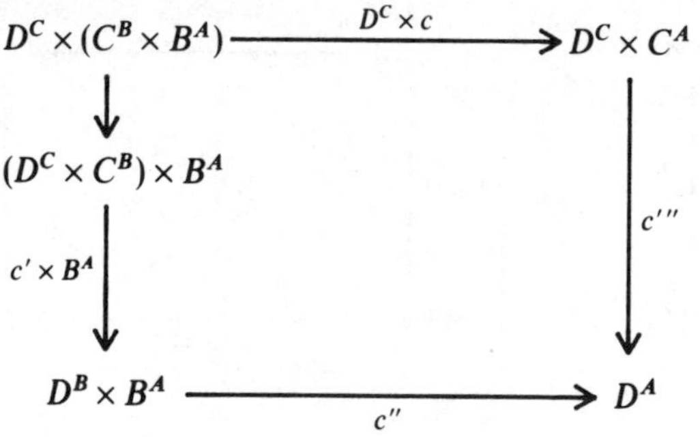

In terms of generalized elements, for any object T and any T-elements $\bar{s}$, $\bar{t}$, and $\bar{r}$ of D^C, C^B, and B^A respectively, we have $c'''(\bar{t}, c(\bar{s}, \bar{r})) = c''(c'(\bar{t}, \bar{s}), \bar{r})$. One could show directly that the transpose of $c'''(\bar{t}, c(\bar{s}, \bar{r}))$ is $ev \circ (D^C \times ev_{B,C}) \circ (D^C \times C^B \times ev)$; and that the transpose of $c''(c'(\bar{t}, \bar{s}), \bar{r})$ is the same.

6.4 Initial objects and pushouts

Suppose that our Cartesian closed category also has an initial object $\emptyset$. The next two theorems show how it is reasonable to think of $\emptyset$ as an empty object.

THEOREM 6.2 For every object A, $\emptyset \times A \cong \emptyset$. By symmetry, $A \times \emptyset \cong \emptyset$. And for any object B, $B^{\emptyset} \cong 1$.

PROOF For each object B there are as many arrows from $\emptyset \times A$ to B as there are from $\emptyset$ to B^A; that is, exactly one. Proof of $B^{\emptyset} \cong 1$ is left to the reader. □

THEOREM 6.3 If an object A has an arrow $f: A \longrightarrow \emptyset$ then $A \cong \emptyset$.

PROOF The graph of f is a monic $\Gamma_f: A \rightarrowtail A \times \emptyset$. By Theorem 6.2 this gives a monic from A to $\emptyset$. But any arrow from A to $\emptyset$ is split epic, with $!_A$ as right inverse. So the monic is iso. □

In any category (whether Cartesian closed or not) a *strict initial object* is an initial object such that every object with an arrow to it is also initial. Clearly, if one initial object in a given category is strict then all in that category are. Theorem 6.3 says that initial objects in a Cartesian closed category are strict.

We return to our Cartesian closed base category:

THEOREM 6.4 If there is an initial object $\emptyset$ and an arrow $1 \longrightarrow \emptyset$ then every object is a zero object, since every object has an arrow to 1. □

Theorem 6.3 says, for any object A, forming products with A preserves initial objects. It also preserves coproducts, coequalizers, and pushouts. We prove this for pushouts, leaving the other cases to the reader.

THEOREM 6.5 For any object A, if the left-hand square is a pushout, so is the right:

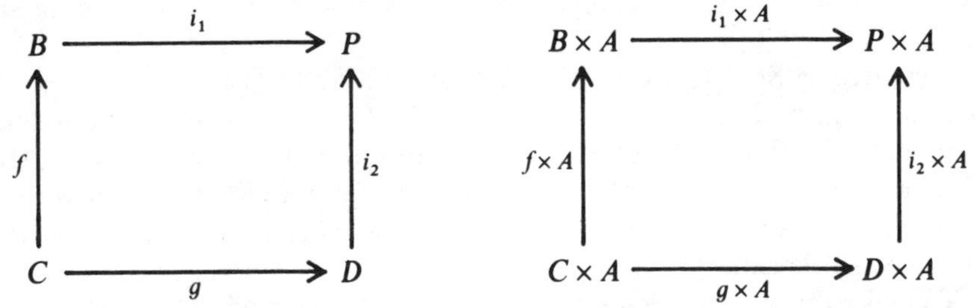

PROOF Let $h: B \times A \longrightarrow T$ and $k: D \times A \longrightarrow T$ have $h \circ (f \times A) = k \circ (g \times A)$. Then the transposes are also equal: $\bar{h} \circ f = \bar{k} \circ g$ (see Exercise 6.2). Thus there exists a unique $u: P \longrightarrow T^A$ with $u \circ i_1 = \bar{h}$ and $u \circ i_2 = \bar{k}$. Taking transposes in those equations gives $\bar{u} \circ (i_1 \times A) = h$ and $\bar{u} \circ (i_2 \times A) = k$, and if any arrow w has those composites with $i_1 \times A$ and $i_2 \times A$, then $\bar{w} = u$ and so $w = \bar{u}$. □

6.5 Intuitive discussion

Consider the definition of f^A. Intuitively, the composite

$$B^A \times A \xrightarrow{ev} B \xrightarrow{f} C$$

takes an arrow g in B^A and an argument y in A, evaluates g at y, and then applies f; that is, it forms $f(g(y))$. So its transpose takes any arrow g in B^A and assigns to it an arrow in C^A, namely the arrow the value of which at any y is $f(g(y))$. But that arrow is just $f \circ g$. So f^A takes any g in B^A to $f \circ g$. This intuitive view may also make the proofs clearer. We wanted to show that f^A('h') equals '$f \circ h$'. But these are names of arrows, so we have simply shown that pairing f^A('h') with any argument y in A and evaluating gives the same result as just applying h and then f to that argument y.

For another example, take any arrow $g: A \longrightarrow A'$ and object B and define an arrow $B^g: B^{A'} \longrightarrow B^A$ that internalizes composition with g on the right. We can describe the transpose of the sought B^g as the arrow which takes any f in $B^{A'}$ and y in A, first applies g to y giving $g(y)$, and then evaluates f at $g(y)$, giving $f(g(y))$. In other words, define B^g as the transpose of

$$B^{A'} \times A \xrightarrow{B^{A'} \times g} B^{A'} \times A' \xrightarrow{ev} B$$

Clearly, exponentiating B preserves identity arrows; that is, $B^{(1_A)} = 1_{(B^A)}$. One can also show that it preserves composition, but reverses it as it reverses the domain and codomain of an arrow. For any $g: A \longrightarrow A'$ and $h: A' \longrightarrow A''$, we have $B^g \circ B^h = B^{(h \circ g)}$. This could be proved along the lines that were used for f^A, or see the proof in Section 6.6. One should also look at the definition of internal composition arrows $c: C^B \times B^A \longrightarrow C^A$ in these terms.

The category **Set** is Cartesian closed. For sets A and B the exponential B^A is the set of all functions from A to B, and $ev: B^A \times A \longrightarrow B$ is the obvious evaluation function. A set is determined by its (global) elements (see Chapter 22) and so the above arguments are rigorous in **Set** if we interpret 'any arrow in B^A' to mean any actual arrow $A \longrightarrow B$ and 'any argument y in A' to mean any (global) element y of A.

To make this intuitive thinking rigorous in all Cartesian closed categories we have to interpret 'any arrow in B^A' and 'any argument y in A' to mean any *generalized elements* of B^A and A, respectively. We have already remarked that Theorem 6.1 fails to imply that $(g \circ f)^A = g^A \circ f^A$ because it only says how $(g \circ f)^A$ acts on global elements of B^A.

The category **Man** is not Cartesian closed. It only includes finite-dimensional manifolds, while the space of functions from one manifold to another is naturally infinite-dimensional. But in Chapter 23 a related Cartesian closed category called **Spaces** is described. Given spaces A and B in this category the global elements of B^A correspond to differentiable maps from A to B. So once we know the maps, that is the arrows from A to B, we know how many points B^A has. But to give the geometric structure of B^A we need more—we need to say how other spaces map into B^A. In other words, we need give all generalized elements $T \longrightarrow B^A$, and that is precisely what the definition of an exponential does in terms of maps from $T \times A$ to B.

Whether or not **Data** is Cartesian closed depends on just what programming language we used to define it. In fact, few if any languages have a genuinely Cartesian closed category of data types, but some powerful languages can be reasonably idealized as though they did. Then, for any data types A and B, the exponential B^A is the type of computable functions from A to B. But, of course, we cannot define a data type by saying how many different values it has; we must say what functions to it are computable. And, again, that is just what the definition of an exponential does. It says that for B^A to be an exponential means that, for any type T, the computable functions $T \longrightarrow B^A$ correspond exactly to the computable functions $T \times A \longrightarrow B$. As we have seen, that defines B^A up to isomorphism in **Data**.

6.6 Indexed families of arrows

By definition, of course, a generalized element of an exponential, $\bar{r}: T \longrightarrow B^A$, corresponds to a uniquely determined arrow $r: T \times A \longrightarrow B$. We will refer to such an arrow as a *T-indexed family of arrows* from A to B, and think of T as a parameter space so that, intuitively, for each value of the parameter t in T there is an arrow $r(t, \_): A \longrightarrow B$.

An arrow $f: B \longrightarrow C$ composes on the left with any T-indexed family $r: T \times A \longrightarrow B$ to give a T-indexed family $f \circ r: T \times A \longrightarrow C$ of arrows from A to C. We have $f^A(\bar{r}) = \overline{f \circ r}$. The proof is in the following diagram:

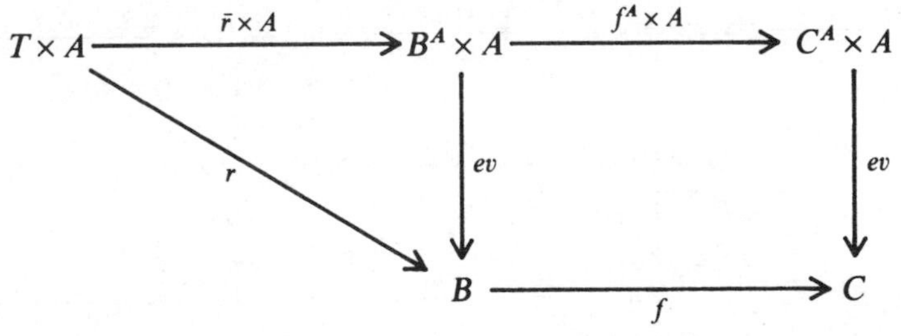

The triangle and square are the definitions of $\bar{r}$ and f^A respectively, while $\overline{f \circ r}$ is certainly the transpose of $f \circ r$. This is the generalized element version of Theorem 6.1. Consider any $f: B \longrightarrow C$ and $g: C \longrightarrow D$. For any object T and $\bar{r} \in_T B^A$ we have

$$(g^A \circ f^A)(\bar{r}) = \overline{g \circ f \circ r} = (g \circ f)^A(\bar{r})$$

and this proves that $(g^A \circ f^A) = (g \circ f)^A$.

On the other hand, given any arrow $g: A' \longrightarrow A$, define its composite on the right with $r: T \times A \longrightarrow B$ to be $r \circ (T \times g): T \times A' \longrightarrow B$. Call this $r \circ_T g$. Calculating the transpose of $B^g(\bar{r})$ directly by taking its product with A and composing with ev proves that $\overline{B^g(\bar{r})} = r \circ_T g$. (Use $(B^{A'} \times g) \circ (\bar{r} \times A) = (\bar{r} \times A') \circ (T \times g)$ to move r from one side of g to the other.) This is the generalized element version of Theorem 6.7, and the reader can use it for a quick proof that $B^g \circ B^h = B^{(h \circ g)}$.

Given two T-indexed families of arrows, $r: T \times A \longrightarrow B$ and $s: T \times B \longrightarrow C$, one from A to B and one from B to C, we can define a *T-indexed composite* of r and s, written as $s \circ_T r$, from A to C. Intuitively, for each value of the parameter t in T, we want to compose $r(t, \_)$ with $s(t, \_)$, so that each y in A goes to $s(t, r(t, y))$; that is, we must pass both the value of $r(t, y)$ and the parameter t to s. Spelling it out precisely, we define $s \circ_T r: T \times A \longrightarrow C$ to be this composite:

$$T \times A \xrightarrow{\langle p_1, r \rangle} T \times B \xrightarrow{s} C$$

Exercise 6.15 shows how the earlier notation $r \circ_T g$ can be seen as a case of this T-indexed composition. Now we can give our most general result on internal composition:

THEOREM 6.6 Take any $r: T \times A \longrightarrow B$ and $s: T \times B \longrightarrow C$, and the arrow $c: C^B \times B^A \longrightarrow C^A$ as defined in Section 6.3. Then $c(\bar{s}, \bar{r}) = \overline{s \circ_T r}$.

PROOF First recall that $\langle \bar{s}, \bar{r} \rangle$ can be factored as $(\bar{s} \times \bar{r}) \circ \Delta_T$. So we can calculate the transpose of $c(\bar{s}, \bar{r})$ as the composite along the top and right of the following diagram:

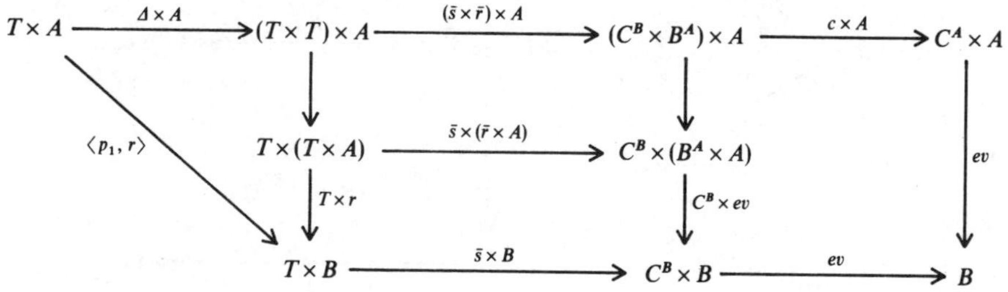

To verify that the triangle commutes, project on to each of T and B. Each square commutes by definition of $\bar{r}$, $\bar{s}$, or c, or by naturality of some isomorphism, and $ev \circ (\bar{s} \times B) = s$. □

Exercise 6.14 shows that T-indexed composition is associative, which gives a generalized element proof that internal composition is associative in the sense given in Section 6.3.

Exercises

In these exercises (except for Exercise 6.4) our base category is Cartesian closed.

6.1 Show that if I, e is an exponential of B by A, and $i: J \xrightarrow{\sim} I$ is iso, then $J, e \circ (i \times A)$ is also an exponential of B by A. Given two exponentials I, e and J, f, both of B by A, show the unique $i: J \longrightarrow I$ such that $f = e \circ (i \times A)$ is iso.

6.2 Given $f: C \longrightarrow B$ and $h: B \times A \longrightarrow T$, show that the transpose of $h \circ (f \times A)$ is $\bar{h} \circ f$. Conclude that for any $k: B \longrightarrow T^A$ the transpose of $k \circ f$ is $\bar{k} \circ (f \times A)$.

6.3 Given $f: C \longrightarrow T$ and $h: B \longrightarrow C^A$, show that the transpose of $f^A \circ h$ is $f \circ \bar{h}$. Conclude that for any $k: B \times A \longrightarrow C$ the transpose of $f \circ k$ is $f^A \circ \bar{k}$.

6.4 Show that every arrow from a strict initial object is monic. In any category with a strict initial object we say that two monics $s: S \rightarrowtail A$ and $t: T \rightarrowtail A$ are *disjoint* if their intersection is initial. Show, in any category with a strict initial object, that if s and t are disjoint then so are $f^{-1}(s)$ and $f^{-1}(t)$ for any $f: B \longrightarrow A$.

6.5 Show, for every object A, that $1^A \cong 1$. Show that $A^1 \cong A$.

6.6 Show that if there is an initial object $\emptyset$ then, for every object A, $\emptyset^A$ is a subterminator. Show that if A has a global element, $x: 1 \longrightarrow A$, then $\emptyset^A \cong \emptyset$. [Hint: given x, every object B has at least the arrow $x \circ !_B$ to A.] There is no general proof that $\emptyset^A \cong \emptyset$, and in fact $\emptyset^\emptyset \cong 1$.

6.7 Products preserve coproducts, and in fact all colimits. Suppose that $B + C$, i_1, i_2 is a coproduct diagram for B and C. Show that $(B + C) \times A$, $i_1 \times A, i_2 \times A$ is a coproduct diagram for $B \times A$ and $C \times A$. This result is often abbreviated as $(B + C) \times A \cong (B \times A) + (C \times A)$.

Show that products preserve coequalizers, and in fact preserve all colimits of diagrams. [Compare with Theorem 6.5.]

6.8 Exponentiation preserves products, and in fact all limits. Consider a product diagram $B \times C$, p_1, p_2. Show that $(B \times C)^A, p_1^A, p_2^A$ is a product diagram for B^A and C^A. Show that for any $\bar{h}: T \longrightarrow B^A$ and $\bar{k}: T \longrightarrow C^A$ the induced arrow $\langle \bar{h}, \bar{k} \rangle: T \longrightarrow (B \times C)^A$ is the transpose of $\langle h, k \rangle: T \times A \longrightarrow B \times C$. [Use Exercise 6.3.]

This is often abbreviated as $(B \times C)^A \cong B^A \times C^A$. It internalizes the fact that an arrow from A to $B \times C$ corresponds to a pair of one arrow from A to B and one from A to C.

Show that, if e is an equalizer for f and g, then e^A is an equalizer for f^A and g^A. In fact, show that exponentiation preserves limits of any diagram.

6.9 Show that $(C^B)^A$, $ev_{B,C} \circ ev_{(A,C^B)}$ is an exponential of C by $A \times B$. Thus show that $C^{(A \times B)} \cong (C^B)^A \cong (C^A)^B$.

6.10 Suppose that $A + C, i_1, i_2$ is a coproduct diagram. Show, for any B, that $B^{(A+C)}$, $B^{(i_1)}$, $B^{(i_2)}$ is a product diagram for B^A and B^C. [Use Exercise 6.3.] In short, $B^{(A+C)} \cong B^A \times B^C$.

6.11 Show, for any object A and monic $f: B \rightarrowtail C$, that f^A is monic.

6.12 For any object T and epic $g: A \longrightarrow A'$, show that $g \times T: A \times T \longrightarrow A' \times T$ is epic. [Use Exercise 6.2.] Then show that for any object B, B^g is monic. [Use the description of B^g in terms of T-indexed composition.]

6.13 Any $f: A \longrightarrow B$ gives a 1-indexed family $f \circ u_A: 1 \times A \longrightarrow B$, with u_A the unitary isomorphism, and of course 'f' is the transpose of this 1-indexed family. Given another arrow $g: B \longrightarrow C$ show that $(g \circ u_B) \circ_1 (f \circ u_A) = (g \circ f) \circ u_A$. Now use Theorem 6.6 to show that $c(`g', `f') = `g \circ f'$.

6.14 For any $r: T \times A \longrightarrow B$ and $s: T \times A \longrightarrow C$ show that the following two arrows are equal:

$$T \times A \xrightarrow{\langle p_1, s \circ \langle p_1, r \rangle \rangle} T \times C \qquad T \times A \xrightarrow{\langle p_1, r \rangle} T \times B \xrightarrow{\langle p_1, s \rangle} T \times C$$

Use this to prove that T-indexed composition is associative; that is, $(t \circ_T s) \circ_T r = t \circ_T (s \circ_T r)$. Conclude that internal composition is associative.

6.15 Any arrow $f: B \longrightarrow C$ gives a T-indexed family $f_T: T \times B \longrightarrow C$, namely the composite $f \circ p_1$. Intuitively, f_T is the T-indexed family the value of which anywhere in T is f.

Consider the name of f, 'f': $1 \longrightarrow C^B$, and $!_T: T \longrightarrow 1$. Show that 'f' $\circ\, !_T$ equals 'f_T'. This is another sense in which f_T is the T-indexed family the value of which anywhere in T is f.

Show that ordinary composition of any $r: T \times A \longrightarrow B$ with $f: B \longrightarrow C$ gives the same result as T-indexed composition of r with f_T; that is, $f \circ r = f_T \circ_T r$. Show that the composite on the right of any $g: A' \longrightarrow A$ with r is just the same as T-indexed composition of g_T with r; that is, $r \circ_T g$ as defined in Section 6.6 is just $r \circ_T g_T$.

6.16 Use Exercise 6.14 to prove that f^A equals

$$B^A \xrightarrow{\sim} 1 \times B^A \xrightarrow{`f' \times B^A} C^B \times B^A \xrightarrow{c} C^A$$

that is, for any $\bar{r} \in_T B^A$ we have $f^A(\bar{r}) = c(\bar{f}_T, \bar{r})$. Show that B^g is

$$B^A \xrightarrow{u} B^A \times 1 \xrightarrow{B^A \times `g'} B^A \times A^{A'} \xrightarrow{c''''} B^{A'}$$

with the obvious isomorphism u and internal composition c''''. In an equation, $B^g(\bar{r}) = c''''(\bar{r}, \overline{g_T})$.

We could also prove that f^A and B^g have these factorizations by calculating transposes.

6.17. Suppose that the right-hand square below is a pushout. Then show that the left-hand square is a pullback:

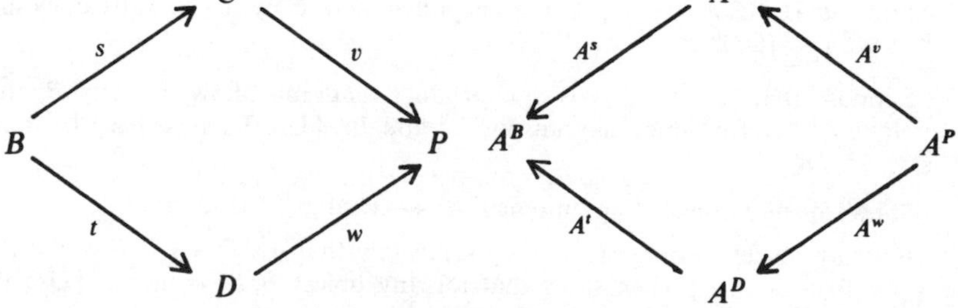

Consider series of arrows like this:

$$\dfrac{\dfrac{\dfrac{h\colon T \longrightarrow A^C \quad k\colon T \longrightarrow A^D}{\bar{h}\colon T \times C \longrightarrow A \quad \bar{k}\colon T \times D \longrightarrow A}}{\bar{u}\colon T \times P \longrightarrow A}}{u\colon T \longrightarrow A^P} \qquad A^s \circ h = A^t \circ k$$

Give the appropriate conditions on the arrows, and show that they are met. Recall that $T \times \_$ preserves pushouts.

7

Product operators and others

This chapter deals with a formal issue. With products, for example, once we know that objects A and B have at least one product we pick one and call it $A \times B$. Similarly, if B has at least one exponential by A we pick one and call it B^A. As long as we deal with only a finite number of objects at a time this can be formalized in first order logic, with no axiom of choice, using no actual terms $A \times B$ or B^A. But if, for a given A, we want to once and for all pick one exponential B^A for each B, that involves an axiom of choice. And even if only a finite number of choices are made there is a kind of awkwardness in them. Therefore we adopt a device for avoiding them.

7.1 Extending the language

Our approach is to extend the language of category theory. We already have operators Dom(_), Cod(_), $1_{\_}$, and $\_\circ\_$, for domain, codomain, identity arrow, and composition. From now on when we speak of a *category with all binary products* we will mean a category described in an extended language that also has an operator $\_\times\_$ and two operators $p_{1,\_,\_}$ and $p_{2,\_,\_}$. The operator $\_\times\_$ applies to a pair of objects to give an object; and each of $p_{1,\_,\_}$ and $p_{2,\_,\_}$ applies to a pair of objects to give an arrow.

Then the category axioms are extended by a new axiom:

For any objects A and B:

$\mathrm{Dom}(p_{1,A,B}) = A \times B$ and $\mathrm{Dom}(p_{2,A,B}) = A \times B$
$\mathrm{Cod}(p_{1,A,B}) = A$ and $\mathrm{Cod}(p_{2,A,B}) = B$
And for any object T and arrows f and g, if
$\mathrm{Dom}(f) = T$ and $\mathrm{Dom}(g) = T$ and $\mathrm{Cod}(f) = A$ and $\mathrm{Cod}(g) = B$,
there exists a unique arrow u with $\mathrm{Dom}(u) = T$ and
$\mathrm{Cod}(u) = A \times B$ and $p_{1,A,B} \circ u = f$ and $p_{2,A,B} \circ u = g$.

Of course, $A \times B, p_{1,A,B}, p_{2,A,B}$ will generally not be the only product diagram for A and B; and nothing distinguishes it from any other except that it is referred to by the terms '$A \times B$', '$p_{1,A,B}$', and '$p_{2,A,B}$'. We axiomatize a selection of one product diagram for each pair of objects, but this selection is arbitrary.

A *category with all finite products* is a category with all binary products, plus a constant 1 that satisfies another axiom:

For every object A there is a unique arrow from A to 1.

For a *category with all finite limits* we keep the operators and axioms for all finite products and add a partial operator $e_{\_,\_}$ which applies to a pair of arrows to give an arrow. The term $e_{f,g}$ is only defined when $\mathrm{Dom}(f) = \mathrm{Dom}(g)$ and $\mathrm{Cod}(f) = \mathrm{Cod}(g)$. There is then a new axiom:

For all f and g, if $\mathrm{Dom}(f) = \mathrm{Dom}(g)$ and $\mathrm{Cod}(f) = \mathrm{Cod}(g)$ then $\mathrm{Cod}(e_{f,g}) = \mathrm{Dom}(f)$ and $f \circ e_{f,g} = g \circ e_{f,g}$, and for any arrow h if $\mathrm{Cod}(h) = \mathrm{Dom}(f)$ and $f \circ h = g \circ h$ there exists a unique u with $\mathrm{Dom}(u) = \mathrm{Dom}(h)$ and $\mathrm{Cod}(u) = \mathrm{Dom}(e_{f,g})$ and $e_{f,g} \circ u = h$.

Theorem 4.5 showed how to define a selected pullback for each corner of arrows explicitly using the product and equalizer operators.

By a *Cartesian closed category* we will mean one described in an extension of the language for a category with all finite limits, extended by the obvious operators $\_^{\_}$ and $ev_{\_,\_}$, each of which applies to a pair of objects. The axioms make B^A, $ev_{A,B}$ an exponential of B by A.

Any reader who prefers to avoid partially defined operators can replace $f \circ g$ and $e_{f,g}$ by ternary relations $C(g, f; h)$ and $Eq(f, g; e)$.

Exercises

7.1 To avoid the existential quantifier in the axiom for a terminal object, 'For every A *there exists* an arrow . . .' introduce an operator $!_{\_}$, applying to an object and giving an arrow, and take axioms:

For every object A: $\mathrm{Dom}(!_A) = A$ and $\mathrm{Cod}(!_A) = 1$.

For every arrow f: If $\mathrm{Cod}(f) = 1$, then $f = !_{\mathrm{Dom}(f)}$.

Show that we can reduce all the axioms in this chapter to universally quantified conditionals, where each antecedent is a conjunction of equations and so is each consequent. For example, for products add an operator $u_{\_,\_,\_,\_}$ which applies to two objects and two arrows, so that $u_{A,B,f,g}$ is defined when $\mathrm{Dom}(f) = \mathrm{Dom}(g)$ and $\mathrm{Cod}(f) = A$ and $\mathrm{Cod}(g) = B$. Add axioms saying that u is the obvious arrow from $\mathrm{Dom}(f)$ to $A \times B$.

Part II

THE CATEGORY OF CATEGORIES

8

Functors and categories

As we begin to compare categories with one another we must always specify which category an object or arrow is in. We will speak of an object A of a category **A**, or an arrow $f: B \longrightarrow C$ of a category **B**.

8.1 Functors

A *functor* **F** from a category **A** to a category **B**, written $\mathbf{F}: \mathbf{A} \longrightarrow \mathbf{B}$, assigns to each object A of **A** an object $\mathbf{F}A$ of **B**, and to each arrow f of **A** an arrow $\mathbf{F}f$ of **B**, meeting the following conditions:

It preserves domains and codomains: given $f: A \longrightarrow B$ of **A** we have $\mathbf{F}f: \mathbf{F}A \longrightarrow \mathbf{F}B$.
It preserves identities: for any A of **A**, $\mathbf{F}(1_A) = 1_{\mathbf{F}A}$.
It preserves composition: if f and g are composable in **A** then $\mathbf{F}(g \circ f) = \mathbf{F}g \circ \mathbf{F}f$, where the second composite is formed in **B**.

It follows that applying **F** to all the objects and arrows of any commutative diagram in **A** gives a commutative diagram in **B**.

If one thinks vividly of the category **A** as a network of arrows between objects, then a functor $\mathbf{F}: \mathbf{A} \longrightarrow \mathbf{B}$, as Mac Lane says, produces a 'picture' of **A** in **B**, and the picture of any commutative diagram also commutes.

Functors compose. Given functors $\mathbf{F}: \mathbf{A} \longrightarrow \mathbf{B}$ and $\mathbf{G}: \mathbf{B} \longrightarrow \mathbf{C}$ define the composite $\mathbf{G} \circ \mathbf{F}: \mathbf{A} \longrightarrow \mathbf{B}$ by the obvious rules $(\mathbf{G} \circ \mathbf{F})A = \mathbf{G}(\mathbf{F}A))$ and $(\mathbf{G} \circ \mathbf{F})f = \mathbf{G}(\mathbf{F}f))$. Clearly $\mathbf{G} \circ \mathbf{F}$ is a functor. This composition is associative, and the reader can define an identity functor $\mathbf{1}_\mathbf{A}$ for every category **A**. So it is natural to talk about the category of categories, **CAT**, with categories as objects and functors as arrows. We will do this freely, and discuss foundations in Chapter 12.

A functor $\mathbf{F}: \mathbf{A} \longrightarrow \mathbf{B}$ is *faithful* iff: for any pair of objects A and A' of **A**, for any arrows f and g both from A to A' if $\mathbf{F}f = \mathbf{F}g$, then $f = g$. In other words, for a given A and A', **F** is one-to-one on arrows from A to A'. That does not mean that **F** is one-to-one on arrows altogether, since a faithful functor need not be one-to-one on objects.

A functor $\mathbf{F}: \mathbf{A} \longrightarrow \mathbf{B}$ is *full* iff the following holds. For any objects A and A' of **A** and arrow $g: \mathbf{F}A \longrightarrow \mathbf{F}A'$ of **B**, there is some $f: A \longrightarrow A'$ with $\mathbf{F}f = g$. For a given A and A', **F** is onto from arrows $A \longrightarrow A'$ to arrows $\mathbf{F}A \longrightarrow \mathbf{F}A'$.

Of course, **F** may still not be onto for arrows altogether, since it need not be onto for objects.

Given any two categories **A** and **B** and any object B of **B**, there is a functor from **A** to **B**, taking every object of **A** to B, and every arrow to 1_B.

Suppose that **A** has products. Then for any object A of **A** there is a functor $\_\times A: \mathbf{A} \longrightarrow \mathbf{A}$. It takes every object B of **A** to $B \times A$ and every arrow $f: B \longrightarrow C$ to $f \times A: B \times A \longrightarrow C \times A$.

If **A** is Cartesian closed then for any object A of **A** there is a functor $\_^A: \mathbf{A} \longrightarrow \mathbf{A}$ that takes every object B of **A** to B^A and every $f: B \longrightarrow C$ to $f^A: B^A \longrightarrow C^A$.

8.2 Preserving structures

Since a functor $\mathbf{F}: \mathbf{A} \longrightarrow \mathbf{B}$ preserves composition and identities, it preserves inverses; that is, if $g = f^{-1}$ in **A** then $\mathbf{F}g = (\mathbf{F}f)^{-1}$ in **B**. But it does not follow that **F** preserves all the categorical structures defined in Part I.

We say that $\mathbf{F}: \mathbf{A} \longrightarrow \mathbf{B}$ *preserves* a product diagram P, p_1, p_2 for A and A' in **A**, if $\mathbf{F}P, \mathbf{F}p_1, \mathbf{F}p_2$ is a product diagram for $\mathbf{F}A$ and $\mathbf{F}A'$. We say that **F** *preserves all finite products* if, for every terminal object P of **A**, $\mathbf{F}P$ is terminal, and **F** preserves every product diagram in **A**.

If **A** and **B** have all finite products in the sense of Chapter 7, a product-preserving functor **F** need not take the selected product for A and A' to the selected one for $\mathbf{F}A$ and $\mathbf{F}A'$. But it will have $\mathbf{F}(A \times A') \cong \mathbf{F}A \times \mathbf{F}A'$.

We say that **F** preserves the equalizer e of f and g if $\mathbf{F}e$ is an equalizer for $\mathbf{F}f$ and $\mathbf{F}g$. Again, we do not require an equalizer-preserving functor to take selected equalizers to selected ones.

Given any diagram D in **A** with an object A_i at each vertex i and an arrow f_e at each edge e, and a limit $L, l_i: L \longrightarrow A_i$, we say that **F** preserves that limit if $\mathbf{F}L, \mathbf{F}l_i: \mathbf{F}L \longrightarrow \mathbf{F}A_i$ is a limit for the diagram in **B** with $\mathbf{F}A_i$ at each vertex i and $\mathbf{F}f_e$ at each edge e. By Section 4.6, if **F** preserves equalizers and all (or all finite) products, then it preserves all (or all finite) limits.

The definitions of preserving colimits and exponentials are analogous.

For example, if **A** has all finite products then, for any object A of **A**, the functor $\_\times A: \mathbf{A} \longrightarrow \mathbf{A}$ preserves equalizers and pullbacks, but not terminal objects or products. If **A** is Cartesian closed then $\_\times A$ also preserves all colimits, as noted in Chapter 6. The functor $\_^A: \mathbf{A} \longrightarrow \mathbf{A}$ preserves terminal objects, products, equalizers, and pullbacks.

8.3 Constructing categories from categories

Other functors are associated with categories constructed from categories. For example, any two categories **A** and **B** have a product category, $\mathbf{A} \times \mathbf{B}$. An object of $\mathbf{A} \times \mathbf{B}$ is a pair $\langle A, B \rangle$, where A is an object of **A** and B an object of

B. An arrow is a pair $\langle f, g\rangle:\langle A, B\rangle \rightarrowtail \langle A', B'\rangle$, where $f: A \longrightarrow A'$ is an arrow of **A** and $g: B \longrightarrow B'$ an arrow of **B**. The identity arrow for $\langle A, B\rangle$ is $\langle 1_A, 1_B\rangle$ and composition is defined component-wise, so $\langle f, g\rangle \circ \langle f', g'\rangle$ is $\langle f \circ f', g \circ g'\rangle$. There is a functor $\mathbf{p}_1: \mathbf{A} \times \mathbf{B} \longrightarrow \mathbf{A}$ defined by $\mathbf{p}_1\langle A, B\rangle = A$ and $\mathbf{p}_1\langle f, g\rangle = f$, and an obvious $\mathbf{p}_2$.

Given any category **T** and functors $\mathbf{H}: \mathbf{T} \longrightarrow \mathbf{A}$ and $\mathbf{K}: \mathbf{T} \longrightarrow \mathbf{B}$ the functor $\langle \mathbf{H}, \mathbf{K}\rangle: \mathbf{T} \longrightarrow \mathbf{A} \times \mathbf{B}$ defined by $\langle \mathbf{H}, \mathbf{K}\rangle T = \langle \mathbf{H}T, \mathbf{K}T\rangle$ and $\langle \mathbf{H}, \mathbf{K}\rangle f = \langle \mathbf{H}f, \mathbf{K}f\rangle$ is the unique functor with the required composites. So $\mathbf{A} \times \mathbf{B}, \mathbf{p}_1, \mathbf{p}_2$ is a product diagram for **A** and **B** in **CAT**.

Any category with one object, and the identity arrow for that object as sole arrow, is a terminal object in **CAT**, so **CAT** has all finite products.

By a *subcategory* of a category **A** we mean any monic functor to **A**, but we will often construct a subcategory $\mathbf{i}: \mathbf{S} \rightarrowtail \mathbf{A}$ by saying that certain objects and arrows of **A** are the objects and arrows of **S**. Then we define **i** to be the obvious inclusion, taking each object or arrow of **S** to itself as an object or arrow of **A**. Such a construction makes **S** a category and **i** a functor iff the following conditions are met. For every object A in **S** the arrow 1_A (as defined in **A**) is also in **S**, and for any composable pair of arrows f and g in **S** the composite $g \circ f$ (as defined in **A**) is also in **S**. A *full subcategory* is a full monic **i**; that is, a full subcategory is determined by the objects of **A** in it, since all **A** arrows between them are in it.

For any parallel pair of functors, $\mathbf{F}, \mathbf{G}: \mathbf{A} \longrightarrow \mathbf{B}$ define a category **E**, the objects of which are those objects A of **A** with $\mathbf{F}A = \mathbf{G}A$, and the arrows of which are those arrows f of **A** with $\mathbf{F}f = \mathbf{G}f$. The reader can verify that the inclusion $\mathbf{e}: \mathbf{E} \rightarrowtail \mathbf{A}$ is an equalizer for **F** and **G**. So **CAT** has all finite limits.

Every category **A** has an *arrow category*, $\mathbf{A}^2$. The objects of $\mathbf{A}^2$ are the arrows of **A**. Given two $\mathbf{A}^2$ objects $f: A \longrightarrow B$ and $g: A' \longrightarrow B'$, an $\mathbf{A}^2$ arrow from f to g is a commutative square of **A**:

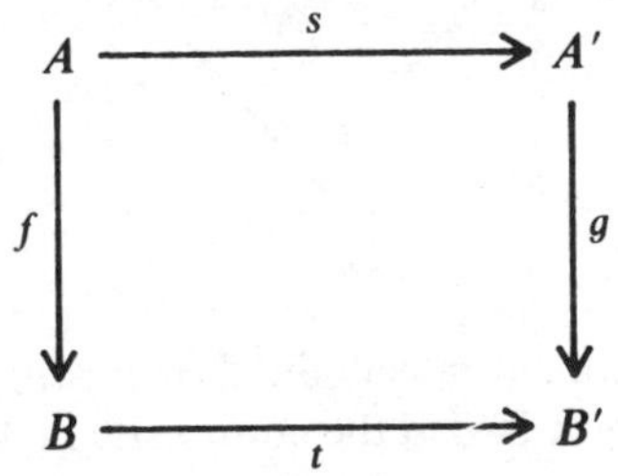

We will write $\langle s, t\rangle: f \longrightarrow g$. (Strictly, s and t alone do not determine the $\mathbf{A}^2$ arrow, since they may be the top and bottom of many different commutative squares in **A**.) Given another $\mathbf{A}^2$ arrow $\langle u, w\rangle: g \longrightarrow h$, define $\langle u, w\rangle \circ \langle s, t\rangle$ to be $\langle u \circ s, w \circ t\rangle: f \longrightarrow h$. That is, we compose commutative squares by pasting them together:

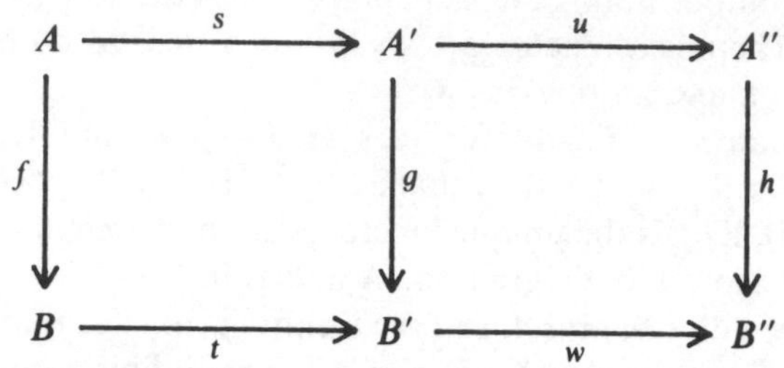

Any $\mathbf{A}^2$ object $f: A \longrightarrow B$ has an identity arrow $\langle 1_A, 1_B \rangle$ in $\mathbf{A}^2$.

There are three important functors between any category $\mathbf{A}$ and its arrow category. To define $\mathbf{id}: \mathbf{A} \longrightarrow \mathbf{A}^2$ set $\mathbf{id}(A) = 1_A$, and for any $\mathbf{A}$ arrow $f: A \longrightarrow B$ set $\mathbf{id}(f) = \langle f, f \rangle : 1_A \longrightarrow 1_B$. There is a 'domain functor' $\mathbf{D}: \mathbf{A}^2 \longrightarrow \mathbf{A}$ defined in this way: for any $\mathbf{A}^2$ object $f: A \longrightarrow B$, $\mathbf{D}f = A$. For any $\mathbf{A}^2$ arrow $\langle s, t \rangle$, $\mathbf{D}\langle s, t \rangle = s$. And there is the obvious 'co-domain functor' $\mathbf{C}: \mathbf{A}^2 \longrightarrow \mathbf{A}$ with $\mathbf{C}f = B$ and $\mathbf{C}\langle s, t \rangle = t$.

For any category $\mathbf{A}$ and object A of $\mathbf{A}$ there is a *slice category* $\mathbf{A}/A$, also called a *comma category*. The objects of $\mathbf{A}/A$ are arrows of $\mathbf{A}$ with codomain A. Given $\mathbf{A}/A$ objects $f: B \longrightarrow A$ an $g: C \longrightarrow A$, an $\mathbf{A}/A$ arrow is a commutative triangle of $\mathbf{A}$:

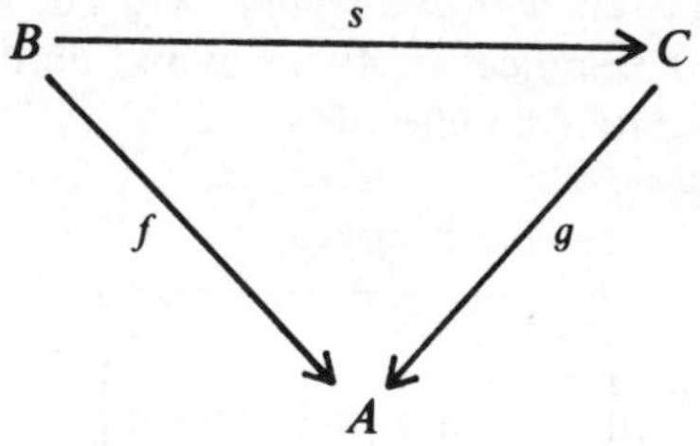

We write $s: f \longrightarrow g$. Composition is defined by pasting the triangles together, so the composite $t \circ s$ from $\mathbf{A}$ is also the composite in $\mathbf{A}/A$. Identity arrows are obvious.

There is a functor $\Sigma_A: \mathbf{A}/A \longrightarrow \mathbf{A}$ that takes any $\mathbf{A}/A$ object $f: B \longrightarrow A$ to B. For any $s: f \longrightarrow g$, $\Sigma_A s$ is s thought of as an arrow in $\mathbf{A}$.

If the category $\mathbf{A}$ has binary products there is a functor $A^*: \mathbf{A} \longrightarrow \mathbf{A}/A$. For any object B of $\mathbf{A}$, let A^*B be $p_1: A \times B \longrightarrow A$; and for any $f: B \longrightarrow C$ let A^*f be

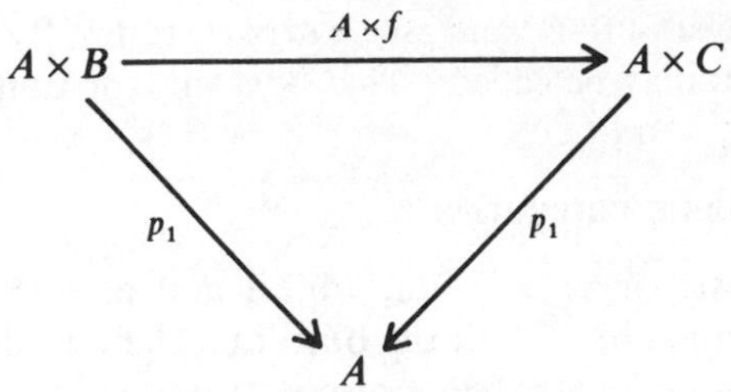

Given any category $\mathbf{A}$ with all finite products there is a category $\mathbf{Grp_A}$ with the groups of $\mathbf{A}$ as objects and their homomorphisms as arrows. There is a faithful *underlying object functor*, $\mathbf{U}\colon \mathbf{Grp_A} \longrightarrow \mathbf{A}$ defined by setting $\mathbf{U}\langle G, e_G, m_G, -_G^{-1}\rangle = G$, and for a homomorphism f from $\langle G, e_G, m_G, -_G^{-1}\rangle$ to $\langle H, e_H, m_H, -_H^{-1}\rangle$ let $\mathbf{U}f$ just be the arrow $f\colon G \longrightarrow H$. This functor preserves the terminal object and products, and if $\mathbf{A}$ has equalizers then $\mathbf{U}$ preserves the equalizers from $\mathbf{Grp_A}$. It does not preserve the initial object, i.e. the zero group 1, unless the terminator 1 of $\mathbf{A}$ is also initial.

Every category $\mathbf{A}$ has a *dual* or *opposite* category $\mathbf{A}^{\mathrm{op}}$. The objects of $\mathbf{A}^{\mathrm{op}}$ are the objects of $\mathbf{A}$, and the arrows of $\mathbf{A}^{\mathrm{op}}$ are the arrows of $\mathbf{A}$ but with domain and codomain reversed. So every arrow $f\colon A \longrightarrow A'$ of $\mathbf{A}$ appears as an arrow $f^{\mathrm{op}}\colon A' \longrightarrow A$ of $\mathbf{A}^{\mathrm{op}}$. The superscript 'op' on 'f^{op}' is only a reminder to think of f as an arrow of $\mathbf{A}^{\mathrm{op}}$. Since arrows

$$A \xrightarrow{f} B \xrightarrow{g} C$$

of $\mathbf{A}$ appear as arrows

$$C \xrightarrow{g^{\mathrm{op}}} B \xrightarrow{f^{\mathrm{op}}} A$$

of $\mathbf{A}^{\mathrm{op}}$, the order of composability is reversed. So we define composition in $\mathbf{A}^{\mathrm{op}}$ as $f^{\mathrm{op}} \circ g^{\mathrm{op}} = (g \circ f)^{\mathrm{op}}$, where the right-hand side uses composition in $\mathbf{A}$. The category axioms are easily verified. Every category is the dual of its dual: $\mathbf{A} = (\mathbf{A}^{\mathrm{op}})^{\mathrm{op}}$.

All of this is the objective form of the process of forming dual statements, explained in Chapter 2. An arrow is monic in $\mathbf{A}$ iff it is epic in $\mathbf{A}^{\mathrm{op}}$. Any P, p_1, p_2 is a product diagram in $\mathbf{A}$ iff it is a coproduct diagram in $\mathbf{A}^{\mathrm{op}}$, and so on for all the dual concepts described in Chapter 2. If a statement holds in $\mathbf{A}$ then its dual holds in $\mathbf{A}^{\mathrm{op}}$. Every category is dual to its own dual, so if a statement holds in all categories so does its dual.

A functor $\mathbf{G}\colon \mathbf{A}^{\mathrm{op}} \longrightarrow \mathbf{B}$ is often called a *contravariant functor* from $\mathbf{A}$ to $\mathbf{B}$. That is, $\mathbf{G}$ assigns to each object A of $\mathbf{A}$ an object $\mathbf{G}A$ of $\mathbf{B}$, and to each $f\colon A \longrightarrow A'$ of $\mathbf{A}$, an arrow $\mathbf{G}f\colon \mathbf{G}A' \longrightarrow \mathbf{G}A$, satisfying the conditions $\mathbf{G}(1_A) = 1_{(\mathbf{G}A)}$ and $\mathbf{G}(g \circ f) = (\mathbf{G}f) \circ (\mathbf{G}g)$. For example, given any object B of a Cartesian closed category $\mathbf{A}$, there is a contravariant functor B^{-} from $\mathbf{A}$ to

A, taking any **A** object A to B^A and any **A** arrow f to B^f. By way of contrast, a functor $\mathbf{F}:\mathbf{A}\longrightarrow\mathbf{B}$ may be called a *covariant* functor from **A** to **B**.

8.4 Aspects of finite categories

We will use the category **1** with one object and one arrow, which is the identity arrow for the object. Call the object 0. There is also the category **2** defined in Exercise 1.5 with two objects, 0 and 1, and just one arrow $\alpha:0\longrightarrow 1$ in addition to the identity arrows. The fact that one object of **2** is called by the same name as the object of **1** means nothing. They are just objects in different categories.

There are two functors from **1** to **2**. Define $0:\mathbf{1}\longrightarrow\mathbf{2}$ by $0(0) = 0$ and define $1:\mathbf{1}\longrightarrow\mathbf{2}$ by $1(0) = 1$. The actions of 0 and 1 on the arrow in **1** are determined, since functors preserve identity arrows.

In general, for any category **A**, each object A of **A** determines a unique functor which we will also call $A:\mathbf{1}\longrightarrow\mathbf{A}$, taking the object 0 of **1** to A. The objects of **A** correspond exactly to the **1**-elements of **A**. Since **1** is a terminator in **CAT**, these are the global elements of **A**. We can also say that the identity arrows of **A** correspond to the global elements, if we are thinking of the identity arrow in **1**. But we already knew that objects and identity arrows correspond exactly to each other!

Each arrow f of **A** determines a unique functor from **2** to **A**, which we call $f:\mathbf{2}\longrightarrow\mathbf{A}$, taking the non-identity arrow of **2** to the arrow f. In short, the arrows of **A** correspond exactly to the **2**-elements of **A**.

It is possible to have functors $\mathbf{F}:\mathbf{A}\longrightarrow\mathbf{B}$ and $\mathbf{G}:\mathbf{A}\longrightarrow\mathbf{B}$ which have the same effect on all objects A of **A**, but not on all arrows (see Exercise 8.11). On the other hand, if they agree on all arrows that includes identities, and so they must agree on objects, and so $\mathbf{F} = \mathbf{G}$.

We call an object I of any category **C** a *generator* if, for every parallel pair of arrows of **C**, $f, g: C\longrightarrow C'$, either $f = g$ or there is some $x: I\longrightarrow C$ with $f\circ x \neq g\circ x$. In other words, I-elements suffice to distinguish arrows. Thus, in **CAT**, **1** is not a generator but **2** is. (One version of the axioms for a category of sets will say that it is a topos and that the terminator 1 is a generator.)

There is also a category **3** with objects and arrows as shown below:

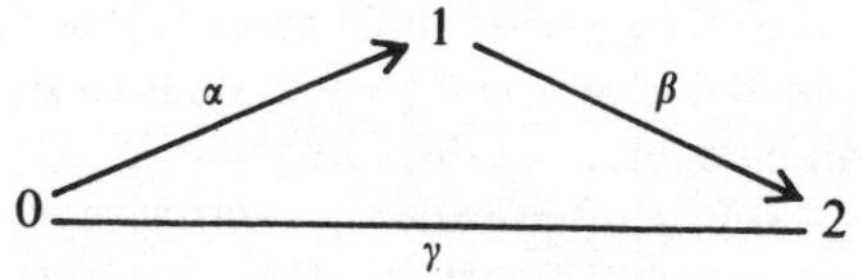

plus identity arrows for each object. The composite $\beta\circ\alpha$ can only be γ, and then the category axioms are easily verified. For any category **A**, commutative triangles correspond to **3**-elements of **A**.

Exercises

8.1 A functor is monic iff it is one-to-one on arrows; and iso iff it is one-to-one and onto on arrows.

8.2 Show that any functor $\mathbf{F}: \mathbf{A} \longrightarrow \mathbf{B}$ such that every arrow of **B** either is in the image of the functor or has an inverse in that image is epic. Define a category **I** with two objects 0 and 1, and two arrows $f: 0 \longrightarrow 1$ and $g: 1 \longrightarrow 0$ in addition to the identities, with composition defined by $f = g^{-1}$. Define a functor from **2** to **I** which is monic and epic, but not iso.

8.3 Notice that the empty category is initial in **CAT**. Show that **CAT** has coproducts, by considering for any **A** and **B** the category formed of one copy of **A** and a separate copy of **B**.

8.4 Given a parallel pair of functors $\mathbf{F}, \mathbf{G}: \mathbf{A} \longrightarrow \mathbf{B}$ show that the objects A of **A** such that $\mathbf{F}A = \mathbf{G}A$, together with the arrows f such that $\mathbf{F}f = \mathbf{G}f$, form a subcategory of **A**. Show that the inclusion of this subcategory into **A** is an equalizer for **F** and **G** in **CAT**.

Show that if an arrow f in the equalizer has an inverse f^{-1} in **A**, then f^{-1} is also in the equalizer. Show that this is not truc for all subcategories (either use one of the examples from the introduction, or construct a finite category illustrating the fact). Conclude that not every monic in **CAT** is an equalizer (this also follows from Exercise 8.2).

8.5 Show that a functor preserves split monics and split epics. Show that functors do not always preserve monics or epics. [Use one of the examples from the introduction, or construct a finite category with a non-monic arrow and a non-epic arrow and suitable functors.]

8.6 Let **N** be a category with one object, and one arrow f_n for each natural number n including 0, with composition defined by $f_n \circ f_m = f_{(n+m)}$. Show that the functor assigning f_1 to the non-identity arrow of **2** is a coequalizer for 0 and $1: \mathbf{1} \longrightarrow \mathbf{2}$, thus epic, although far from onto for arrows.

8.7 If **A** has a terminal object 1 then **A**/1 is isomorphic to **A**.

8.8 For any category **A** and object A of **A**, show that $\mathbf{A}/A$ is a subcategory of $\mathbf{A}^2$. In fact, it is a pullback:

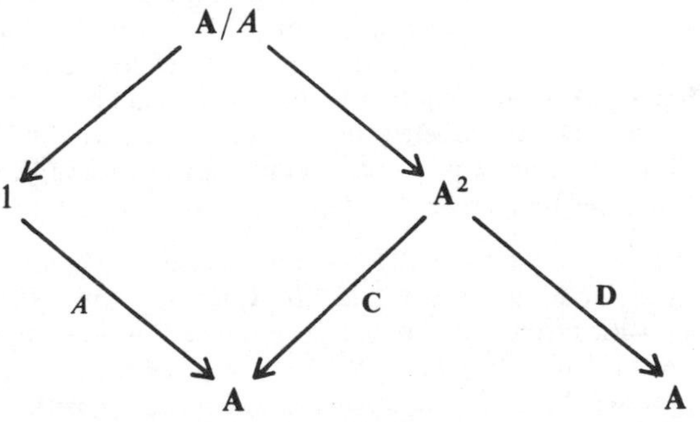

while Σ_A is the composite along the right.

8.9 If $f: \mathbf{2} \longrightarrow \mathbf{A}$ corresponds to an arrow f of $\mathbf{A}$, then the composite $f \circ 0$ corresponds to the domain of f, and $f \circ 1$ to the co-domain. For any $t: \mathbf{3} \longrightarrow \mathbf{A}$ the functor $t \circ \gamma: \mathbf{2} \longrightarrow \mathbf{A}$ corresponds to the composite of the arrows of $\mathbf{A}$ corresponding to $t \circ \alpha$ and $t \circ \beta$.

8.10 Show that every relation $\mathbf{r}: \mathbf{R} \rightarrowtail \mathbf{A} \times \mathbf{B}$ has the following properties:

For any objects A of $\mathbf{A}$ and B of $\mathbf{B}$, if A has $\mathbf{r}$ to B then 1_A has $\mathbf{r}$ to 1_B.

For any arrows f, g, of $\mathbf{A}$ and f', g' of $\mathbf{B}$, if f has $\mathbf{r}$ to f' and g has $\mathbf{r}$ to g' then $f \circ f'$ has $\mathbf{r}$ to $g \circ g'$.

Show that an external (i.e. not necessarily represented by a monic in **CAT**) relation of objects and arrows of $\mathbf{A}$ to those of $\mathbf{B}$ is represented by a subcategory of $\mathbf{A} \times \mathbf{B}$ iff it has these properties. Show that any $\mathbf{r}: \mathbf{R} \rightarrowtail \mathbf{A} \times \mathbf{A}$ is an equivalence relation iff it is reflexive, symmetric, and transitive on objects and on arrows.

Equivalence relations in **CAT** have quotients, but the explicit construction of a quotient category is messy and little used. It does not consist only of equivalence classes of objects and arrows, because two arrows that were not composable in $\mathbf{A}$ may have equivalent domain and codomain so that they are composable in the quotient. Think of the kernel pair of the coequalizer in Exercise 8.6. Ehresmann (1965) gives details.

8.11 Suppose that $\mathbf{A}$ has all finite products and that $\mathbf{Grp_A}$ includes non-zero groups, groups not isomorphic to 1. Show that there is a functor from $\mathbf{Grp_A}$ to $\mathbf{Grp_A}$ that agrees with the identity functor on all objects but not on all arrows. [Hint: see Exercise 3.4.]

8.12 Suppose that $\mathbf{A}$ is Cartesian closed. Show, for every group $\langle G, e, m, \_^{-1} \rangle$ in $\mathbf{A}$ and object A of $\mathbf{A}$, that $\langle G^A, e^A, m^A, (\_^{-1})^A \rangle$ is also a group. Show that $\_^A$ gives a functor from $\mathbf{Grp_A}$ to $\mathbf{Grp_A}$. In fact, any product-preserving functor from $\mathbf{A}$ to $\mathbf{A}$ gives a product-preserving functor from $\mathbf{Grp_A}$ to itself.

Show that $\mathbf{Grp_A}$ is not Cartesian closed unless every group in $\mathbf{A}$ is isomorphic to 1.

8.13 Suppose that a category $\mathbf{A}$ with all finite limits also has a selected representative for each equivalence class of monics. For any object A of a category $\mathbf{A}$, define $\mathrm{Sub}(A)$ to be the full subcategory of $\mathbf{A}/A$ the objects of which are all the selected monics to A; that is, the category of selected sub-objects of A with inclusions as arrows.

Show that if you always take the selected monic as a value for the pullback, pulling back along any $f: A \longrightarrow B$ of $\mathbf{A}$ gives a functor $f^{-1}: \mathrm{Sub}(B) \longrightarrow \mathrm{Sub}(A)$. Show, for any $g: B \longrightarrow C$, that $(g \circ f)^{-1} = (f^{-1}) \circ (g^{-1})$. Thus define a contravariant functor from $\mathbf{A}$ to **CAT** that takes each object A to $\mathbf{A}/A$ and each arrow f to f^{-1}.

Without the restriction to selected representatives we could only prove that $(g \circ f)^{-1} \cong (f^{-1}) \circ (g^{-1})$. This would give a contravariant *pseudofunctor* from $\mathbf{A}$ to **CAT** (for details see Bénabou (1985) Section 12 and references there).

8.14 A *preorder* is a collection $\mathbf{X}$ with a reflexive transitive relation $\leq$ on it. So, for every x in $\mathbf{X}$, $x \leq x$; and for every x, y, and z in $\mathbf{X}$, if $x \leq y$ and $y \leq z$ then $x \leq z$. An *order-preserving function* from a preorder $\mathbf{X}$ to another $\mathbf{Y}$ is a function f from $\mathbf{X}$ to $\mathbf{Y}$ such that if $x \leq x'$ in $\mathbf{X}$ then $fx \leq fx'$ in $\mathbf{Y}$.

Show that a preorder $\mathbf{X}$, $\leq$ can be construed as a category, with the elements of $\mathbf{X}$ as objects, and with one arrow $x \longrightarrow y$ if $x \leq y$ and none otherwise. What is composition?

In fact, we formally define a *preorder* to be a category in which, for any objects A and B, there is at most one arrow from A to B. Then define $A \leq B$ to mean that there is an arrow from A to B. Show that a category is a preorder iff every diagram in it commutes. Show that an order-preserving function between preorders is just the same thing as a functor between them. Thus the category of preorders, **Preord**, is a full subcategory of **CAT**.

Show that an order-preserving function is monic in **Preord** iff it is one-to-one on objects, and in that case it is also monic as a functor in **CAT**. Show it is epic in **Preord** iff it is onto for objects, and then it need not be epic as a functor in **CAT**. [Hint: consider the preorder $1 + 1$, which has two elements neither less than the other, and **2**, and the category **E** defined in Exercise 10.6.]

8.15 A *supremum* for a preorder **X** is an object t of **X** such that, for every object x of **X**, $x \leq t$. An *infimum* for **X** is an object i such that for every object x, $i \leq x$. A greatest lower bound for two objects x and y is an object p such that $p \leq x$ and $p \leq y$, and for any object s if $s \leq x$ and $s \leq y$ then $s \leq p$. We abbreviate greatest lower bound to g.l.b. Give a definition of least upper bound for x and y, abbreviated to l.u.b. Then show that these are terminal objects, initial objects, products, and coproducts in **X**, respectively. What does uniqueness up to isomorphism mean in a preorder?

Show that in any preorder **X** every parallel pair of arrows trivially has an equalizer and a coequalizer.

Suppose that a preorder **X** has a g.l.b. for every pair of its objects. Then in **X** define a *material implicate* of y by x to be an object c such that, for any object s, $s \leq c$ iff any g.l.b. of s and x is $\leq y$. Show that a material implicate of y by x is just an exponential of y by x.

8.16 Let **P** be a preorder with all finite products (equivalently, every finite set of objects, including the empty set, has a l.u.b.). Show that a group in **P** must be a zero group. The object G must be terminal.

8.17 A *poset*, or *partially ordered set*, is an antisymmetric preorder; that is, a preorder in which, if $x \leq y$ and $y \leq x$ then $x = y$. Show a poset has at most one supremum, at most one g.l.b. for any given pair of objects x and y, and so on. What does uniqueness up to isomorphism mean in a poset?

A *lattice* is a poset with a supremum t, an infimum i, and a g.l.b. and a l.u.b. for every pair of objects x and y. The g.l.b. of x and y is unique and is called their *meet*, written $x \wedge y$. The l.u.b. is called their *join* and is written $x \vee y$.

A *Heyting algebra* is a Cartesian closed lattice. The material implicate of y by x is often written $x \rightarrow y$.

A *linear order* is a poset such that, for any pair of objects x and y, either $x \leq y$ or $y \leq x$. Show that every linear order with supremum and infimum is a Heyting algebra. [Hint: show that $x \rightarrow y$ is t when $x \leq y$, and equals y otherwise.]

8.18 If you are familiar with them, show that every Boolean algebra, and every poset of open sets of a topological space, is a Heyting algebra. For more on Heyting algebras see Rasiowa and Sikorski (1963), and with emphasis on applications to toposes, Fourman and Scott (1979).

9

Natural transformations

9.1 Definition

Given a parallel pair of functors $\mathbf{F}, \mathbf{G}: \mathbf{A} \longrightarrow \mathbf{B}$, a *natural transformation* from $\mathbf{F}$ to $\mathbf{G}$ is a family of arrows $\nu_A: \mathbf{F}A \longrightarrow \mathbf{G}A$, one for each object A of $\mathbf{A}$, such that for every arrow $f: A \longrightarrow A'$ of $\mathbf{A}$ the following square commutes:

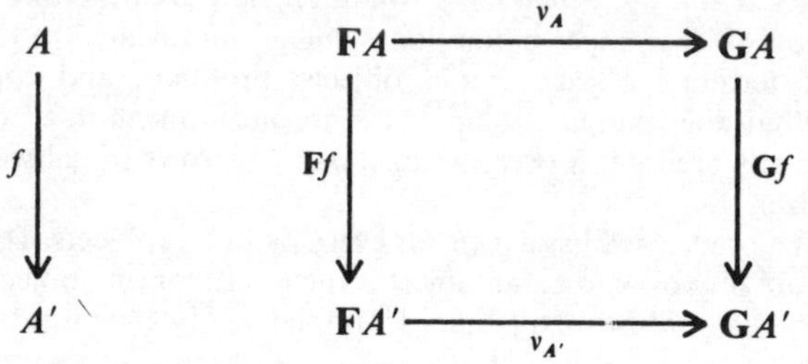

We write $\nu: \mathbf{F} \longrightarrow \mathbf{G}$ for the natural transformation, and call the arrows ν_A the *components* of ν. If the above square appears in a diagram we may say that it commutes by naturality of ν.

For example, let $\mathbf{A}$ be a category with all finite products and take any object A of $\mathbf{A}$. There is a product functor $\_ \times A: \mathbf{A} \longrightarrow \mathbf{A}$ as well as the identity functor $\mathbf{1_A}: \mathbf{A} \longrightarrow \mathbf{A}$ and the projection arrows $p_1: B \times A \longrightarrow B$ are the components of a natural transformation $p_1: \_ \times A \longrightarrow \mathbf{1_A}$. For any $f: B \longrightarrow C$ of $\mathbf{A}$ the following square commutes by definition of $f \times A$:

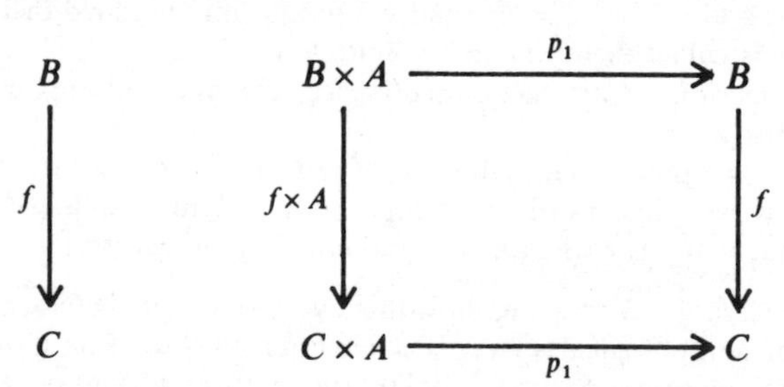

One sense of the word 'natural' here is that the projections form a natural family of arrows in that $p_1: B \times A \longrightarrow B$ 'does the same thing' with B as $p_1: C \times A \longrightarrow C$ 'does with' C. The definition of naturality captures this intuitive idea of 'doing the same thing' with different objects.

Suppose that **A** is Cartesian closed and take any object A of **A**. The functor $\_^A$ composes with $\_ \times A$ to give $(\_^A) \times A: \mathbf{A} \longrightarrow \mathbf{A}$ and the evaluation arrows give a natural transformation $ev_A: (\_^A) \times A \longrightarrow \mathbf{1_A}$. For any arrow $f: B \longrightarrow C$ the naturality square commutes by definition of f^A.

A natural transformation with every component iso is called a *natural isomorphism*. If **A** has all finite products then the natural transformation $p_1: \_ \times 1 \longrightarrow \mathbf{1_A}$ is a natural isomorphism.

9.2 Functor categories

Natural transformations compose. Given functors **F**, **G**, and **H**, all from **A** to **B**, and natural transformations $v: \mathbf{F} \longrightarrow \mathbf{G}$ and $\varphi: \mathbf{G} \longrightarrow \mathbf{H}$, there is a natural $\varphi \circ v: \mathbf{F} \longrightarrow \mathbf{H}$ defined by composing components $\varphi_A \circ v_A$:

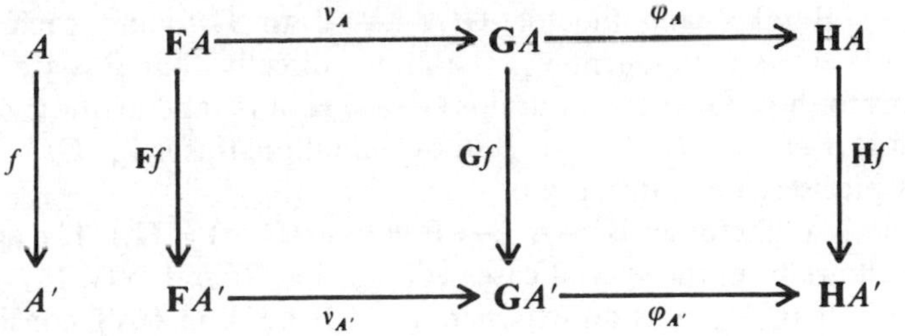

Naturality is clear, and this composition is associative. There is an identity natural transformation $1_\mathbf{F}: \mathbf{F} \longrightarrow \mathbf{F}$, with components $1_{\mathbf{F}A}$. Thus there is a category, $\mathbf{B^A}$, called the *functor category* from **A** to **B**, the objects of which are the functors from **A** to **B** and the arrows of which are the natural transformations between them. For example, the arrow category $\mathbf{A^2}$ is the functor category from **2** to **A**, since an arrow of **A** can be seen as a functor from **2** to **A**, and by definition an arrow of $\mathbf{A^2}$ is a natural transformation between these functors.

In fact, **CAT** is Cartesian closed, with functor categories giving exponentials. The proof is easier using this description of functors from product categories:

THEOREM 9.1 Consider categories **A**, **B**, and **C**. Suppose, for each object B of **B**, that there is a functor $\mathbf{R}_B: \mathbf{A} \longrightarrow \mathbf{C}$ and, for each object A of **A**, a functor $\mathbf{L}_A: \mathbf{B} \longrightarrow \mathbf{C}$, such that for all A and B we have $\mathbf{L}_A(B) = \mathbf{R}_B(A)$. Finally, suppose that for every $f: A \longrightarrow A'$ of **A** and $g: B \longrightarrow B'$ of **B** the functors satisfy the interchange rule $\mathbf{R}_{B'}f \circ \mathbf{L}_{A'}g = \mathbf{L}_A g \circ \mathbf{R}_B f$. Then there is a unique functor $\mathbf{F}: \mathbf{A} \times \mathbf{B} \longrightarrow \mathbf{C}$ with $\mathbf{F}(A, B) = \mathbf{F}_A B$ and $\mathbf{F}(1_A, g) = \mathbf{L}_A g$ and $\mathbf{F}(f, 1_B) = \mathbf{R}_B f$.

PROOF Since every arrow $\langle f, g \rangle$ of $\mathbf{A} \times \mathbf{B}$ factors as $\langle 1_{A'}, g \rangle \circ \langle f, 1_B \rangle$ there is at most one such functor $\mathbf{F}$, and it must have

$$\mathbf{F}(f, g) = \mathbf{F}(1_{A'}, g) \circ \mathbf{F}(f, 1_B) = \mathbf{L}_{A'}g \circ \mathbf{R}_B f.$$

So take that as defining $\mathbf{F}$ on arrows. It remains to show that

$$\mathbf{F}(f' \circ f, g' \circ g) = \mathbf{F}(f', g') \circ \mathbf{F}(f, g).$$

But this follows from the definition of $\mathbf{F}$ by calculation, using the interchange rule. □

A functor from a product category can be called a *bifunctor*. Then Theorem 9.1 says that a bifunctor is a function of pairs of objects and pairs of arrows, functorial in each variable and satisfying the interchange rule.

We apply Theorem 9.1 to obtain a functor $\mathbf{ev}: \mathbf{B}^{\mathbf{A}} \times \mathbf{A} \longrightarrow \mathbf{B}$. For any functor $\mathbf{G}$ from $\mathbf{A}$ to $\mathbf{B}$, let $\mathbf{L}_{\mathbf{G}}: \mathbf{A} \longrightarrow \mathbf{B}$ be $\mathbf{G}$. For any object A of $\mathbf{A}$, let $\mathbf{R}_A: \mathbf{B}^{\mathbf{A}} \longrightarrow \mathbf{B}$ take each functor $\mathbf{G}: \mathbf{A} \longrightarrow \mathbf{B}$ to $\mathbf{G}A$, and each natural $v: \mathbf{G} \longrightarrow \mathbf{H}$ to its component v_A. It follows directly that $\mathbf{R}_A$ is a functor. It remains to show for every natural $v: \mathbf{G} \longrightarrow \mathbf{H}$ of $\mathbf{B}^{\mathbf{A}}$ and arrow $f: A \longrightarrow A'$ of $\mathbf{A}$ that $\mathbf{R}_A v \circ \mathbf{L}_{\mathbf{G}} f = \mathbf{L}_{\mathbf{H}} f \circ \mathbf{R}_{A'} v$. But, by definition, this is $v_A \circ \mathbf{G}f = \mathbf{H}f \circ v_{A'}$, which is precisely the naturality of v.

So there is a functor $\mathbf{ev}: \mathbf{B}^{\mathbf{A}} \times \mathbf{A} \longrightarrow \mathbf{B}$ with $\mathbf{ev}(\mathbf{G}, A) = \mathbf{G}A$. The action on arrows follows from the special cases $\mathbf{ev}(1_{\mathbf{G}}, f) = \mathbf{G}f$ and $\mathbf{ev}(v, 1_A) = v_A$.

To show that $\mathbf{B}^{\mathbf{A}}$, $\mathbf{ev}$ is an exponential of $\mathbf{B}$ by $\mathbf{A}$ in $\mathbf{CAT}$ consider any functor $\mathbf{F}: \mathbf{T} \times \mathbf{A} \longrightarrow \mathbf{B}$. Define the transpose $\bar{\mathbf{F}}: \mathbf{T} \longrightarrow \mathbf{B}^{\mathbf{A}}$ in the following way. For any object T of $\mathbf{T}$, $\bar{\mathbf{F}}\mathrm{T}$ is the functor $\mathbf{F}(\mathrm{T}, \_): \mathbf{A} \longrightarrow \mathbf{B}$. For any arrow $f: T \longrightarrow T'$, $\bar{\mathbf{F}}f$ is the natural transformation the component of which at any object A of $\mathbf{A}$ is $\mathbf{F}(f, 1_A)$. Proof that these are functors and natural transformations, and that $\bar{\mathbf{F}}$ is actually the transpose of $\mathbf{F}$, is left to the reader.

9.3 Equivalence

An *equivalence situation* consists of two functors $\mathbf{F}: \mathbf{X} \longrightarrow \mathbf{A}$ and $\mathbf{G}: \mathbf{A} \longrightarrow \mathbf{X}$ such that there are natural isomorphisms $\varphi: \mathbf{1}_{\mathbf{X}} \longrightarrow \mathbf{GF}$ and $\psi: \mathbf{1}_{\mathbf{A}} \longrightarrow \mathbf{FG}$. Often there will be many different natural isomorphisms for a given $\mathbf{F}$ and $\mathbf{G}$. Then, while $\mathbf{F}$ and $\mathbf{G}$ may not be inverse to one another, they are 'inverses up to natural isomorphisms'.

THEOREM 9.2 In an equivalence situation, both functors are full and faithful.

PROOF For any $h: X \longrightarrow Y$ of $\mathbf{X}$, $h = (\varphi_Y^{-1}) \circ \mathbf{GF}h \circ \varphi_X$. So $\mathbf{F}$ is faithful. For any $k: \mathbf{G}A \longrightarrow \mathbf{G}B$ of $\mathbf{X}$, $k = \mathbf{G}((\psi_B^{-1}) \circ \mathbf{F}k \circ \psi_A)$. So $\mathbf{G}$ is full. By symmetry, both are full and faithful. □

Of course, every object X of $\mathbf{X}$ is isomorphic to one in the image of $\mathbf{G}$, for example $\mathbf{GF}X$, and every object of $\mathbf{A}$ to one in the image of $\mathbf{F}$.

Conversely, any full and faithful functor $\mathbf{H}: \mathbf{A} \longrightarrow \mathbf{B}$ such that every object of $\mathbf{B}$ is isomorphic to an object in the image of $\mathbf{F}$ is called an *equivalence functor*. Fitting an equivalence functor into an equivalence situation requires a selection of isomorphisms:

THEOREM 9.3 Suppose that $\mathbf{F}: \mathbf{X} \longrightarrow \mathbf{A}$ is an equivalence functor. Suppose that for every object A of $\mathbf{A}$ there is a selected object $\mathbf{G}A$ of $\mathbf{X}$ and a selected isomorphism $\psi_A: A \longrightarrow \mathbf{FG}A$. Then $\mathbf{G}$ extends to a functor $\mathbf{G}: \mathbf{A} \longrightarrow \mathbf{X}$ in exactly one way that makes the arrows ψ_A components of a natural isomorphism from $\mathbf{1_A}$ to $\mathbf{FG}$. There is also at least one natural isomorphism from $\mathbf{1_X}$ to $\mathbf{GF}$, so $\mathbf{F}$ and $\mathbf{G}$ form an equivalence situation.

PROOF This will be a corollary of Theorem 10.3 and Exercise 10.15. □

In many categorical contexts no axiom of choice is available, and the selections required in Theorem 9.3 cannot always be made. This is true, for example, of category theory in most toposes. Then some equivalence functors do not fit into equivalence situations. Of course, even then, many particular equivalence functors $\mathbf{F}$ fit into equivalence situations that are definable without arbitrary choices.

Equivalence functors and situations often arise in mathematical practice when the values of the functors, say $\mathbf{F}X$ and $\mathbf{G}A$, have only been defined up to isomorphism. It may follow that $X \cong \mathbf{GF}X$ and $A \cong \mathbf{FG}A$, while the definition simply does not determine whether or not $X = \mathbf{GF}X$.

Exercises

9.1 Let $\mathbf{A}$ be a Cartesian closed category. For any object A of $\mathbf{A}$ consider the functor $(\_ \times A)^A: \mathbf{A} \longrightarrow \mathbf{A}$. Show that there is a natural transformation $\varepsilon: \mathbf{1_A} \longrightarrow (\_ \times A)^A$, where the component at any object B of $\mathbf{A}$ is the transpose of the identity arrow $1_{B \times A}$.

9.2 Show that the natural isomorphisms between functors from $\mathbf{A}$ to $\mathbf{B}$ are precisely the isomorphisms in the functor category $\mathbf{B}^A$; that is, a natural transformation is iso iff each of its components is.

9.3 Let $\mathbf{A}$ have all finite products. Show that there is a functor $\mathbf{X}_{12}: \mathbf{A} \times \mathbf{A} \longrightarrow \mathbf{A}$ that takes any object $\langle A, A' \rangle$ of $\mathbf{A} \times \mathbf{A}$ to $A \times A'$, and any arrow $\langle f, g \rangle$ to $f \times g$. There is also a functor $\mathbf{X}_{21}: \mathbf{A} \times \mathbf{A} \longrightarrow \mathbf{A}$ that takes $\langle A, A' \rangle$ to $A' \times A$ and $\langle f, g \rangle$ to $g \times f$. Show that the twist arrows form a natural isomorphism $tw: \mathbf{X}_{12} \xrightarrow{\sim} \mathbf{X}_{21}$.

Show that there is a functor $\mathbf{X}_{(12)3}: \mathbf{A} \times \mathbf{A} \times \mathbf{A} \longrightarrow \mathbf{A}$ that takes any object $\langle A, A', A'' \rangle$ to $(A \times A') \times A''$ and any arrow $\langle f, g, h \rangle$ to $(f \times g) \times h$. There is also a functor $\mathbf{X}_{1(23)}$ that takes $\langle A, A', A'' \rangle$ to $A \times (A' \times A'')$. Show that the associativity arrows form a natural isomorphism from $\mathbf{X}_{(12)3}$ to $\mathbf{X}_{1(23)}$.

Conclude that the same holds for coproducts, $+_{12}: \mathbf{A} \times \mathbf{A} \longrightarrow \mathbf{A}$ and so on, assuming that $\mathbf{A}$ has all finite coproducts.

9.4 Consider functors $\mathbf{F}: \mathbf{A} \longrightarrow \mathbf{B}$, $\mathbf{G}$ and $\mathbf{H}$ both from $\mathbf{B}$ to $\mathbf{C}$, and $\mathbf{K}: \mathbf{C} \longrightarrow \mathbf{D}$, and a natural transformation $\varphi: \mathbf{G} \longrightarrow \mathbf{H}$. Show that there is a natural transformation $\varphi_{\mathbf{F}}: \mathbf{G} \circ \mathbf{F} \longrightarrow \mathbf{H} \circ \mathbf{F}$, where the component at any object A of $\mathbf{A}$ is the component of φ at $\mathbf{F}A$. Show that there is a natural transformation $\mathbf{K}\varphi: \mathbf{K} \circ \mathbf{G} \longrightarrow \mathbf{K} \circ \mathbf{H}$, where the component at any object B of $\mathbf{B}$ is $\mathbf{K}(\varphi_B)$.

Note that $\varphi_{\mathbf{F}}$ and $\mathbf{K}\varphi$ are both natural isomorphisms if φ is.

9.5 A transformation between bifunctors is natural if it is natural in each variable separately. More precisely, consider a parallel pair of bifunctors $\mathbf{F}, \mathbf{G}: \mathbf{A} \times \mathbf{B} \longrightarrow \mathbf{C}$ and a family of arrows $\nu_{A,B}: \mathbf{F}(A, B) \longrightarrow \mathbf{G}(A, B)$. Suppose, for every object A of $\mathbf{A}$ and arrow $f: B \longrightarrow B'$ of $\mathbf{B}$, that the following square commutes:

$$\begin{array}{ccc} \mathbf{F}(A, B) & \xrightarrow{\nu_{A,B}} & \mathbf{G}(A, B) \\ {\scriptstyle \mathbf{F}(1_A, f)}\downarrow & & \downarrow{\scriptstyle \mathbf{G}(1_A, f)} \\ \mathbf{F}(A, B') & \xrightarrow[\nu_{A,B'}]{} & \mathbf{G}(A, B') \end{array}$$

Suppose that the analogous condition is met for every arrow of $\mathbf{A}$ and object of $\mathbf{B}$. Show that $\nu: \mathbf{F} \longrightarrow \mathbf{G}$ is a natural transformation.

9.6 For functors $\mathbf{F}, \mathbf{G}: \mathbf{A} \longrightarrow \mathbf{B}$ a natural transformation $\nu: \mathbf{F} \longrightarrow \mathbf{G}$ is an arrow of $\mathbf{B}^{\mathbf{A}}$. Thus there is a functor that we also call $\nu: \mathbf{2} \longrightarrow \mathbf{B}^{\mathbf{A}}$, and this has a transpose $\bar{\nu}: \mathbf{2} \times \mathbf{A} \longrightarrow \mathbf{B}$. Twisting and transposing gives $\nu^{\circ}: \mathbf{A} \longrightarrow \mathbf{B}^{2}$. Describe ν°. What arrow does it assign to A, and what square to f? Describe $\bar{\nu}$.

9.7 (Exercise 9.4 revisited.) Given the functors of Exercise 9.4 there are functors $\mathbf{K}^{\mathbf{B}}: \mathbf{C}^{\mathbf{B}} \longrightarrow \mathbf{D}^{\mathbf{B}}$ and $\mathbf{C}^{\mathbf{F}}: \mathbf{C}^{\mathbf{B}} \longrightarrow \mathbf{C}^{\mathbf{A}}$. Show, for any arrow φ of $\mathbf{C}^{\mathbf{B}}$, that $\mathbf{K}^{\mathbf{B}}(\varphi)$ is $\mathbf{K}_{\varphi}$ and $\mathbf{C}^{\mathbf{F}}(\varphi)$ is $\varphi_{\mathbf{F}}$.

9.8 Show that the functor $\mathbf{D}: \mathbf{A}^{2} \longrightarrow \mathbf{A}$ is induced by $0: \mathbf{1} \longrightarrow \mathbf{2}$; that is, up to the isomorphism $\mathbf{A} \cong \mathbf{A}^{1}$, $\mathbf{D}$ is $\mathbf{A}^{0}: \mathbf{A}^{2} \longrightarrow \mathbf{A}^{1}$. Show that $\mathbf{C}$ is induced by $1: \mathbf{1} \longrightarrow \mathbf{2}$, and id by $!_2: \mathbf{2} \longrightarrow \mathbf{1}$.

9.9 Show that a functor category $\mathbf{B}^{\mathbf{A}}$ has a terminal object (or an initial object) iff $\mathbf{B}$ does.

9.10 Limits in functor categories are computed pointwise.

Suppose that $\mathbf{B}$ has binary products. Show that for any $\mathbf{F}$ and $\mathbf{G}: \mathbf{A} \longrightarrow \mathbf{B}$ there is a unique functor $\mathbf{F} \times \mathbf{G}: \mathbf{A} \longrightarrow \mathbf{B}$ such that the following holds. For each object A of $\mathbf{A}$, $(\mathbf{F} \times \mathbf{G})A = \mathbf{F}A \times \mathbf{G}A$, and $p_1: \mathbf{F}A \times \mathbf{G}A \longrightarrow \mathbf{F}A$ and $p_2: \mathbf{F}A \times \mathbf{G}A \longrightarrow \mathbf{G}A$ are components of natural transformations $p_1: \mathbf{F} \times \mathbf{G} \longrightarrow \mathbf{F}$ and $p_2: \mathbf{F} \times \mathbf{G} \longrightarrow \mathbf{G}$. That is, show that this requirement of naturality forces the definition of $(\mathbf{F} \times \mathbf{G})f$ for each $\mathbf{A}$ arrow f. Show that $\mathbf{F} \times \mathbf{G}, p_1, p_2$ is a product diagram for $\mathbf{F}$ and $\mathbf{G}$ in the functor category $\mathbf{B}^{\mathbf{A}}$.

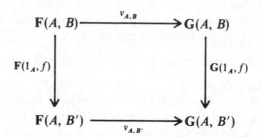

Suppose that $\mathbf{B}$ has equalizers. Given functors $\mathbf{F}$ and $\mathbf{G}$ as before, and a parallel pair of natural transformations $\varphi, \psi: \mathbf{F} \longrightarrow \mathbf{G}$, define a functor $\mathbf{E}: \mathbf{A} \longrightarrow \mathbf{B}$ and natural transformation $e: \mathbf{E} \longrightarrow \mathbf{F}$ forming an equalizer for φ and ψ in $\mathbf{B}^{\mathbf{A}}$. Do this by making each e_A an equalizer for φ_A and ψ_A in $\mathbf{A}$.

Generalize to all limits, and to all co-limits.

9.11 Exercise 9.10 does not generalize to exponentials. Show that $\mathbf{2}$ is Cartesian closed, as is the functor category $\mathbf{2}^{\mathbf{2}} \cong \mathbf{3}$ (in fact, both are Heyting algebras). No functor from $\mathbf{2}$ to $\mathbf{2}$ takes each object A of $\mathbf{2}$ to 0^A. What is the exponential in $\mathbf{2}^{\mathbf{2}}$ of $0 \circ !_{\mathbf{2}}$ by the identity functor?

9.12 Suppose that $\mathbf{A}$ has all finite products. Prove that $\_ \times 1: \mathbf{A} \longrightarrow \mathbf{A}$ is an equivalence functor, and fits into an equivalence situation. [Do this with no special assumption such as $A \times 1 = A$, so that $\_ \times 1$ is not necessarily either one-to-one or onto for objects.]

9.13 Let $\mathbf{A}$ be any category with all finite products. For each group G in $\mathbf{A}$ consider the zero homomorphism $G \longrightarrow 1 \xrightarrow{e} G$. Show that these homomorphisms form a natural transformation from the identity functor for $\mathbf{Grp}_{\mathbf{A}}$ to the functor which takes each group to itself and each homomorphism $f: G \longrightarrow H$ to the zero homomorphism $G \longrightarrow 1 \xrightarrow{e} H$.

10

Adjunctions

Most of this chapter is devoted to proving fundamental results on adjunctions. More specific results and calculations are given in the exercises.

10.1 Universal arrows

Given any functor $\mathbf{F}: \mathbf{X} \longrightarrow \mathbf{A}$ and object A of $\mathbf{A}$, a *universal arrow* from $\mathbf{F}$ to A consists of an object X of $\mathbf{X}$ and an arrow $\varepsilon_A: \mathbf{F}X \longrightarrow A$ such that given any object Y of $\mathbf{X}$ and arrow $f: \mathbf{F}Y \longrightarrow A$ there is a unique arrow $g: Y \longrightarrow X$ of $\mathbf{X}$ that makes the following triangle commute:

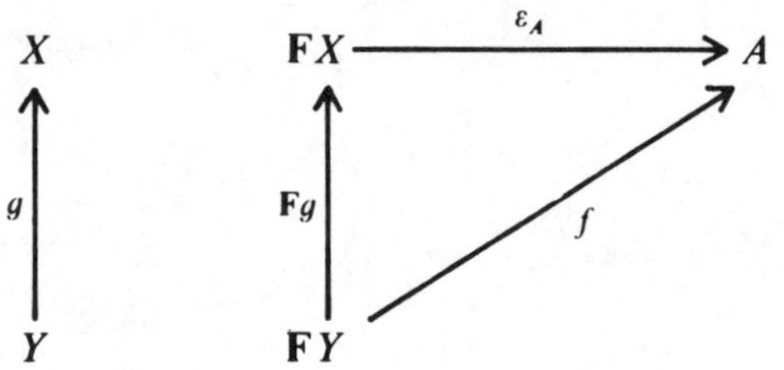

We will say that g is the *adjunct to f with respect to* $\varepsilon_{\mathbf{A}}$, or just that g is adjunct of f when the arrow ε_A is clear from the context. At the same time f is uniquely determined by g since f is $\varepsilon_A \circ \mathbf{F}g$. We also call f the adjunct of g with respect to ε_A.

For example, for any objects A and B of a category $\mathbf{A}$ with all finite products, the pair B^A, $ev_{A,B}$ is universal from the functor $\_ \times A$ to B. In that case adjuncts are called transposes.

A functor may or may not have a universal arrow to a given object, but if it does the universal arrow is determined up to isomorphism.

THEOREM 10.1 If X, ε_A is universal from $\mathbf{F}$ to A and there is an isomorphism $i: Y \longrightarrow X$ then Y, $\varepsilon_A \circ \mathbf{F}i$ is also universal from $\mathbf{F}$ to A. Conversely, if X, ε_A is universal from $\mathbf{F}$ to A and Y, ε'_A is too, and then the unique arrow $i: Y \longrightarrow X$ with $\varepsilon'_A = \varepsilon_A \circ \mathbf{F}i$ is iso.

PROOF The proof follows the familiar argument. □

There is a dual notion. Given $\mathbf{F}$ and A as above, a *universal arrow* from A to $\mathbf{F}$ consists of an object X of $\mathbf{X}$ and an arrow $\eta_A : A \longrightarrow \mathbf{F}X$ with the property suggested by the following diagram:

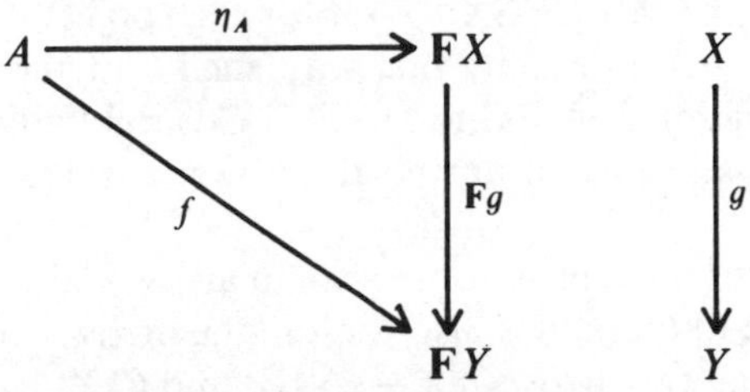

This is the definition of a universal arrow from $\mathbf{F}$ to A, with all arrows in $\mathbf{X}$ and $\mathbf{A}$ reversed. In fact, $\mathbf{F} : \mathbf{X} \longrightarrow \mathbf{A}$ can be reconstrued as a functor $\mathbf{F} : \mathbf{X}^{\mathrm{op}} \longrightarrow \mathbf{A}^{\mathrm{op}}$ (the functor $\mathbf{F}$ is not reversed), and this reconstrual turns a universal arrow to $\mathbf{F}$ into one from $\mathbf{F}$ and vice versa.

10.2 Adjunctions

An *adjunction* $\langle \mathbf{F}, \mathbf{G}, \eta, \xi \rangle$ consists of two functors and two natural transformations as shown:

$$\mathbf{X} \underset{\mathbf{G}}{\overset{\mathbf{F}}{\rightleftarrows}} \mathbf{A} \qquad \eta : \mathbf{1}_{\mathbf{X}} \longrightarrow \mathbf{GF} \qquad \varepsilon : \mathbf{FG} \longrightarrow \mathbf{1}_{\mathbf{A}}$$

such that $\varepsilon_{\mathbf{F}} \circ \mathbf{F}\eta = 1_{\mathbf{F}}$ and $\mathbf{G}\varepsilon \circ \eta_{\mathbf{G}} = 1_{\mathbf{G}}$; that is, for every object X of $\mathbf{X}$ we have $\varepsilon_{\mathbf{F}X} \circ \mathbf{F}\eta_X = 1_{\mathbf{F}X}$ and similarly for every A of $\mathbf{A}$.

Given an adjunction as above, we say that $\mathbf{F}$ is *left adjoint* to $\mathbf{G}$, or $\mathbf{G}$ is *right adjoint* to $\mathbf{F}$, and write $\mathbf{F} \dashv \mathbf{G}$. Then η is called the *unit* of the adjunction and ε the *counit*. The equations on the unit and counit are called the *triangular identities.*

The following Theorems 10.2 and 10.3 characterize an adjunction as a selection of universal arrows. Proofs are given in the next section. An adjunction is described as an isomorphic correspondence between arrows $f : \mathbf{F}X \longrightarrow A$ and their adjuncts $g : X \longrightarrow \mathbf{G}A$ in Section 10.4.

THEOREM 10.2 Given an adjunction as above, every arrow $\eta_X : X \longrightarrow \mathbf{GF}X$ is universal from X to $\mathbf{G}$. Specifically, the adjunct of any $f : X \longrightarrow \mathbf{G}A$ with respect to η_X is $\varepsilon_A \circ \mathbf{F}f$. Dually, every $\varepsilon_A : \mathbf{FG}A \longrightarrow A$ is universal from $\mathbf{F}$ to A, and the adjunct to any $g : \mathbf{F}X \longrightarrow A$ is $\mathbf{G}f \circ \eta_X$.

COROLLARY If $\mathbf{G}$ and $\mathbf{G}'$ are both right adjoint to one $\mathbf{F} : \mathbf{X} \longrightarrow \mathbf{A}$ there is a natural isomorphism $i : \mathbf{G} \longrightarrow \mathbf{G}'$. More specifically, if $\langle \mathbf{F}, \mathbf{G}, \eta, \varepsilon \rangle$ is an adjunction and $\langle \mathbf{F}, \mathbf{G}', \eta', \varepsilon' \rangle$ is another, then there is a unique natural isomorphism i with $\varepsilon' = \varepsilon \circ \mathbf{F}i$. This i is also the unique natural isomorphism satisfying the dual equation for the units. The dual result holds for **F** and **F**′ both left adjoint to **G**.

THEOREM 10.3 Given a functor $\mathbf{F}:\mathbf{X}\longrightarrow\mathbf{A}$ and a selected object $\mathbf{G}A$ of **X** and universal arrow $\mathbf{G}A,\ \varepsilon_A:\mathbf{FG}A\longrightarrow A$ for each object A of **A**, **G** extends to a functor $\mathbf{G}:\mathbf{A}\longrightarrow X$ in exactly one way such that the arrows ε_A form a natural transformation from **FG** to $\mathbf{1_A}$. This **G** is right adjoint to **F**, with ε as counit. Dually, a selection of universal arrows to a functor **G** gives a left adjoint **F** to **G**.

Thus, given an adjunction as above and an arrow $f: X\longrightarrow \mathbf{G}A$ the adjunct $g:\mathbf{F}X\longrightarrow A$ is $\varepsilon_A\circ\mathbf{F}f$; and g is also the unique arrow with $f=\mathbf{G}g\circ\eta_X$. We will often prove that two arrows $f: X\longrightarrow\mathbf{G}A$ and $f': X\longrightarrow\mathbf{G}A$ are equal by finding the adjunct of f one way and of f' the other way, and showing the adjuncts are equal. We did this repeatedly in Chapter 6, where in fact for every object A the functor $\_^A$ was right adjoint to $\_\times A$ with counit ev.

10.3 Proofs

LEMMA 10.4 Given an adjunction as above, we can construe **F** as a functor $\mathbf{F}:\mathbf{X}^{\mathrm{op}}\longrightarrow\mathbf{A}^{\mathrm{op}}$, and similarly for **G**. On this construal **F** is right adjoint to **G** with unit ε and counit η. Thus any theorem on all left adjoints or all counits has a dual theorem on all right adjoints or all units.

PROOF The proof follows by working through the definitions. □

PROOF OF THEOREM 10.2 Naturality of η, specifically $\mathbf{GF}f\circ\eta_X=\eta_{\mathbf{G}A}\circ f$, and the triangular identity show that the following triangle commutes:

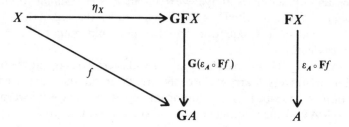

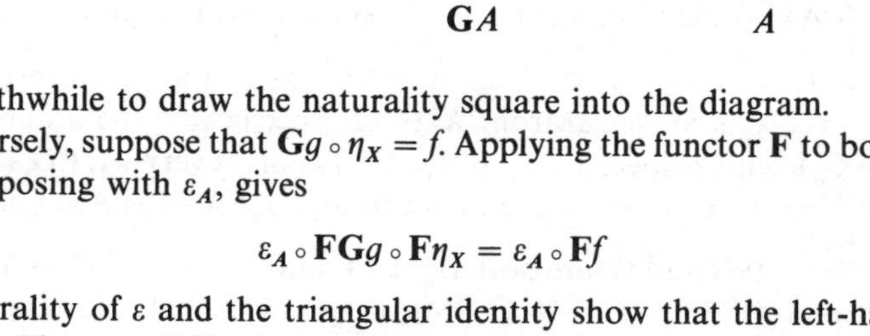

It is worthwhile to draw the naturality square into the diagram.

Conversely, suppose that $\mathbf{G}g\circ\eta_X=f$. Applying the functor **F** to both sides, and composing with ε_A, gives

$$\varepsilon_A\circ\mathbf{FG}g\circ\mathbf{F}\eta_X=\varepsilon_A\circ\mathbf{F}f$$

But naturality of ε and the triangular identity show that the left-hand side equals g. Thus $\varepsilon_A\circ\mathbf{F}f$ is the one and only arrow that makes the triangle commute, the adjunct of f. □

PROOF OF COROLLARY By Theorems 10.1 and 10.2, for each object A of $\mathbf{A}$ there is a unique isomorphism $i_A: \mathbf{G}'A \longrightarrow \mathbf{G}A$ with $\varepsilon'_A = \varepsilon_A \circ \mathbf{F}i_A$. We only need show that these form the components of a natural transformation; that is, for every $h: A \longrightarrow B$ of $\mathbf{A}$, $\mathbf{G}h \circ i_A = i_B \circ \mathbf{G}'h$. By Theorem 10.2 it suffices to show that both sides give the same result when plugged into $\varepsilon_B \circ \mathbf{F}(\_)$. Naturality of ε and the definition of i_A shows that the left-hand side gives $h \circ \varepsilon'_A$. Naturality of ε' and the definition of i_B show that the right-hand side gives the same. □

LEMMA 10.5 Let a functor $\mathbf{F}: \mathbf{X} \longrightarrow \mathbf{A}$ have a selected universal arrow $\mathbf{G}A$, ε_A to each object A of $\mathbf{A}$, when '$\mathbf{G}A$' names some object of $\mathbf{X}$ and shows that this object depends on A. Then there is just one way to extend $\mathbf{G}$ to a functor $\mathbf{G}: \mathbf{A} \longrightarrow \mathbf{X}$ so that the universal arrows become components of a natural transformation $\varepsilon: \mathbf{FG} \longrightarrow \mathbf{1_A}$.

PROOF For any arrow $f: B \longrightarrow C$ of $\mathbf{A}$, define $\mathbf{G}f$ as the unique arrow of $\mathbf{X}$ that makes the following square commute:

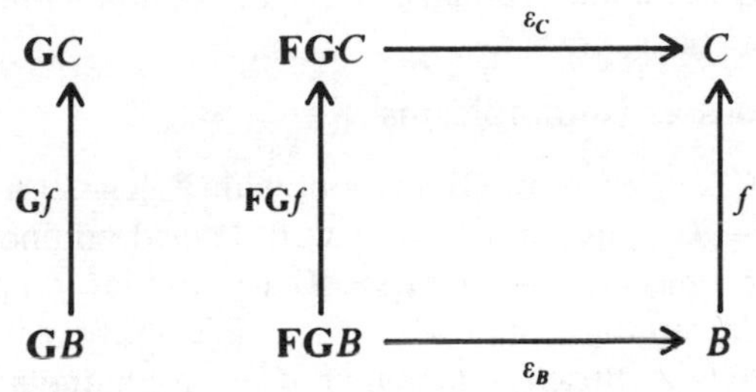

that is, $\mathbf{G}f$ is adjunct to $f \circ \varepsilon_B$. The rest of the proof copies the proof that $\_^A$ is a functor in Section 6.2. □

For every object X of $\mathbf{X}$, define $\eta_X: X \longrightarrow \mathbf{GF}X$ to be the adjunct to $1_{\mathbf{F}X}$ relative to $\varepsilon_{\mathbf{F}X}$, and thus the unique arrow that makes the following diagram commute:

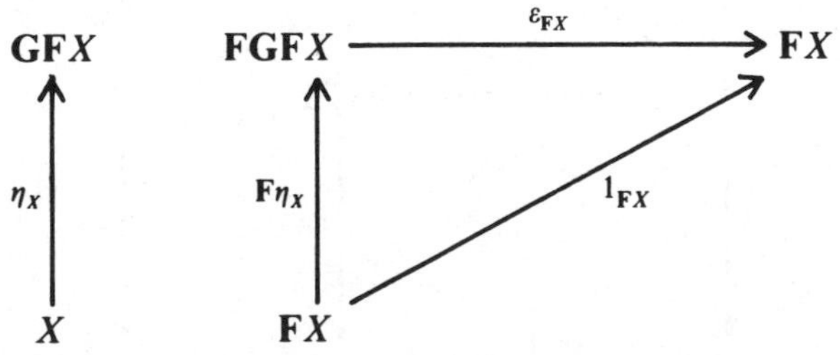

LEMMA 10.6 The arrows η_X form a natural transformation $\eta: 1_{\mathbf{X}} \longrightarrow \mathbf{GF}$.

PROOF Show, for every $f: X \longrightarrow Y$ of $\mathbf{X}$, that $\mathbf{GF}f \circ \eta_X = \eta_Y \circ f$, by showing that both sides give the same result when plugged into $\varepsilon_{\mathbf{F}Y} \circ \mathbf{F}(\_)$, namely $\mathbf{F}f$.

The calculation is very similar to the proof of the corollary to Theorem 10.2, using the definition of η_X in place of the triangular identity. □

LEMMA 10.7 The natural transformations η and ε satisfy the triangular identities.

PROOF One of the identities is the definition of η. To prove the other, $\mathbf{G}\varepsilon_A \circ \eta_{\mathbf{G}A} = 1_{\mathbf{G}A}$, again plug both sides into $\varepsilon_A \circ \mathbf{F}(\_)$ to show that they have the same adjunct relative to ε_A. □

PROOF OF THEOREM 10.3 Lemmas 10.5–10.7 show that a functor $\mathbf{F}:\mathbf{X} \longrightarrow \mathbf{A}$ and selected universal arrows from it give a unique right adjoint that makes the arrows a natural transformation. The other claim follows as the dual to this one. □

A functor **F** may or may not have a right (or left) adjoint, but if it has one the adjoint is determined only up to a natural isomorphism. An adjunction $\langle \mathbf{F}, \mathbf{G}, \eta, \varepsilon \rangle$ is fully determined by **F** and ε, or by **G** and η. Even given both functors **F** and **G** there may be many units and co-units that make $\mathbf{F} \dashv \mathbf{G}$.

10.4 Adjunctions as isomorphisms

Suppose that $\langle \mathbf{F}, \mathbf{G}, \eta, \varepsilon \rangle$ is an adjunction, with $\mathbf{F}:\mathbf{X} \longrightarrow \mathbf{A}$. To be precise, an arrow $f: X \longrightarrow \mathbf{G}A$ may not have a well defined adjunct because there may be another object A' with $\mathbf{G}A = \mathbf{G}A'$, so that f has one adjunct $g: \mathbf{F}X \longrightarrow A$ and another $g': \mathbf{F}X \longrightarrow A'$. We have let the notation $f: X \longrightarrow \mathbf{G}A$ specify A. Strictly, the pair f, A has a unique adjunct X, g.

Define a category $(\mathbf{X}, \mathbf{G})$ called the *comma category* of **X** over **G**. An object of $(\mathbf{X}, \mathbf{G})$ consists of an **X** arrow f and an **A** object A as shown:

$$X \xrightarrow{f} \mathbf{G}A \qquad A$$

An arrow consists of a commutative square of **X** and an **A** arrow as shown:

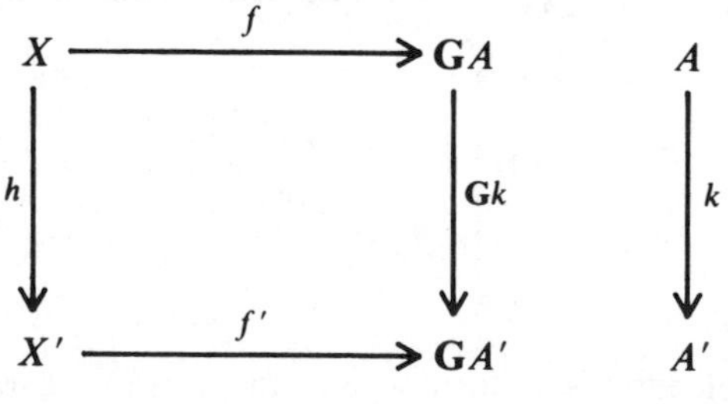

We call this arrow $\langle h, k \rangle$. Composition is defined by the obvious pasting, $\langle h, k \rangle \circ \langle h' \circ k' \rangle = \langle h \circ h', k \circ k' \rangle$. Arrows $\langle 1_X, 1_A \rangle$ are identities.

There is a *base functor* $\mathbf{b}:(\mathbf{X}, \mathbf{G}) \longrightarrow \mathbf{X} \times \mathbf{A}$ which forgets the arrow. It takes an object $\langle f: X \longrightarrow \mathbf{G}A, A\rangle$ to $\langle X, A\rangle$, and an arrow $\langle h, k\rangle$ to the $\mathbf{X} \times \mathbf{A}$ arrow $\langle h, k\rangle$.

For the functor $\mathbf{F}: \mathbf{X} \longrightarrow \mathbf{A}$ there is an analogous comma category of $\mathbf{F}$ over $\mathbf{A}$, $(\mathbf{F}, \mathbf{A})$. An object is an $\mathbf{X}$ object X and an $\mathbf{A}$ arrow $g: \mathbf{F}X \longrightarrow A$, and so on. This also has a base functor to $\mathbf{X} \times \mathbf{A}$.

THEOREM 10.8 Given an adjunction $\langle \mathbf{F}, \mathbf{G}, \eta, \varepsilon\rangle$, there is a unique functor isomorphism $\mathbf{I}$ that takes each $f: \mathbf{F}X \longrightarrow A$ to its adjunct and commutes with base functors as shown:

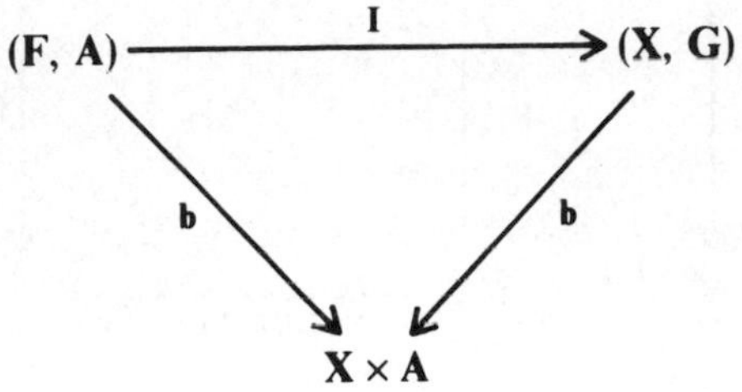

PROOF Consider any arrow of $(\mathbf{F}, \mathbf{A})$:

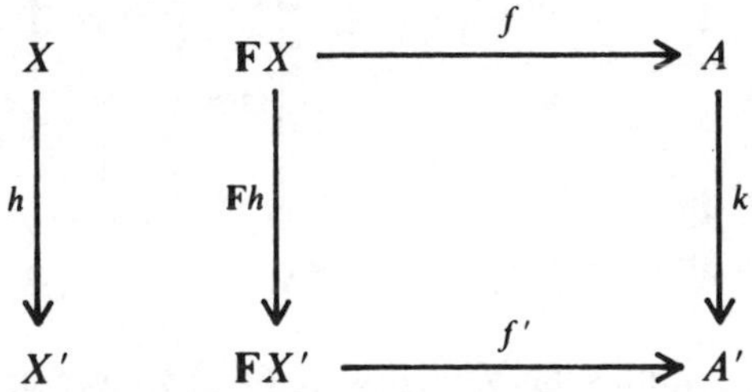

If $\mathbf{I}$ is to take the top and bottom arrows to their adjuncts and commute with the base functors, then it must take this arrow to

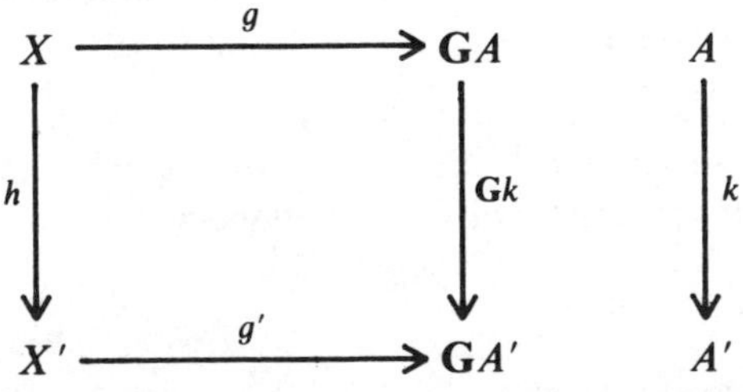

with g adjunct to f, and g' adjunct to f'. To show that $\mathbf{I}$ is a well defined functor we must show that the lower square commutes if the upper one does. To show that $\mathbf{I}$ has an inverse is to show that the upper square commutes if the lower one does. These calculations form Exercise 10.9. □

THEOREM 10.9 Conversely, an isomorphism $\mathbf{I}:(\mathbf{F},\mathbf{A})\overset{\sim}{\longrightarrow}(\mathbf{X},\mathbf{G})$ that commutes with base functors gives an adjunction $\langle \mathbf{F},\mathbf{G},\eta,\varepsilon\rangle$, where for every $\mathbf{X}$ object X the unit η_X is $\mathbf{I}(X, 1_{\mathbf{F}X})$, and for every $\mathbf{A}$ object A the counit is $\mathbf{I}^{-1}(1_{\mathbf{G}A}, A)$. Then $\mathbf{I}$ and its inverse take arrows to their adjuncts.

PROOF It suffices to show that, for every $\mathbf{X}$ object X, the arrow $\mathbf{I}(1_{\mathbf{F}X}, \mathbf{F}X)$ is universal from X to $\mathbf{G}$. For any arrow $g: X \longrightarrow \mathbf{G}A$ of $\mathbf{A}$ consider the following diagrams:

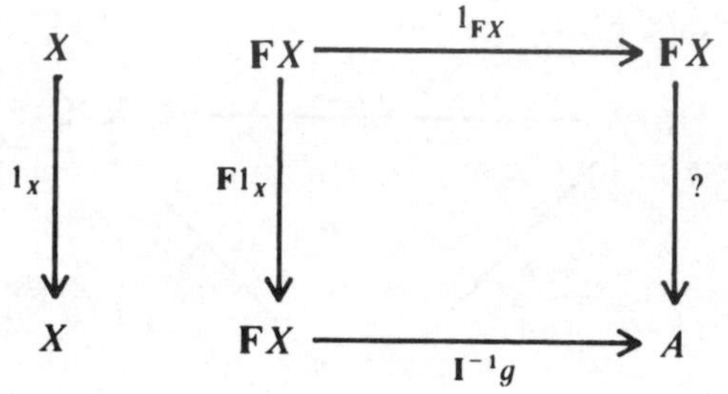

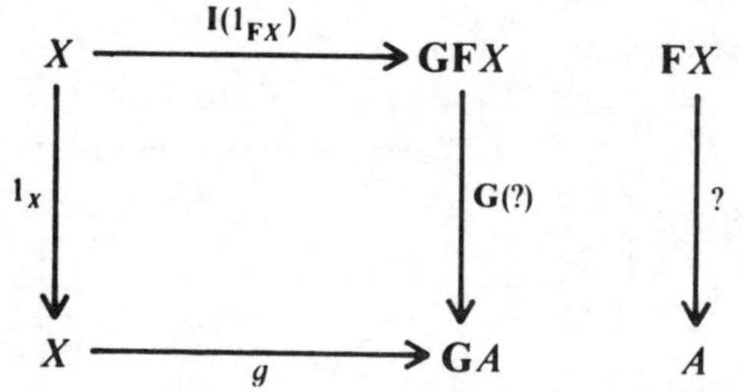

If either of these commutes it is an arrow in the obvious comma category, and the other corresponds to it under $\mathbf{I}$ and so also commutes. But $\mathbf{I}^{-1}g$ is the only arrow that can go in place of ? to make the upper square commute, and so it is the only one that makes the lower one commute. Thus $\mathbf{I}^{-1}g$ is the adjunct of g relative to $\mathbf{I}(X, 1_{\mathbf{F}X})$. □

The constructions of an isomorphism from an adjunction and of an adjunction from an isomorphism are inverse to one another. Adjunctions correspond exactly to isomorphisms of comma categories that commute with base functors.

10.5 Adjunctions compose

Consider adjoint functors $\mathbf{F} \dashv \mathbf{G}$ and $\mathbf{F}' \dashv \mathbf{G}'$ as shown:

$$\mathbf{A} \underset{\mathbf{G}}{\overset{\mathbf{F}}{\leftrightarrows}} \mathbf{B} \underset{\mathbf{G}'}{\overset{\mathbf{F}'}{\leftrightarrows}} \mathbf{C}$$

Given any **C** object C and **A** arrow as shown:

$$C \qquad \mathbf{FF}'C \xrightarrow{\;f\;} A$$

form its adjunct under $\mathbf{F} \dashv \mathbf{G}$

$$C \qquad \mathbf{F}'C \xrightarrow{\;g\;} \mathbf{G}A \qquad A$$

and the adjunct under $\mathbf{F}' \dashv \mathbf{G}'$

$$C \xrightarrow{\;h\;} \mathbf{G}'\mathbf{G}A \qquad A$$

By the same reasoning that was used to show that **I** was a functor in Theorem 10.8, both steps take commutative squares to commutative squares, and each has an inverse. Thus the two give an isomorphism from $(\mathbf{FF}', \mathbf{A})$ to $(\mathbf{C}, \mathbf{G}'\mathbf{G})$ that commutes with base functors. So $\mathbf{FF}' \dashv \mathbf{G}'\mathbf{G}$.

Exercises

10.1 Take any category **A** and the diagonal functor $\Delta : \mathbf{A} \longrightarrow \mathbf{A} \times \mathbf{A}$. Show that P, p_1, p_2 is a product for A and B in **A** iff

$$P \qquad \langle P, P \rangle \xrightarrow{\langle p_1, p_2 \rangle} \langle A, B \rangle$$

is a universal arrow from Δ to the object $\langle A, B \rangle$ of $\mathbf{A} \times \mathbf{A}$. (What is an $\mathbf{A} \times \mathbf{A}$ arrow from $\langle T, T \rangle$ to $\langle A, B \rangle$?) Conclude that Theorems 2.1 and 2.2 are special cases of Theorem 10.1. Conclude that co-products are universal arrows to Δ.

10.2 Given a functor $\mathbf{F} : \mathbf{X} \longrightarrow \mathbf{A}$, define a category $(\mathbf{F}, A)$ in the following way. An object is an **X** object X and an **A** arrow $f : \mathbf{F}X \longrightarrow A$. An arrow from X, f to Y, g is an arrow $h : X \longrightarrow Y$ that makes the following diagram commute:

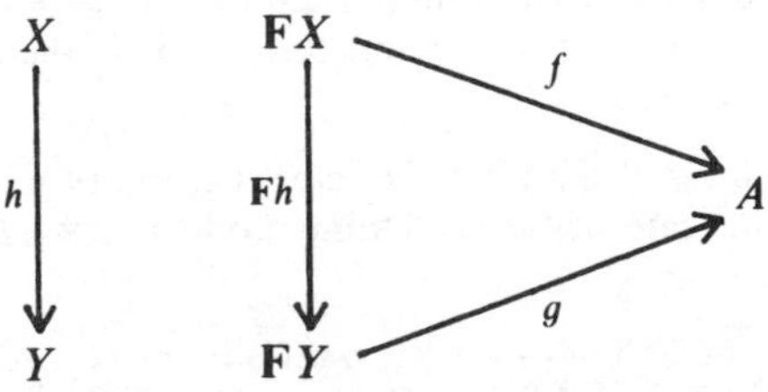

with the obvious composition and identities, taken from **X**. Show that a universal arrow from **F** to A is a terminal object in this category. Conclude that Theorem 10.1 is a special case of Theorems 1.1 and 1.2.

10.3 Suppose that a functor $\mathbf{F}$ is an isomorphism. Show that $\mathbf{F} \dashv \mathbf{F}^{-1}$, and that $\mathbf{F}^{-1} \dashv \mathbf{F}$, both with identity natural transformations as unit and counit.

10.4 Show that a functor $A: \mathbf{1} \longrightarrow \mathbf{A}$ is right adjoint to $!_{\mathbf{A}}$ iff A is a terminal object. In this case the functors uniquely determine the unit and counit. What are they? Describe left adjoints to $!_{\mathbf{A}}$.

10.5 Show that a category $\mathbf{A}$ has all binary products iff there is a right adjoint $\_\times\_: \mathbf{A} \times \mathbf{A} \longrightarrow \mathbf{A}$ to Δ. By Exercise 10.1 we can take $\langle p_1, p_2 \rangle$ as the counit. Then what is the unit?

Describe a left adjoint to Δ, and its unit and counit.

10.6 Define a category $\mathbf{E}$ with two objects 0 and 1, their identity arrows, a parallel pair of arrows e_1 and e_2 from 0 to 1, and no other arrows. Show that a parallel pair f, g in any category $\mathbf{A}$ corresponds to a functor $(f, g): \mathbf{E} \longrightarrow \mathbf{A}$.

Define $\Delta: \mathbf{A} \longrightarrow \mathbf{A}^{\mathbf{E}}$, taking each object A of $\mathbf{A}$ to the constant functor which takes both objects of $\mathbf{E}$ to A and all arrows to $\mathbf{1}_A$. Show that Δ is the transpose of $\mathbf{p}_1: \mathbf{A} \times \mathbf{E} \longrightarrow \mathbf{A}$ and that, up to the isomorphism $\mathbf{A} \cong \mathbf{A}^{\mathbf{1}}$, it is also $\mathbf{A}^{(!_{\mathbf{E}})}: \mathbf{A}^{\mathbf{1}} \longrightarrow \mathbf{A}^{\mathbf{E}}$.

Show that a cone over a parallel pair f, g is a natural transformation from some ΔT to (f, g), and that a limit is a universal arrow from Δ to (f, g). Therefore $\mathbf{A}$ has equalizers iff Δ has a right adjoint $\mathbf{Eq}: \mathbf{A}^{\mathbf{E}} \longrightarrow \mathbf{A}$.

State the dual result for coequalizers.

10.7 Generalize Exercises 10.5 and 10.6 to all limits and colimits. A diagram itself can be defined as a functor. A diagram of two objects of $\mathbf{A}$ is a functor from the category $\mathbf{1} + \mathbf{1}$ with two objects, their identity arrows, and no others. (Show that this category is the coproduct of $\mathbf{1}$ with itself in $\mathbf{CAT}$.) Show that $\mathbf{A} \times \mathbf{A}$ is isomorphic to $\mathbf{A}^{(\mathbf{1}+\mathbf{1})}$. Exercise 10.6 showed that a parallel pair in $\mathbf{A}$ is a functor to $\mathbf{A}$ from $\mathbf{E}$. Define a category $\mathbf{C}$ with three objects and two non-identity arrows, so that a corner of arrows in any category $\mathbf{A}$ is a functor to $\mathbf{A}$ from $\mathbf{C}$.

Given a category $\mathbf{D}$, with objects i thought of as vertices and arrows e thought of as edges, define a *diagram of shape* $\mathbf{D}$ in a category $\mathbf{A}$ to be a functor $\mathbf{d}: \mathbf{D} \longrightarrow \mathbf{A}$. For each object i of $\mathbf{D}$ the object $\mathbf{d}i$ of $\mathbf{A}$ corresponds to A_i of Section 4.5, and $\mathbf{d}e$ corresponds to f_e. For every object T of $\mathbf{A}$ define the *constant diagram* $\Delta T: \mathbf{D} \longrightarrow \mathbf{A}$ to be the functor that takes each i to T and each e to 1_T. Define a functor $\Delta: \mathbf{A} \longrightarrow \mathbf{A}^{\mathbf{D}}$ that takes each object of $\mathbf{A}$ to its constant functor.

Show that a cone over the diagram $\mathbf{d}$ is a natural transformation from some constant functor ΔT to $\mathbf{d}$, and that a limit is a universal arrow from some ΔT to $\mathbf{d}$. Conclude that every diagram of shape $\mathbf{D}$ in $\mathbf{A}$ has a limit, iff there is a right adjoint to Δ. Dualize all this for colimits.

10.8 Show, for any object A of a Cartesian closed category $\mathbf{A}$, that $\_ \times \mathbf{A} \dashv \_^{A}$ with unit ev. Show that, in any category with all finite products, $\_ \times A$ has a left adjoint iff A is a terminal object.

10.9 Consider a functor $\mathbf{F}: \mathbf{X} \longrightarrow \mathbf{A}$ with right adjoint $\mathbf{G}$, and arrows $k: A \longrightarrow A'$ and $j: X' \longrightarrow X$. Let $f: \mathbf{F}X \longrightarrow A$ have adjunct $g: X \longrightarrow \mathbf{G}A$. Show that the adjunct of $k \circ f$ is $\mathbf{G}k \circ g$. Show that the adjunct of $g \circ j$ is $f \circ \mathbf{F}j$.

10.10 Let $\mathbf{F}: \mathbf{X} \longrightarrow \mathbf{A}$ have right adjoint $\mathbf{G}$. Show that for any epic $f: X \twoheadrightarrow Y$ of $\mathbf{X}$, $\mathbf{F}f$ is also epic. Show (or conclude by duality) that for any monic $f: A \rightarrowtail B$ of $\mathbf{A}$, $\mathbf{G}f$ is monic.

10.11 Again with $\mathbf{F} \dashv \mathbf{G}$ as above, show that if 1 is initial in **X** then **F**1 is initial in **A**. Show that **G** preserves terminal objects.

Show that **F** preserves coproducts, coequalizers, and in fact all colimits. Show (or conclude) that **G** preserves all limits.

Compare Exercises 10.9–10.11 with the results in Chapter 6.

10.12 Let **A** be a category with products. Show that for any **A** object A the functor A^* (defined in Section 8.3) is right adjoint to Σ_A.

10.13 This is another proof that adjoints compose. Suppose that the functors of Section 10.5 fit into adjunctions $\langle \mathbf{F}, \mathbf{G}, \eta, \varepsilon \rangle$ and $\langle \mathbf{F}', \mathbf{G}', \eta', \varepsilon' \rangle$. Then show that $\mathbf{FF}' \dashv \mathbf{G}'\mathbf{G}$ with unit $\mathbf{G}'\eta_{\mathbf{F}'} \circ \eta'$ and counit $\varepsilon \circ \mathbf{F}\varepsilon'_{\mathbf{G}}$. [Either show that all components of the unit are universal arrows, or show that the unit and counit satisfy the triangular identities.]

10.14 Given functors $\mathbf{G}: \mathbf{A} \longrightarrow \mathbf{B}$ and $\mathbf{F}': \mathbf{C} \longrightarrow \mathbf{B}$ define the *comma category of* **G** *over* $\mathbf{F}'$, $(\mathbf{G}, \mathbf{F}')$, in the following way. An object consists of an **A** object A, a **C** object C, and a **B** arrow $f: \mathbf{G}A \longrightarrow \mathbf{F}'C$. An arrow consists of an **A** arrow h and a **C** arrow k that make a commutative square in **B**, $\mathbf{F}'\,k \circ f' \circ \mathbf{G}h$. $(\mathbf{G}, \mathbf{F}')$ has an obvious base functor to $\mathbf{A} \times \mathbf{C}$. Use this to factor the construction in Section 10.5 into two isomorphisms of comma categories.

10.15 Suppose that $\mathbf{F}: \mathbf{X} \longrightarrow \mathbf{A}$ is full and faithful, and there is given a selected object $\mathbf{G}A$ and isomorphism $\psi_A: A \longrightarrow \mathbf{FG}A$ for each object A of **A**. Show that ψ_A^{-1} is universal from **F** to A. Conclude that **G** extends to a functor that is right adjoint to **F** with counit ψ^{-1}.

Show, for every X of **X**, that the unit η_X is iso. [Hint: show that $\mathbf{F}\eta_{\mathbf{X}} = \psi_{\mathbf{F}X}$.) Conclude that **F** and **G** make up an equivalence situation. In fact $\langle \mathbf{F}, \mathbf{G}, \eta, \psi^{-1} \rangle$ is an adjunction.

Consider any adjunction $\mathbf{H} \dashv \mathbf{K}$ with unit η and counit ε that are both natural isomorphisms. Show that also $\mathbf{K} \dashv \mathbf{H}$ with unit ε^{-1} and counit η^{-1}. Such an adjunction is called an *adjoint equivalence*.

10.16 For any object B of a Cartesian closed category **A**, consider the contravariant exponentiation functor $B^{-}: \mathbf{A}^{\mathrm{op}} \longrightarrow \mathbf{A}$. Show that B^{-} is right adjoint to itself considered as $B^{-}: \mathbf{A} \longrightarrow \mathbf{A}^{\mathrm{op}}$. Consider a series of arrows:

$$
\begin{array}{cl}
C \longrightarrow B^A & \text{of } \mathbf{A} \\
\hline
C \times A \longrightarrow B & \text{of } \mathbf{A} \\
\hline
A \longrightarrow B^C & \text{of } \mathbf{A} \\
\hline
B^C \longrightarrow A & \text{of } \mathbf{A}^{\mathrm{op}}
\end{array}
$$

Therefore Exercises 6.10 and 6.17 are special cases of Exercises 10.11.

10.17 Show that the comma categories of Section 10.4 and Exercise 10.14 can be defined by pullbacks as in the following diagram:

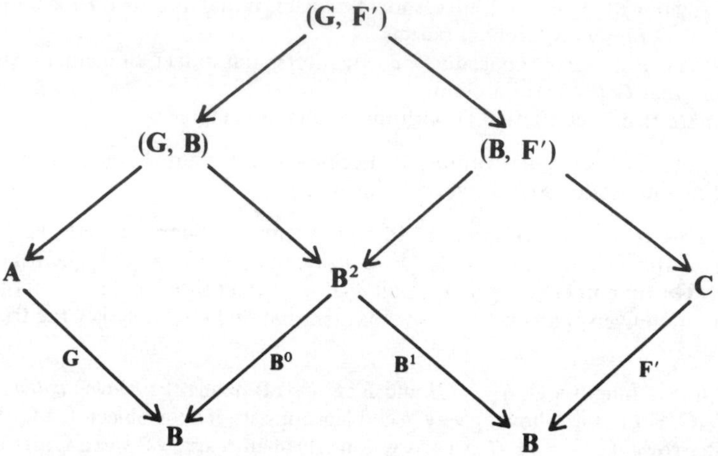

How are the base functors defined in terms of functors in this diagram?

10.18 Exponentiation preserves pullbacks, and so it preserves comma categories and adjunctions. Show that $(\mathbf{G}, \mathbf{B})^{\mathbf{D}} \cong (\mathbf{G}^{\mathbf{D}}, \mathbf{B}^{\mathbf{D}})$ for any category **D** and, up to the obvious isomorphisms, that $\mathbf{b}^{\mathbf{D}}$ is the base functor. So, if an isomorphism **I** commutes with base functors then $\mathbf{I}^{\mathbf{D}}$ does also. In other words, if $\mathbf{G} \dashv \mathbf{F}$ then $\mathbf{G}^{\mathbf{D}} \dashv \mathbf{F}^{\mathbf{D}}$.

10.19 Show, up to the isomorphism $(\mathbf{B}^{\mathbf{A}})^{\mathbf{D}} \cong (\mathbf{B}^{\mathbf{D}})^{\mathbf{A}}$, that exponentiating $\Delta : \mathbf{B} \longrightarrow \mathbf{B}^{\mathbf{D}}$ by **A** gives $\Delta : \mathbf{B}^{\mathbf{A}} \longrightarrow (\mathbf{B}^{\mathbf{A}})^{\mathbf{D}}$. Do this directly from the definition of Δ as transpose of p_1, and use Exercises 10.7 and 10.18 for a functorial proof of Exercise 9.10.

10.20 Compare our definition of adjunction with, for example, Mac Lane (1971, p. 78) or Barr and Wells (1985, p. 50). Show that a natural isomorphism $I : \hom_{\mathbf{A}}(\mathbf{F}\_, \_) \xrightarrow{\sim} \hom_{\mathbf{B}}(\_, \mathbf{G}\_)$ determines an isomorphism $\mathbf{I} : (\mathbf{F}, \mathbf{A}) \xrightarrow{\sim} (\mathbf{X}, \mathbf{G})$ over $\mathbf{A} \times \mathbf{B}$ and vice versa. [Assume that **X** and **A** are locally small, as defined in Chapter 12, so that the sets of arrows do exist.]

11

Slice categories

When dealing with sets we often work with sets of sets. In geometry we often deal with spaces that vary over another space (as with fibre bundles or sheaves). And high-level programming languages have dependent types, data types which themselves depend on parameters in other data types. All of these are represented categorically by indexed families of objects.

11.1 Indexed families of objects

For any object B of a category $\mathbf{A}$, an arrow $g: C \longrightarrow B$ to B can be seen as a B-indexed family of objects. Then the slice category $\mathbf{A}/B$ appears as the category of B-indexed families of objects and arrows. In particular, if $\mathbf{A}$ has a terminator 1 then a 1-indexed family of objects is just an object A of $\mathbf{A}$, with its unique $!_A: A \longrightarrow 1$. Thus $\mathbf{A} \cong \mathbf{A}/1$, and we will not distinguish objects from 1-indexed families.

The general idea of indexed families is easily motivated for categories $\mathbf{A}$ with all finite limits. Given $g: C \longrightarrow B$, for any global element $y: 1 \longrightarrow B$ there is a *fibre* of g over y, defined by the following pullback:

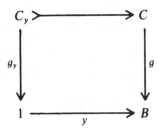

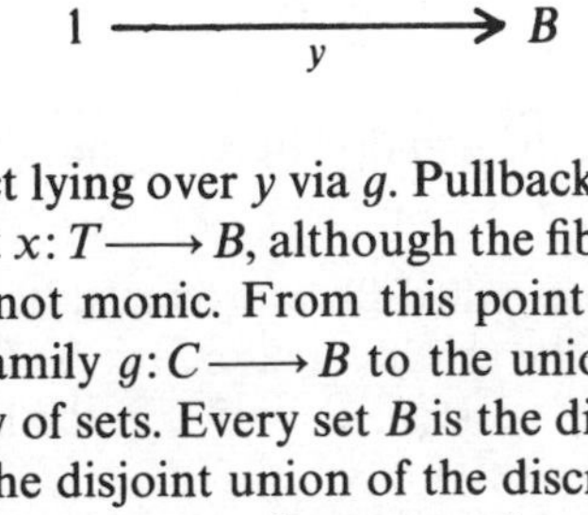

Think of C_y as the object lying over y via g. Pullback gives a fibre g_x of g over any generalized element $x: T \longrightarrow B$, although the fibre will not generally be a sub-object of C if x is not monic. From this point of view, $\Sigma_B: \mathbf{A}/B \longrightarrow \mathbf{A}$ takes each B-indexed family $g: C \longrightarrow B$ to the union of its fibres, C.

Let $\mathbf{A}$ be the category of sets. Every set B is the disjoint union of its global elements, so C will be the disjoint union of the discrete fibres C_y over global elements of B. But if $\mathbf{A}$ is a category of spaces and smooth maps, then not only will the fibres be spaces, but they will be smoothly related to each other over the indexing space B. In a category of data types and computable functions, the fibres will be data types, computably dependent on the indexing type B.

For any category $\mathbf{A}$, each slice $\mathbf{A}/B$ has a terminal object, 1_B, whether or not $\mathbf{A}$ has a terminal object. Intuitively, 1_B is a B-indexed family of singletons, in the sense that it is terminal among B-indexed families, but also in the sense that the fibre of 1_B over any index $x: T \longrightarrow B$ is just x itself. (Note that every category has pullbacks along any identity arrow.)

THEOREM 11.1 If $\mathbf{A}$ has all finite limits, then so does each slice $\mathbf{A}/B$.

PROOF The product of $\mathbf{A}/B$ objects, $f: A \longrightarrow B$ and $g: C \longrightarrow B$, is the following pullback in $\mathbf{A}$:

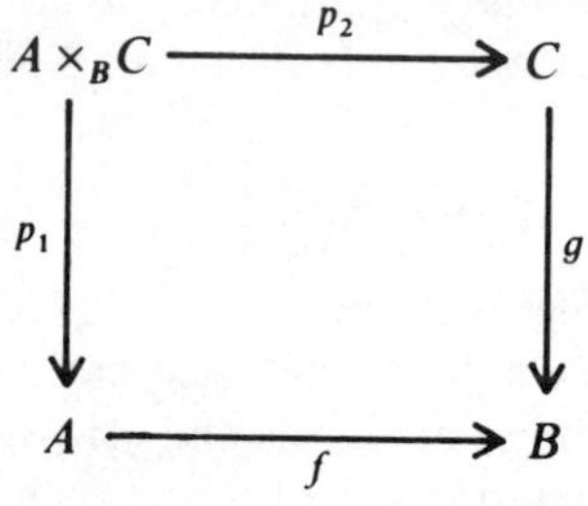

with its arrow $A \times_B C \longrightarrow B$ obtained by composing either way around. The reader can verify directly that $A \times_B C, p_1, p_2$ is a product diagram for f and g in $\mathbf{A}/B$.

For any $\mathbf{A}/B$, and arrows $k: f \longrightarrow g$ and $k': f \longrightarrow g$, the following diagram is an equalizer:

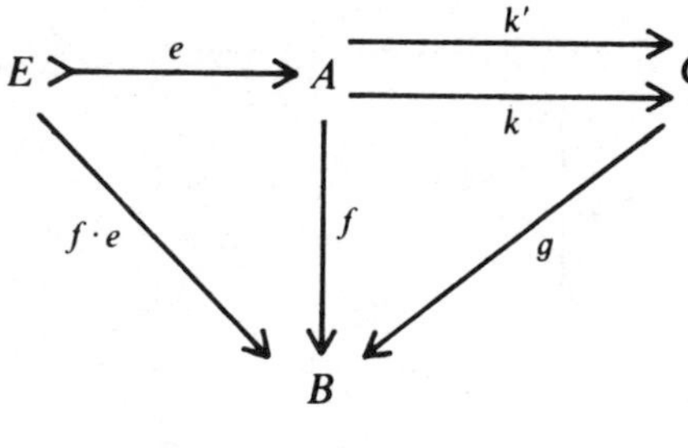

where E, e is an equalizer for k and k' in $\mathbf{A}$. Again, the verification is almost immediate. □

Pullbacks are sometimes called fibre products, because $A \times_B C$ is not the whole product $A \times C$ but the product taken fibre by fibre. The fibre over any $x: T \longrightarrow B$ contains those pairs $\langle h, k \rangle$ with h in the fibre of f over x and k in the fibre of g; that is, $f(h) = g(k) = x$.

Notice that Σ_B preserves equalizers but generally not products or the terminator.

Colimits in slice categories are discussed in Exercises 11.3–11.5, but for now we can say that if **A** has binary products then every $\Sigma_B: \mathbf{A}/B \longrightarrow \mathbf{A}$ preserves all colimits. This, and much more, follows from the next theorem:

THEOREM 11.2 If **A** has binary products, then for every **A** object B, the functor Σ_B is left adjoint to B^*.

PROOF Given objects $g: C \longrightarrow B$ of $\mathbf{A}/B$ and A of **A**, an arrow $\langle h, k\rangle: g \longrightarrow B^* A$ must have $h = g$, and so it is determined by any

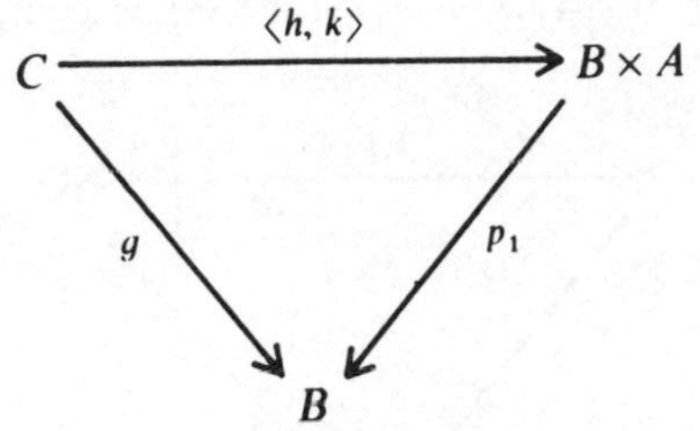

$k: C \longrightarrow A$. In short, $k: \Sigma_B g \longrightarrow A$ is adjunct to $\langle g, k \rangle$. The reader should calculate the unit and counit of the adjunction, and verify that they are universal arrows. □

Intuitively, $B^* A$ is a constant B-indexed family, the fibre of which everywhere is A. Precisely, the fibre over any $x: T \longrightarrow B$ is $T^* A$. The theorem says that a B-indexed family of arrows from $g: C \longrightarrow B$ to $B^* A$ can be completely determined by ignoring the indexing and giving an arrow k from the union of the indexed objects, C, to one copy of the fibre A. The product arrow $\langle g, k\rangle$ restores the indexing.

T-indexed families of arrows from A to B, as defined in Section 6, are adjuncts under $\Sigma_T \dashv T^*$ to arrows $T^* A \longrightarrow T^* B$. T-indexed composition corresponds to composition in $\mathbf{A}/T$, and the arrow f_T of Exercise 6.15 is the adjunct of $T^* f$. Now, by passing to $\mathbf{A}/T$ we gain the non-constant T-indexed families of objects, $k: C \longrightarrow T$ (see Exercise 11.6).

11.2 Internal products

If **A** is Cartesian closed, any B-indexed family of objects f has a B-indexed product $\Pi_B f$ defined by the following pullback:

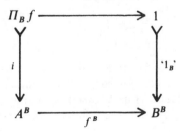

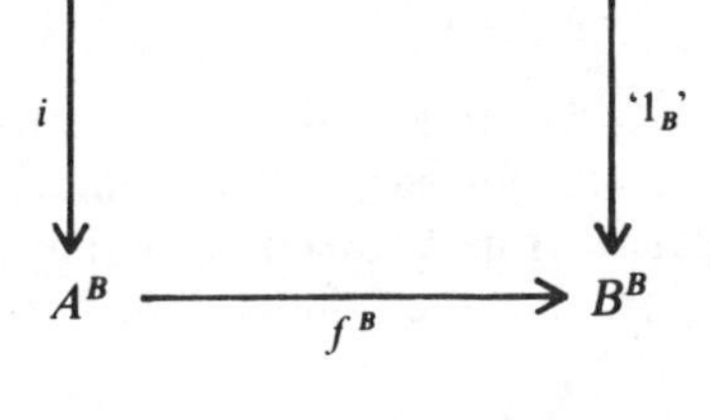

Intuitively, $\Pi_B f$ represents arrows from B to A the composites of which with f are 1_B. Such an arrow maps each index x in B to an element of the fibre A_x over x, so an element of $\Pi_B f$ is a B-tuple of values, with a value in A_x for each index x. If $\mathbf{A}$ is the category of sets, then $\Pi_B f$ is (up to isomorphism) the usual Cartesian product of the fibres A_y. If $\mathbf{A}$ is a category of spaces and smooth maps, then an element of $\Pi_B f$ will have values varying smoothly in the indices, while if $\mathbf{A}$ models data types and computable functions the values will vary computably.

There is a functor $\Pi_B: \mathbf{A}/B \longrightarrow \mathbf{A}$ that takes each $f: A \longrightarrow B$ to $\Pi_B f$. Given an $\mathbf{A}/B$ arrow

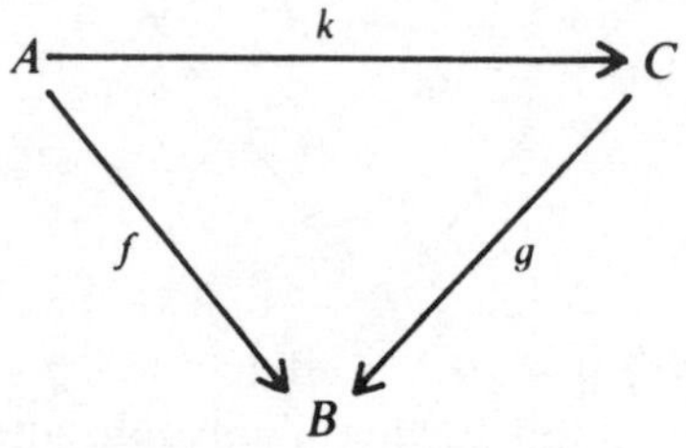

define $\Pi_B k: \Pi_B f \longrightarrow \Pi_B g$ by pulling the following triangle back along '1_B'. This is a functor by the pullback theorems in Chapter 4.

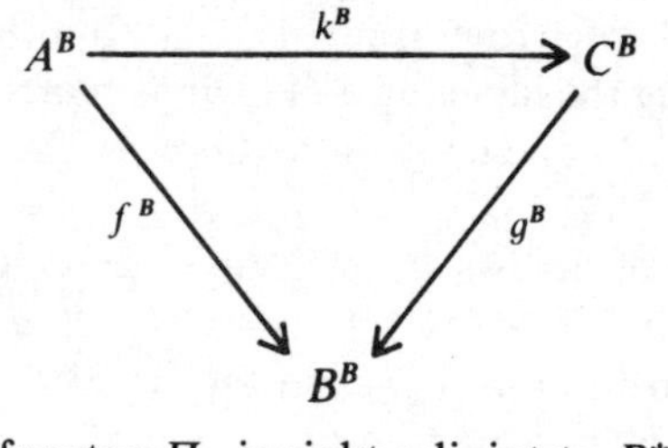

THEOREM 11.3 The functor Π_B is right adjoint to B^*.

PROOF Use the following series of arrows to establish an isomorphism of comma categories, $(B^*, \mathbf{A}/A) \cong (\mathbf{A}, \Pi_B)$:

$$\begin{array}{ll} t: T \longrightarrow \Pi_B f & \text{in } \mathbf{A} \\ \hline s: T \longrightarrow A^B & \text{such that } f^B \circ s = \text{`}1_B\text{'} \circ !_T \\ \hline \bar{s}: T \times B \longrightarrow A & \text{such that } f \circ \bar{s} = p_2 \\ \hline \bar{s}: B^* T \longrightarrow f & \text{in } \mathbf{A}/B \end{array}$$

For the first step, since $!_T$ is unique, the arrow t to the pullback is fully determined by s. The next step forms an adjunct under $\_ \times B \dashv \_^B$. The last is by definition of B^* and arrows in $\mathbf{A}/B$. To show that this is an isomorphism of comma categories, show that the following square in $\mathbf{A}$:

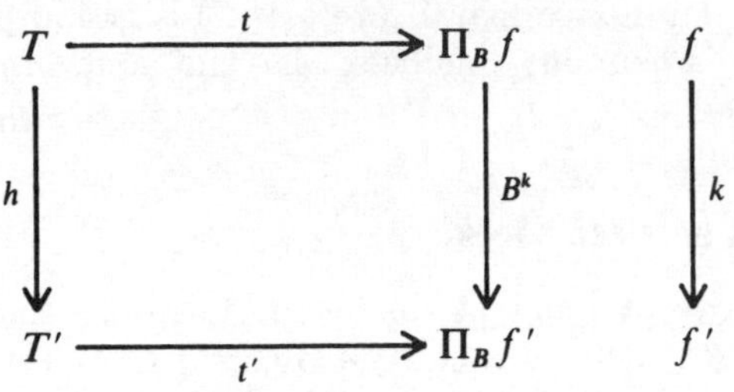

commutes iff this one in $\mathbf{A}/B$ does:

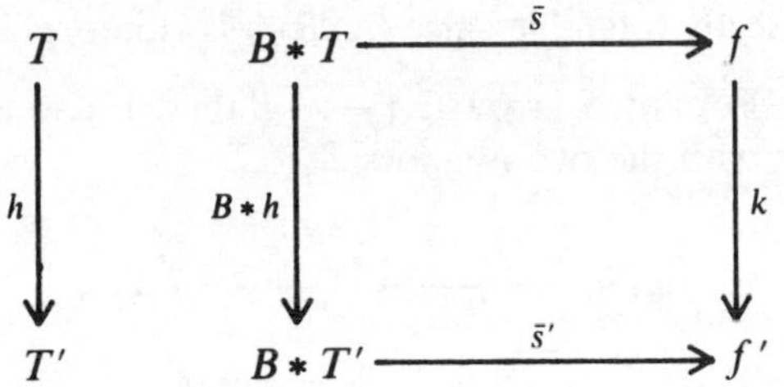

for any $f: A \longrightarrow B$, $f': A' \longrightarrow B$, and $k: f \longrightarrow f'$. This is a direct calculation using the steps in the series (see Exercise 11.8). □

There is a converse:

THEOREM 11.4 For any category $\mathbf{A}$ with all finite limits, if every slice functor A^* has a right adjoint $\Pi_A: \mathbf{A}/A \longrightarrow \mathbf{A}$ then $\mathbf{A}$ is Cartesian closed.

PROOF For each $\mathbf{A}$ object A, the functor $\_ \times A$ factors as $\Sigma_A \circ A^*$. But each factor has right adjoint, so by composition of adjoints $\_ \times A$ has right adjoint $\Pi_A \circ A^*$. □

If $\mathbf{A}$ is Cartesian closed it does not follow that all its slices are (see Exercise 11.9). But even if $\mathbf{A}$ and its slices are not Cartesian closed, each slice functor A^* on $\mathbf{A}$ preserves whatever exponentials there are in $\mathbf{A}$.

THEOREM 11.5 For any category $\mathbf{A}$ with binary products, suppose that B^C is an exponential of B by C. Then, for each $\mathbf{A}$ object A, $A^*(B^C)$ is an exponential of A^*B by A^*C in $\mathbf{A}/A$.

PROOF For any $\mathbf{A}/A$ object $g: C \longrightarrow A$, consider series of arrows:

$$\begin{array}{ll} g \times_A A^*C \longrightarrow A^*B & \text{in } \mathbf{A}/A \\ \hline \Sigma_A(g \times_A A^*C) \longrightarrow B & \text{in } \mathbf{A} \\ \hline (\Sigma_A g) \times C \longrightarrow B & \\ \hline \Sigma_A g \longrightarrow B^C & \\ \hline g \longrightarrow A^*(B^C) & \text{in } \mathbf{A}/A \end{array}$$

For the second step notice that $\Sigma_A(g \times_A A^* C)$ is isomorphic to $(\Sigma_A g) \times C$, which is clear when the pullback is laid out. In fact, this was Exercise 4.3. □

11.3 Functors between slices

For any category **A**, and **A** arrow $f\colon A \longrightarrow B$, there is a functor $\Sigma_f\colon \mathbf{A}/A \longrightarrow \mathbf{A}/B$ that takes each $\mathbf{A}/A$ object $g\colon C \longrightarrow A$ to $f \circ g$, and each $\mathbf{A}/A$ arrow $h\colon g \longrightarrow g'$ to $h\colon f \circ g \longrightarrow f \circ g'$. The notation Σ_f is ambiguous, since f is also an object of $\mathbf{A}/B$ and there is a slice category $(\mathbf{A}/B)/f$. But the ambiguity is perfectly tolerable since $(\mathbf{A}/B)/f$ is isomorphic to $\mathbf{A}/A$:

THEOREM 11.6 For any **A** arrow $f\colon A \longrightarrow B$ there is a functor isomorphism **I** that commutes with the two functors Σ_f:

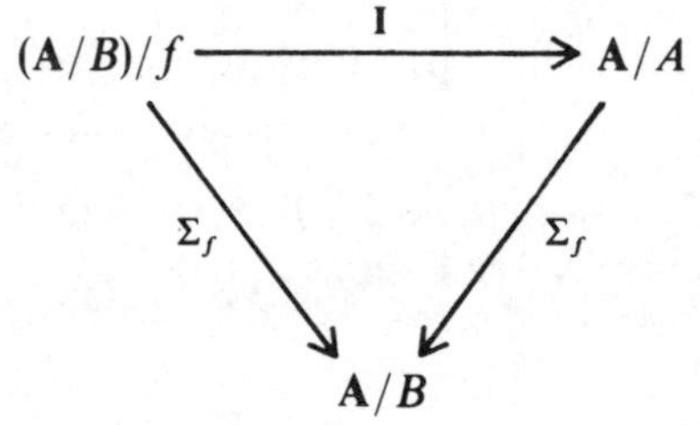

PROOF An object of $(\mathbf{A}/B)/f$ is an arrow of $\mathbf{A}/B$ with codomain f. So let **I** take the $(\mathbf{A}/B)/f$ arrow on the left to its top arrow, an object of $\mathbf{A}/A$:

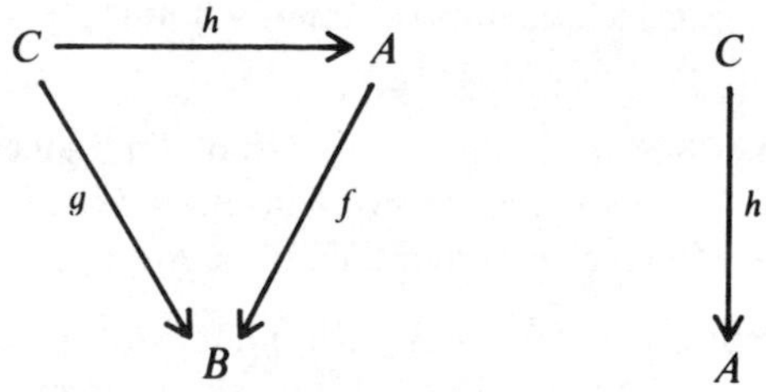

Note that h fully determines the $(\mathbf{A}/B)/f$ object, since $g = f \circ h$. The functor Σ_f defined on $(\mathbf{A}/B)/f$ takes the triangle to its domain g, while Σ_f defined on $\mathbf{A}/A$ takes h and composes it with f to give g. The reader can complete the proof, letting **I** take each $(\mathbf{A}/B)/f$ arrow $k\colon h \longrightarrow h'$ to itself, construed as an $\mathbf{A}/A$ arrow. (Draw the diagrams.) □

Since $\Sigma_f\colon \mathbf{A}/A \longrightarrow \mathbf{A}/B$ is (up to isomorphism) a slice functor over $\mathbf{A}/B$, Theorem 11.2 shows that if $\mathbf{A}/B$ has binary products (that is, if **A** has all pullbacks over B) then Σ_f has a right adjoint f^*. If $\mathbf{A}/B$ is Cartesian closed then f^* has a right adjoint Π_f. A category is called *locally Cartesian closed* if each of its slices is Cartesian closed. The fundamental theorem of topos theory will use the following case of these results:

THEOREM 11.7 If a category **A** is locally Cartesian closed, then so is every slice of **A**. For each arrow $f: A \longrightarrow B$ of **A** there are functors $\Sigma_f \dashv f^* \dashv \Pi_f$, each right adjoint to the one before, and f^* preserves exponentials. □

The functor f^* is defined by products in $\mathbf{A}/B$ so, to describe it in the category **A**, f^* takes an arrow B and forms its pullback along f. The fibre of f^*g over any $x: T \longrightarrow A$ is the fibre of g over $f(x)$. Intuitively, the fibre of $\Sigma_f h$ over an index Y in B is the union of all the fibres h_x for x an index in A with $f(x) = y$; while the fibre of $\Pi_f h$ is the product of those same fibres h_x.

Exercises

11.1 Show that for any category **A** and object B of **A**, an arrow $h: f \longrightarrow g$ of $\mathbf{A}/B$ is monic iff h is monic as an arrow of **A**.

11.2 For any category **A** and **A** object B, show that Σ_B has a left adjoint iff B is terminal, and that Σ_B is then an isomorphism. Show that Σ_B preserves binary products iff B is subterminal.

11.3 Show that if **A** has an initial object $\emptyset$ then $\emptyset \longrightarrow B$ is initial in $\mathbf{A}/B$. Also show that if objects A and C have a coproduct in **A** then any $f: A \longrightarrow B$ and $g: C \longrightarrow B$ have the following coproduct diagram in $\mathbf{A}/B$:

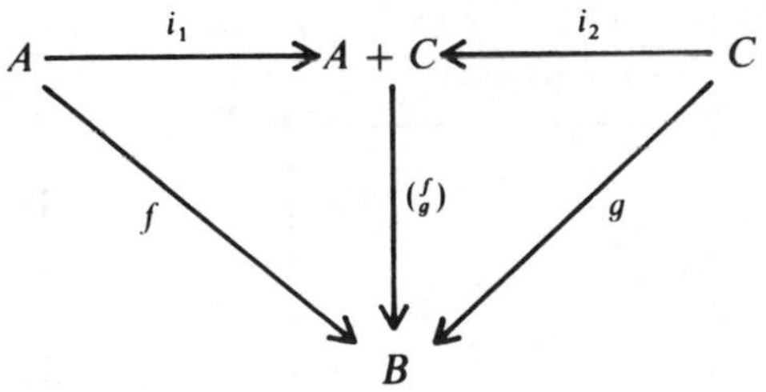

11.4 For any parallel pair $h, k: f \longrightarrow g$ of $\mathbf{A}/B$, if h and k have a coequalizer Q, q in **A** show there is a unique u making the following a coequalizer in $\mathbf{A}/B$:

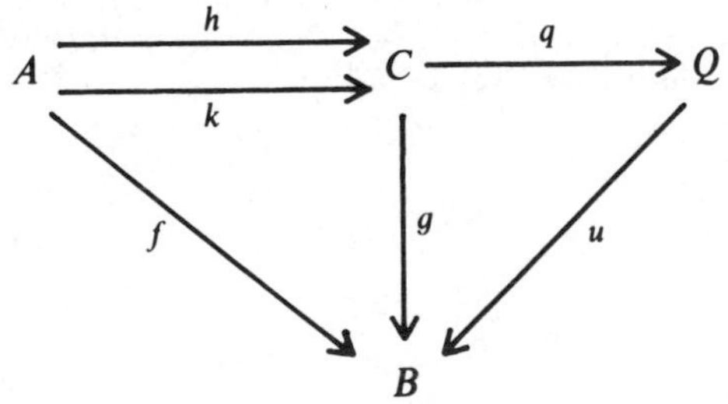

11.5 We say that a functor $\mathbf{F}: \mathbf{A} \longrightarrow \mathbf{B}$ *creates* colimits if it meets the following condition. For any diagram D in $\mathbf{A}$, if the image of D under $\mathbf{F}$ has a colimit L with arrows l_i, then there is a unique cocone C, c_i on D in $\mathbf{A}$, such that $\mathbf{F}C = L$ and $\mathbf{F}c_i = l_i$, and this cocone is a colimit. Show that if $\mathbf{F}$ creates colimits then it preserves all colimits that exist in $\mathbf{B}$. (A colimit might exist in $\mathbf{A}$ but not in $\mathbf{B}$ and, of course, $\mathbf{F}$ would not preserve it.)

Interpret Exercise 11.3 as showing that Σ_B creates initial objects and coproducts; and Exercise 11.4 as showing that it creates coequalizers. Show that Σ_B creates all colimits (finite or infinite).

11.6 Take any category $\mathbf{A}$ with binary products and any $\mathbf{A}$ object T. Define a category with the same objects as $\mathbf{A}$, where the arrows are T-indexed families of arrows, as in Section 6.6. Show that it is isomorphic to the full subcategory of $\mathbf{A}/T$, the objects of which are the constant families T^*A for objects A of $\mathbf{A}$. In particular, show that T-indexed composition corresponds to composition in $\mathbf{A}/T$.

11.7 If $\mathbf{A}$ has binary products, show that, for any $\mathbf{A}$ objects A and T, a T-element of A is adjunct to a global element of T^*A in $\mathbf{A}/T$. In particular, the generic element of A is adjunct to $\Delta: A \longrightarrow A \times A$ in $\mathbf{A}/A$. In fact, Δ is often called the generic element of A. Show, if $\mathbf{A}$ has all finite limits, for every T, that every global element $x: 1_T \longrightarrow T^*A$ is $x^*(\Delta)$. Thus, if we can prove the generic element Δ of A has some property which is preserved by pullback, then every generalized element of A also has it.

This form of generic element is often applied in toposes (see Chapters 18 and 19).

11.8 Show that the arrows $i: \Pi_B f \rightarrowtail A^B$ are the components of a monic natural transformation $i: \Pi_B \rightarrowtail (\Sigma_{\_})^B$. In fact, the naturality squares are pullbacks.

Show that the two squares at the end of the proof of Theorem 11.3 both commute iff the following square commutes:

$$\begin{array}{ccccc} T & \xrightarrow{\;i \circ t\;} & A^B & \qquad & f \\ {\scriptstyle h}\downarrow & & \downarrow{\scriptstyle k^B} & & \downarrow{\scriptstyle k} \\ T' & \xrightarrow[\;i \circ t'\;]{} & A'^B & & f' \end{array}$$

11.9 Show that $\mathbf{CAT}/\mathbf{N}$ is not Cartesian closed, for $\mathbf{N}$ the coequalizer defined in Exercise 8.6. [One proof: Show that the unique arrow from $f_1: \mathbf{2} \longrightarrow \mathbf{N}$ to the terminal object $1_{\mathbf{N}}$ is epic, but the pullback of $f_2: \mathbf{2} \longrightarrow \mathbf{N}$ along f_1 is not epic. Interpret this as a product in $\mathbf{CAT}/\mathbf{N}$ and invoke Exercise 6.12.]

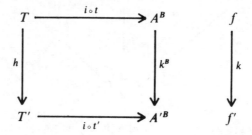

12

Mathematical foundations

The axioms for, say, a Cartesian closed category are a sufficient foundation by themselves for the theory of a single such category. No stronger background theory or metatheory is needed for the theorems of Chapter 6. Similarly, the axioms in Chapter 22 suffice as a formal foundation for a category of sets, and those in Chapter 23 for a category of smooth spaces. But the past few chapters have worked with many categories at once, and this requires some further foundation. In what follows, two styles of mathematical foundations for general category theory are described. The first, found in most of the literature, interprets category theory in some axiomatic set theory. The second, still being developed, axiomatizes a category of categories in categorical terms.

12.1 Set-theoretic foundations

Set-theoretic foundations for category theory are largely concerned with questions of size. If you say that a category must have a set of objects and a set of arrows then there is no category **Set** containing all sets and functions. **Set** is too large to be a set itself. Similar problems arise with the category of groups and group homomorphisms, topological spaces and continuous maps, and so on. The literature contains several solutions to these problems, all using the idea of a *universe*.

The original foundation used by Eilenberg and Mac Lane was Gödel–Bernays (GB) set theory. GB begins with ordinary sets and adds *classes*, which have sets as members. There is a class U of all sets, also called the universe, and for every property of sets there is a class P of all sets having that property. If there is actually a set of all sets with the property, then P is just that set. But if P is too large to be a set, then it is a *proper class*, and it cannot itself be a member of any class. There is no set of all sets, by Russell's paradox, and so the universe U is a proper class.

Then define a category as a class Ar of arrows and a class Ob of objects together with functions:

$$\mathrm{Dom}\colon \mathrm{Ar} \longrightarrow \mathrm{Ob}$$

$$\mathrm{Cod}\colon \mathrm{Ar} \longrightarrow \mathrm{Ob}$$

$$\mathrm{id}\colon \mathrm{Ob} \longrightarrow \mathrm{Ar}$$

$$m\colon \mathrm{Ar} \times_{\mathrm{Ob}} \mathrm{Ar} \longrightarrow \mathrm{Ar}$$

where Ar $\times_{\mathrm{Ob}}$ Ar is the set of pairs of arrows $\langle f, g \rangle$ with $\mathrm{Dom}(f) = \mathrm{Cod}(g)$. (Verify that this is the pullback of Dom and Cod.) The functions must satisfy the equations in Section 1.1, writing 1_A for $\mathrm{id}(A)$ and $f \circ g$ for $m(\langle f, g \rangle)$.

If Ar and Ob are both sets we speak of a *small category*. Then the ordered five-tuple $\langle$Ob, Ar, Dom, Cod, Id$\rangle$ is itself a set. If Ob or Ar is a proper class we speak of a *large category*. In fact, if Ob is a proper class then Ar must be, because of identity arrows, so we can define a large category as one with a proper class of arrows. Similarly, a group with a set of elements is called a small group, and one with a proper class of elements is called large. A topological space with a set of points is a small topological space, and so on. Thus there is a large category **Set** of all sets and functions, one **Grp** of all small groups and homomorphisms, one **Top** of all small topological spaces and continuous maps, and so on. There is a large category **Cat** of all small categories. But no large category or large group or space is a member of any class at all, and there is no large category of all classes or all large groups or whatever, including that there is no large category of all large categories.

A *locally small* category is one such that, for any two objects A and B, there is a set of all arrows from A to B. The large categories named above are all locally small.

A *well-powered* category is one in which every object A has a set of representative sub-objects: that is, there is a set of monics to A such that every monic to A is equivalent to one of those in the set. A category is *co-well-powered* if its dual is well powered.

There are practical reasons for keeping track of what is small and what is not. Consider **Set**. Every set of sets in **Set** has a union in **Set** (even a disjoint union, a coproduct) and a Cartesian product. Also **Set** has equalizers and coequalizers, and so by Section 4.6 every diagram $\mathbf{D} \longrightarrow \mathbf{Set}$ has a limit and a colimit if **D** is a small category. But it may not if **D** is large—a proper class of non-empty sets has neither a disjoint union nor a product in **Set**.

Notions such as well-poweredness help to reduce large diagrams to equivalent small ones. Consider any object A in a well-powered category with limits for all small diagrams, and any class S_i of monics to A, $i \in I$. Even if I is a proper class, this family of monics has an intersection since there is some set J with $J \subseteq I$ such that every S_i is equivalent as a sub-object to some S_j with $j \in J$. The S_j collectively form a small diagram, a set of arrows all with codomain A. That small diagram has a limit, the intersection over A of the S_j, and this limit is also the intersection of all the S_i. Variants on this idea are widely used, notably in Freyd's powerful adjoint functor theorems (see Mac Lane 1971).

The lack of a category of all classes or all large groups is not especially painful, but other restrictions in GB are. There are no functor categories between large categories, such as $\mathbf{Set}^{\mathbf{Grp}}$ would be, because a single functor from **Grp** to **Set** is already a proper class, and so no class contains even one such functor. This foundation is strong enough to produce large categories, but not to do anything with them.

Grothendieck's algebraic geometry required more flexibility, especially with functor categories, so he offered a more powerful foundation. A *Grothendieck universe* is a set V which satisfies the ZF axioms—more precisely, a set V such that:

The set of natural numbers is in V.
If $y \in x$ and $x \in V$ then $y \in V$.
If $x \in V$ and $y \in V$ then $\{x, y\} \in V$.
If $x \in V$ then the power set of x is in V and the union of all members of x is also in V.
If $x \in V$ and $f: x \longrightarrow y$ is a surjection, then $y \in V$.

The ZF axioms cannot prove that such a set V exists, since that would prove the consistency of ZF, so Grothendieck adopted a further axiom:

Every set x is a member of some universe V.

In particular, every universe V is a member of some larger universe V'. For any universe V define a *V-small* category to be a category as above, but with Ob and Ar members of V. (It follows that Dom, Cod, Id, and the ordered five-tuple of them all are also members of V.) Similarly, define V-locally-small, V-well-powered, and so on.

One can perform mathematics within a universe V as if in ZF set theory, since V satisfies the axioms of ZF. But at the same time every subset of V is actually a set and, in fact, a member of any universe V' with $V \in V'$. Since V' also satisfies the ZF axioms you can work in it much more freely than with the proper classes of GB. If we define **Set** to be the category of all sets and functions in V, and **Grp** to be the category all groups and homomorphisms in V, then neither is V-small but both are V'-small for any V' with $V \in V'$. The functor category $\mathbf{Set}^{\mathbf{Grp}}$ is also V'-small and so is much more. The category of all V'-small categories is V''-small whenever $V' \in V''$. The most powerful constructions anyone has had occasion to use in category theory involve at most a few successive universes.

Size considerations are still important. The V'-small category **Set** defined above has all limits and colimits for diagrams defined on V-small categories **D**, but not for all on V'-small categories. Therefore, shifting from one universe to a larger one requires some attention to detail, but it is not difficult in practice. In fact, the major drawback of Grothendieck universes is that it seems extravagant to postulate a vast proliferation of universes, all supporting the same mathematics for all practical purposes.

Mac Lane noticed that a single Grothendieck universe suffices for many purposes, and proposed a foundation extending ZF by a weaker axiom than Grothendieck's:

There is a universe V.

A second universe V' with $V \in V'$ would prove the consistency of ZF plus Mac Lane's axiom, so its existence does not follow from ZF plus Mac Lane's axiom. This foundation does not make the categories **Set** and **Grp** defined above members of any universe, but they are still sets, as is the functor category $\mathbf{Set}^{\mathbf{Grp}}$ and more.

Grothendieck's axiom is probably the most popular foundation among working category theorists, when they feel they must choose one. Mac Lane's seems the most elegant. They all work: so would other strategies, including various ways of truncating the subject to fit within ordinary ZF. The most extensive discussion of set-theoretic foundations to date is in Mac Lane (1969).

12.2 Axiomatizing the category of categories

The other approach describes categories in terms of the functors between them. For example, objects and arrows of a category **A** would be defined as functors to **A** from **1** and **2** respectively. Then the opening of Chapter 1, 'A category has objects and arrows . . . ', becomes shorthand for 'We begin by looking at the functors to a category from **1** and **2**'.

More precisely, such a foundation begins with an axiom that says that categories and functors collectively form a category; that is, functors have domains and codomains, and compose and so on. Then there are special axioms on categories and functors such as:

(Cat_1) There is a terminal category **1**. Every pair of categories **A** and **B** has a product diagram $\mathbf{A} \times \mathbf{B}$, p_1, p_2 and an exponential $\mathbf{B}^{\mathbf{A}}$, $ev_{\mathbf{A},\mathbf{B}}$. Every parallel pair of functors $\mathbf{F}, \mathbf{G}: \mathbf{A} \longrightarrow \mathbf{B}$ has an equalizer $\mathbf{e}: \mathbf{E} \longrightarrow \mathbf{A}$. There is also an initial category, and these are coproducts and coequalizers.

(Cat_2) There is a category **2** with exactly two global elements $0: \mathbf{1} \longrightarrow \mathbf{2}$ and $1: \mathbf{1} \longrightarrow \mathbf{2}$. These two are disjoint as subcategories of **2**. The functors 1_2, $0 \circ !_2$, and $1 \circ !_2$ are the only functors from **2** to **2**. And, given any parallel pair of functors $\mathbf{F}, \mathbf{G}: \mathbf{A} \longrightarrow \mathbf{B}$, either $\mathbf{F} = \mathbf{G}$ or there is some $f: \mathbf{2} \longrightarrow \mathbf{A}$ such that $\mathbf{F} \circ f \neq \mathbf{G} \circ f$.

If objects and arrows of a category **A** are defined as **1**-elements and **2**-elements of **A** respectively, then a functor $\mathbf{F}: \mathbf{A} \longrightarrow \mathbf{B}$ takes them to objects and arrows of **B** (as arrows in any category act on generalized elements). Define the domain of an arrow $f: \mathbf{2} \longrightarrow \mathbf{A}$ to be the composite $f \circ 0: \mathbf{1} \longrightarrow \mathbf{A}$. Then to say that a functor **F** preserves the domain of f means that $\mathbf{F}(f \circ 0) = (\mathbf{F} \circ f) \circ 0$, which is simply associativity of composition of functors. Similarly, **F** automatically preserves codomains. Given an object of **A**, $A: \mathbf{1} \longrightarrow \mathbf{A}$, define the identity arrow on A to be $A \circ !_2: \mathbf{2} \longrightarrow \mathbf{A}$. Then functors preserve identity arrows; that is, $\mathbf{F}(A \circ !_2) = (\mathbf{F} \circ A) \circ !_2$.

Given an arrow $f: \mathbf{2} \longrightarrow \mathbf{A}$ we may call $f \circ 0$ the 'domain of f as an arrow of $\mathbf{A}$', just to distinguish it from the domain of f as a functor which, of course, is $\mathbf{2}$.

Define the category $\mathbf{3}$ as the pushout

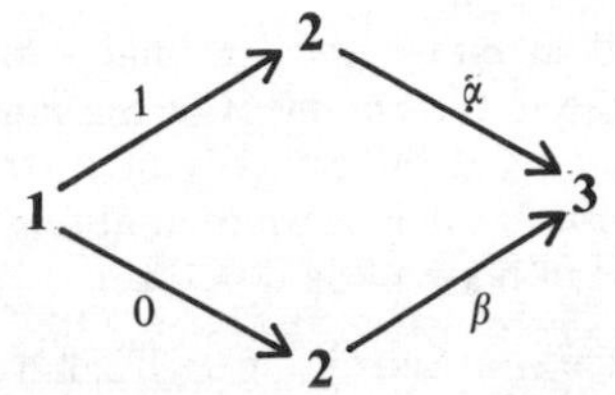

Since a pushout square commutes, $\beta \circ 0 = \alpha \circ 1$. The domain of β as an arrow of $\mathbf{3}$ is the codomain of α. It follows from Cat_{1-2} that the domain of α is distinct from its codomain, and that both are distinct from the codomain of β (see Exercise 12.12). So $\mathbf{3}$ has at least five arrows: α, β, and the three identity arrows. Then an axiom says that:

(Cat_3) The category $\mathbf{3}$ has an arrow, $\gamma: \mathbf{2} \longrightarrow \mathbf{3}$, the domain of which is the domain of α, while its codomain is the codomain of β.

Take any category $\mathbf{A}$ and arrows $f: \mathbf{2} \longrightarrow \mathbf{A}$ and $g: \mathbf{2} \longrightarrow \mathbf{A}$, where the domain of g as an arrow of $\mathbf{A}$ is the codomain of f; that is, $f \circ 1 = g \circ 0$. Since $\mathbf{3}$ is a pushout there is a unique functor $t: \mathbf{3} \longrightarrow \mathbf{A}$ with $t \circ \alpha = f$ and $t \circ \beta = g$. Then define the composite of f and g as arrows of $\mathbf{A}$ to be $t \circ \gamma$. We call the composite $g \circ_{\mathbf{A}} f$. The functor t preserves domains and codomains, so the domain of $g \circ_{\mathbf{A}} f$ as an arrow of $\mathbf{A}$ is the domain of f, and the codomain is the codomain of g. Define a *commutative triangle* in $\mathbf{A}$ to be a $\mathbf{3}$-element t of $\mathbf{A}$, and call the arrows $t \circ \alpha$, $t \circ \beta$, and $t \circ \gamma$ its *sides*. Then a functor $\mathbf{F}: \mathbf{A} \longrightarrow \mathbf{B}$ takes commutative triangles in $\mathbf{A}$ to commutative triangles in $\mathbf{B}$. In other words, any functor preserves composition of arrows.

In this approach, categorical structures are described by generalized elements. Objects are global elements, arrows are $\mathbf{2}$-elements, commutative triangles are $\mathbf{3}$ elements, commutative squares are $\mathbf{2} \times \mathbf{2}$ elements, and so on. Functors automatically preserve all these. This is typical of categorical methods: any structure in any category which can be identified with a generalized element is preserved by all arrows of the category.

Further axioms are needed to prove the usual theorems of category theory, but no one version has established itself as authoritative. The original source is Lawvere (1966), which remains important, although Isbell (1967) pointed out some technical flaws. Blanc and Donnadieu (1976) and McLarty (1991) have followed up different strands in Lawvere's paper. The reader is referred to these, plus Bénabou's (1985) use of fibred categories.

Exercises

In Exercises 12.1–12.8 it is shown that there is a series of four adjoint functors between **Set** and **Cat**,

$$\mathbf{Comp} \dashv \mathbf{Disc} \dashv \mathbf{U} \dashv \mathbf{Indisc}$$

which cannot be extended at either end. Formally, this supposes that we have categories **Set** and **Cat**. For example, **Set** might be the category sets and functions in a Grothendieck universe V, and **Cat** the category of V-small categories and functors. Alternatively, they might rest on categorical foundations, as we will see how to do in Chapter 20, with any topos in place of **Set**.

12.1 Define a functor $\mathbf{U}\colon \mathbf{Cat} \longrightarrow \mathbf{Set}$, taking each small category to its set of objects, and each functor to its function on objects.

12.2 Define a *discrete category* to be a small category with only identity arrows. Show that every set S gives a discrete category with $\text{Ob} = \text{Ar} = S$ and Dom and Cod both the identity function 1_S. Use this construction to define a functor $\mathbf{Disc}\colon \mathbf{Set} \longrightarrow \mathbf{Cat}$. Show that **Disc** makes **Set** a full subcategory of **Cat**.

12.3 For any small category **A**, define an inclusion functor $\mathbf{Disc}(\mathbf{U}(\mathbf{A})) \rightarrowtail \mathbf{A}$ that inserts the objects of **A** into **A**. Show that this functor is universal from **Disc** to **A**, and conclude that **Disc** is left adjoint to **U**.

12.4 Define an *indiscrete category* to be a small category **A** such that for each pair of objects A and A' of **A** there is exactly one arrow from A to A'. In other words, every object of **A** is terminal. Show that every set S gives an indiscrete category with $\text{Ob} = S$ and $\text{Ar} = S \times S$ and $\text{Dom} = p_1$ and $\text{Cod} = p_2$. Define a full monic functor $\mathbf{Indisc}\colon \mathbf{Set} \longrightarrow \mathbf{Cat}$ that takes each set to its indiscrete category. Thus **Set** appears as a full subcategory of **Cat** in a second way.

12.5 For each small category **A**, show that there is a unique functor $\mathbf{A} \longrightarrow \text{Indisc}(\mathbf{U}(\mathbf{A}))$ that takes each object of **A** to itself as an object of $\mathbf{Indisc}(\mathbf{U}(\mathbf{A}))$, and show that this functor is universal from **A** to **Indisc**. Conclude that **U** is left adjoint to **Indisc**. Show that **Indisc** does not preserve the coproduct $1 + 1$, and so is not left adjoint to any functor.

12.6 Exercises 12.4 and 12.5 began with a definition of **Indisc** and deduced that it was left adjoint to **U**. Show how this could be reversed, to deduce the definition of **Indisc** from the supposition that it is left adjoint to **U**. [Hint: An object of $\mathbf{Indisc}(S)$ corresponds to a functor $\mathbf{1} \longrightarrow \mathbf{Indisc}(S)$, and an arrow corresponds to a functor $\mathbf{2} \longrightarrow \mathbf{Indisc}(S)$. What must the adjuncts of these be if **Indisc** is right adjoint to **U**?]

12.7 Let a pair of arrows, f_1 and f_2, of a category **A** be *linked* if they have the same domain, or the same codomain. Then let an object A of **A** be *connected to* an object A' if there is a finite sequence of arrows $f_1, \ldots, f_n$, each linked to the next and with $A = \text{Dom}(f_1)$ and $A' = \text{Cod}(f_n)$. Show that every object is connected to itself, that if A is connected to A' then A' is connected to A, and that if A is connected to A' and A' to A'' then A is connected to A''. Show that if A is connected to A' in **A** then, for any functor $\mathbf{F}\colon \mathbf{A} \longrightarrow \mathbf{B}$, $\mathbf{F}A$ is connected to $\mathbf{F}A'$ in **B**. In short, 'connected to' is an equivalence relation, preserved by all functors.

Define a *component* of a category **A** to be a non-empty, maximal connected full subcategory of **A**; that is, a full subcategory consisting of some object A and all the objects connected to it. Show that **A** is the coproduct of its components.

12.8 Define a functor **Comp**: **Cat** ⟶ **Set** left adjoint to **Disc**. [Hint: **Comp** takes each category to its set of components. What does it do with functors?]

Show that **Comp** preserves products but not equalizers, and so has no left adjoint. [Hint: consider functors from **2** to the category **E** defined in Exercise 10.6.]

Exercises 12.9–12.12 take the viewpoint of Section 12.2. For these exercises, give all proofs using axioms Cat_{1-3} and the definitions in Section 12.2.

12.9 Show (as a trivial exercise in the definitions) that an object of a product category $\mathbf{A} \times \mathbf{B}$ corresponds to a pair $\langle A, B \rangle$, with A an object of **A** and B an object of **B**, and similarly for arrows. Since $\mathbf{A} \times \mathbf{B}$ is only defined up to isomorphism, there is no way to be more definite about the nature of the objects and arrows of $\mathbf{A} \times \mathbf{B}$. Describe the objects and arrows of an equalizer for a parallel pair of functors $\mathbf{F}, \mathbf{G}: \mathbf{A} \longrightarrow \mathbf{B}$.

12.10 Spell out Cat_2 by the definitions as saying that **2** has two objects and three arrows. Note that the identity functor for **2** is the sole non-identity arrow of **2**. Prove the following from the axioms. The product $\mathbf{2} \times \mathbf{2}$ has four objects and nine arrows, with domain and co-domain relations in **2** as shown (omitting identity arrows):

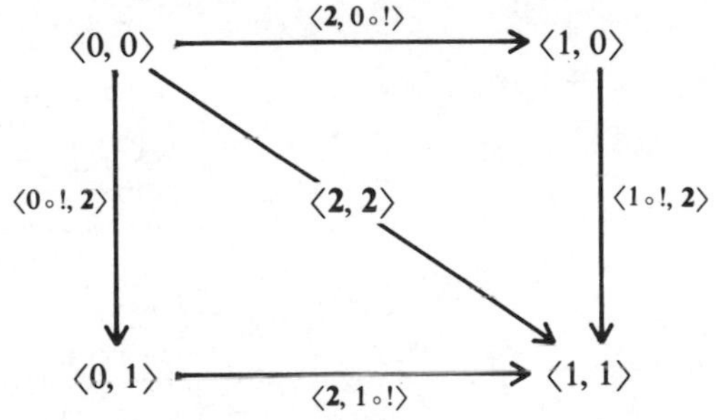

where **2** represents the identity functor on **2**.

Show that the composite of $\langle 1 \circ !, \mathbf{2} \rangle$ and $\langle \mathbf{2}, 0 \circ ! \rangle$ as arrows of $\mathbf{2} \times \mathbf{2}$ is $\langle \mathbf{2}, \mathbf{2} \rangle$. [Hint: $\langle \mathbf{2}, \mathbf{2} \rangle$ is the only arrow of $\mathbf{2} \times \mathbf{2}$ with domain $\langle 0, 0 \rangle$ and codomain $\langle 1, 1 \rangle$.] So the diagram is a commutative square in $\mathbf{2} \times \mathbf{2}$. In general, define a *commutative square* in a category **A** as a functor $\mathbf{S}: \mathbf{2} \times \mathbf{2} \longrightarrow \mathbf{A}$ and say that its *sides* are the composites of **S** with the four non-identity arrows of $\mathbf{2} \times \mathbf{2}$ other than $\langle \mathbf{2}, \mathbf{2} \rangle$. What functor does the diagram correspond to?

12.11 Define the *arrow category* on **A** as the exponential of **A** by **2**, $\mathbf{A}^{\mathbf{2}}$. Show that its objects correspond to arrows of **A** and its arrows to commutative squares. [The above axioms *do not* suffice to prove that if arrows f, g, h, and k of **A** have $g \circ_{\mathbf{A}} f = k \circ_{\mathbf{A}} h$ then there is necessarily a commutative square in **A** with those arrows as its sides. Thus they do not suffice to characterize the arrows of $\mathbf{A}^{\mathbf{2}}$ or other exponential categories very well (see McLarty 1991).]

12.12 Show from Cat_{1-2} and the pushout defining **3** that there is a functor $t: \mathbf{3} \longrightarrow \mathbf{2}$ with $t \circ \alpha \neq t \circ \beta$, and conclude that $\alpha \neq \beta$. Use this reasoning to show that neither α nor β is an identity arrow, and that the three identity arrows listed before Cat_3 are all distinct. Using Cat_3, show that none of those five arrows equals γ.

Part III

TOPOSES

13

Basics

13.1 Definition

A *topos* is a Cartesian closed category with a *sub-object classifier*. A sub-object classifier is an object Ω and a global element $t: 1 \longrightarrow \Omega$ such that for any monic $s: S \rightarrowtail A$ there is a unique arrow $\chi_s: A \longrightarrow \Omega$ that makes the following a pullback:

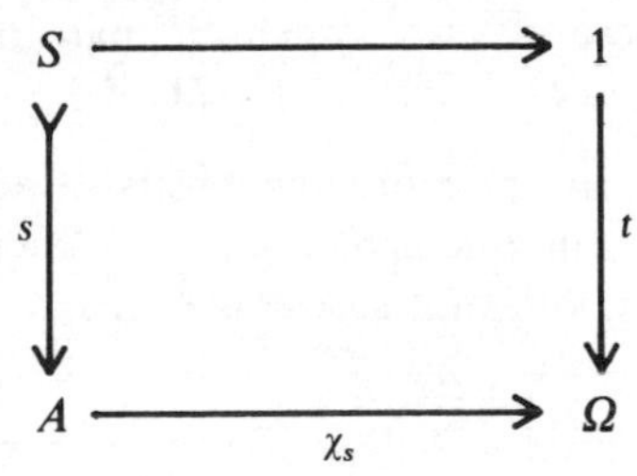

The arrow χ_s is called the *classifying arrow* for s, or the *characteristic arrow* for s.

While every monic has a unique classifying arrow, one arrow may classify many monics. In fact, $\chi_s = \chi_{s'}$ iff $s \equiv s'$ as sub-objects of A, since pullbacks are defined up to isomorphism.

Since t is monic every arrow to Ω classifies a sub-object of its domain, the pullback of t. So arrows from A and Ω correspond exactly to equivalence classes of sub-objects of A. Given an arrow $w: A \longrightarrow \Omega$ we often write $w^*: W \rightarrowtail A$ for the selected pullback of t along w. (By the conventions in Chapter 7 there is a selected one.)

Throughout the rest of this chapter we assume that we are working with a topos **E**.

13.2 The sub-object classifier

THEOREM 13.1 All monics are equalizers, and all monic epics are iso.

PROOF A monic s is an equalizer for χ_s and $t \circ !_A$, and an epic equalizer in any category is iso. □

Think of Ω as the object of truth values, with t as *true*. Then $t \circ !_A$ takes all of A to *true*. The classifying arrow of any $s: S \rightarrowtail A$ is the unique arrow such that, for any T-element x of A, $x \in s$ iff $\chi_s(x) = t \circ !_T$. In short, $x \in s$ iff $\chi_s(x)$ is *true*. Conversely, an arrow to Ω is fully determined by the part of its domain that it takes to *true*; that is, by the sub-object of its domain that it classifies.

We will often use the abbreviation t_T for the composite $t \circ !_T$, and even omit the subscript T when it is clear from the context. Thus, given any $h: A \longrightarrow \Omega$ and an equation $h = t$, this 't' must be short for t_A. In an expression $\langle h, t \rangle: T \longrightarrow \Omega \times \Omega$, '$t$' must be short for t_T. Note that t_T classifies the maximal subobject $1_T: T \rightarrowtail T$.

13.3 Conjunction and intersection

The arrow t is sometimes called the *generic sub-object* because it is a sub-object of Ω itself, and every sub-object is a pullback of it along exactly one arrow. There is also a generic pair of sub-objects, namely $t \times \Omega: 1 \times \Omega \longrightarrow \Omega \times \Omega$ and $\Omega \times t: \Omega \times 1 \longrightarrow \Omega \times \Omega$.

THEOREM 13.2 Given any pair of sub-objects of an object A, $r: R \rightarrowtail A$ and $s: S \rightarrowtail A$, there is a unique arrow $u: A \longrightarrow \Omega \times \Omega$ that makes both the squares below pullbacks, and that arrow is $\langle \chi_r, \chi_s \rangle$:

$$\begin{array}{ccc} R & \longrightarrow & 1 \times \Omega \\ {\scriptstyle r}\downarrow & & \downarrow{\scriptstyle t \times \Omega} \\ A & \xrightarrow{u} & \Omega \times \Omega \end{array} \qquad \begin{array}{ccc} S & \longrightarrow & \Omega \times 1 \\ {\scriptstyle s}\downarrow & & \downarrow{\scriptstyle \Omega \times t} \\ A & \xrightarrow{u} & \Omega \times \Omega \end{array}$$

PROOF Consider the following diagram:

$$\begin{array}{ccccc} R & \longrightarrow & 1 \times \Omega & \longrightarrow & 1 \\ {\scriptstyle r}\downarrow & & \downarrow{\scriptstyle t \times \Omega} & & \downarrow{\scriptstyle t} \\ A & \xrightarrow[u]{} & \Omega \times \Omega & \xrightarrow[p_1]{} & \Omega \end{array}$$

The left-hand square is a pullback iff the outer rectangle is; that is, iff $p_1 \circ u = \chi_r$. Similarly, $p_2 \circ u = \chi_s$. □

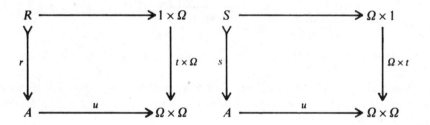

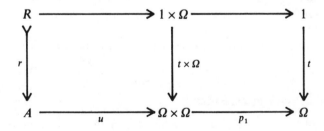

THEOREM 13.3 The generic pair of sub-objects has intersection $\langle t, t\rangle : 1 \longrightarrow \Omega \times \Omega$.

PROOF If $h: T \longrightarrow 1 \times \Omega$ and $k: T \longrightarrow \Omega \times 1$ have $(t \times \Omega) \circ h = (\Omega \times t) \circ k$ then $h = \langle !_T, t_T\rangle$ and $k = \langle t_T, !_T\rangle$, and these factor through $\langle t, t\rangle$. □

Since $\langle t, t\rangle$ is monic it has a classifier $\wedge : \Omega \times \Omega \longrightarrow \Omega$, called the *conjunction* arrow. Given any h and k from A to Ω, we may abbreviate $\wedge \circ \langle h, k\rangle$ by $h \wedge k$. Conjunction is a truth function with the familiar property: for any object T and $u \in_T \Omega$ and $v \in_T \Omega$, $u \wedge v = t_T$ iff $u = t_T$ and $v = t_T$.

THEOREM 13.4 Suppose that $v: A \longrightarrow \Omega$ and $w: A \longrightarrow \Omega$ classify $v^*: V \rightarrowtail A$ and $w^*: W \rightarrowtail A$ respectively. Then $v \wedge w$ classifies $V \cap W$. In other words, for any $r: R \rightarrowtail A$ and $s: S \rightarrowtail A$, $\chi_{r \cap s} = \chi_r \wedge \chi_s$.

PROOF Since $\langle v, w\rangle$ classifies the pair of sub-objects V and W, Theorem 4.10 shows that $\wedge \circ \langle v, w\rangle$ classifies their intersection. □

So, for any $x \in_T \Omega$, $(\chi_s \wedge \chi_r)(x) = t$ iff $\chi_s(x) = t$ and $\chi_r(x) = t$.

Familiar properties of intersections allow us to conclude that for any T-elements u, v, and w of Ω, $u \wedge u = u$, and $u \wedge v = v \wedge u$ and $(u \wedge v) \wedge w = u \wedge (v \wedge w)$. For example, $u^* \cap u^*$ equals u^*, so both $u \wedge u$ and u classify u^*, so $u \wedge u = u$. We also have, for any $u \in_T \Omega$, $u \wedge t = u$, since both sides classify u^*.

13.4 Order and implicates

Define $\leq_1$ as the equalizer of $\wedge$ and p_1:

$$\Omega_1 \overset{\leq_1}{\rightarrowtail} \Omega \times \Omega \underset{p_1}{\overset{\wedge}{\rightrightarrows}} \Omega$$

For any $\langle v, w\rangle : T \longrightarrow \Omega \times \Omega$, write $v \leq_1 w$ to abbreviate $\langle v, w\rangle \in \leq_1$, saying that $\langle v, w\rangle$ factors through $\leq_1$. Then the definition of $\leq_1$ as an equalizer can be restated in the form

$$v \leq_1 w \text{ iff } v = v \wedge w$$

That definition and known properties of conjunction imply, for any T and T-elements u, v, and w of Ω:

$$u \leq_1 t \quad \text{and} \quad u \leq_1 u$$
$$\text{if } u \leq_1 v \quad \text{and } v \leq_1 w \quad \text{then } u \leq_1 w$$
$$\text{if } u \leq_1 v \quad \text{and } v \leq_1 u \quad \text{then } u = v$$

Thus $\leq_1$ is an order relation on Ω. Intuitively, $u \leq_1 v$ if v is at least as true as u.

THEOREM 13.5 Let v and w classify $v^*: V \rightarrowtail T$ and $w^*: W \rightarrowtail T$. Then $v \leq_1 w$ iff $v^* \subseteq w^*$ as sub-objects of T.

PROOF $v^* \subseteq w^*$ is equivalent to $v^* \equiv v^* \cap w^*$. But that is equivalent to $v = v \wedge w$ and so to $v \leq_1 w$. □

The monic $\leq_1: \Omega_1 \rightarrowtail \Omega \times \Omega$ is a relation from Ω to Ω in the sense of Chapter 5. It *internalizes* the relation $v^* \subseteq w^*$ in the sense that a pair $\langle v, w \rangle$ has $v^* \subseteq w^*$ iff it factors through $\leq_1$. Its classifying arrow is written $\rightarrow: \Omega \times \Omega \longrightarrow \Omega$ and is called the *material conditional* arrow.

Write $h \rightarrow k$ to abbreviate $\rightarrow \circ \langle h, k \rangle$. The definition can be restated as follows. For any T and T-elements u and v of Ω, $(u \rightarrow v) = t$ iff $u \leq_1 v$; that is, $u \rightarrow v$ is *true* iff v is at least as true as u.

For any sub-objects $q: Q \rightarrowtail A$ and $r: R \rightarrowtail A$, define $Q \Rightarrow R$ to be the sub-object classified by $\chi_q \rightarrow \chi_r$. Call $Q \Rightarrow R$ the (*material*) *implicate* of R by Q.

THEOREM 13.6 For any sub-objects Q, R, and $s: S \rightarrowtail A$, we have $(S \cap Q) \subseteq R$ iff $S \subseteq (Q \Rightarrow R)$.

PROOF First, $S \subseteq (Q \Rightarrow R)$ is equivalent to $\rightarrow \circ \langle \chi_q, \chi_r \rangle \circ s = t_s$. This is equivalent to $(\chi_q \circ s) \leq_1 (\chi_r \circ s)$ and so to $(S \cap Q) \subseteq (S \cap R)$, which is easily equivalent to $(S \cap Q) \subseteq R$. □

Since $(Q \Rightarrow R) \subseteq (Q \Rightarrow R)$ we have $(Q \Rightarrow R) \wedge Q \subseteq R$. So the theorem says that $Q \Rightarrow R$ is the largest sub-object of A the intersection of which with Q is contained in R.

13.5 Power objects

Since arrows from any A to Ω classify sub-objects of A, the exponential Ω^A is the object of sub-objects of A or the *power object* of A. The membership relation over A is internalized by the following pullback:

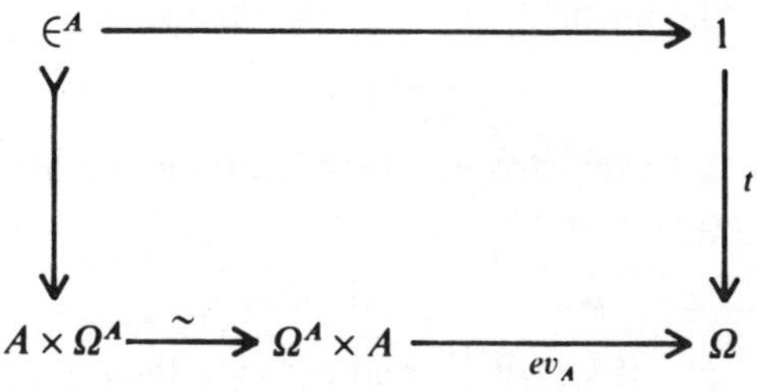

For any $\langle x, h \rangle: T \longrightarrow A \times \Omega^A$ we abbreviate the confusing notation $\langle x, h \rangle \in \in^A$ by $x \in^A h$. Either expression says that $\langle x, h \rangle$ factors through $\in^A$

or, equivalently, $ev_A \circ \langle h, x \rangle = t$ (notice how the twist arrow has disappeared).

THEOREM 13.7 Take any $x: 1 \longrightarrow A$ and $w: A \longrightarrow \Omega$. Then $x \in^A$ 'w' iff $x \in w^*$.

PROOF To say $x \in^A$ 'w' means that the outer quadrilateral commutes:

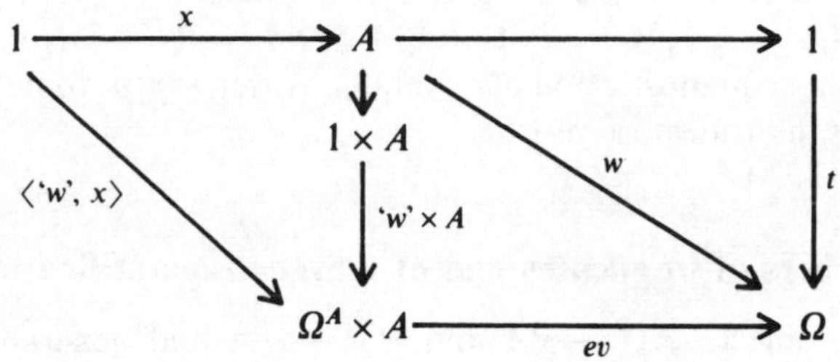

But the two left-hand triangles commute trivially, so the outer quadrilateral commutes iff $w \circ x = t$. □

13.6 Universal quantification

For any E, the name 't_E': $1 \longrightarrow \Omega^E$ represents the maximal sub-object $1_E: E \rightarrowtail E$. And since every arrow from 1 is monic, 't_E' itself has a classifying arrow $\forall_E: \Omega^E \longrightarrow \Omega$; that is, $\forall_E$ takes a sub-object of E to *true* iff it is the maximal sub-object.

Given any relation $r: R \rightarrowtail B \times A$, define the *universal quantification* of r over A to be the following pullback along the transpose of the classifying arrow:

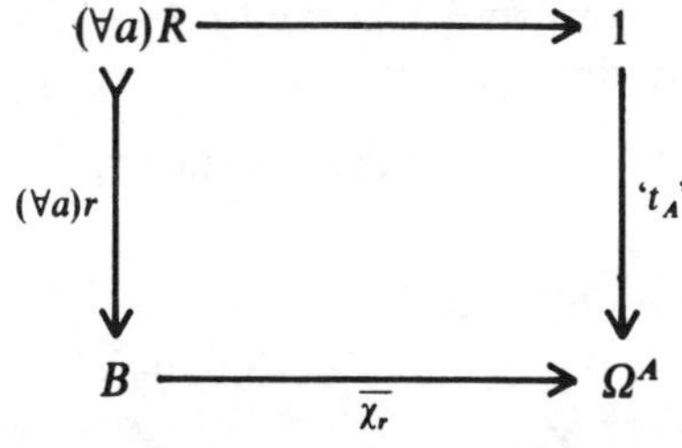

Equivalently, $(\forall a)R$ is the sub-object classified by $\forall_A \circ \overline{\chi_r}$. The following theorem says that $(\forall a)r$ is the largest sub-object of B that is r-related to all of A, and so explains the term 'universal quantification'.

THEOREM 13.8 For any $s: S \rightarrowtail B$, $S \subseteq (\forall a)R$ iff $S \times A \subseteq R$.

PROOF We have $S \subseteq (\forall a)R$ iff $\chi_r \circ s = `t_A' \circ !_S$. Transposing both sides gives $\chi_r \circ (s \times A) = t_{S \times A}$. Exercise 13.4 completes the proof. □

Of course, we might want to quantify over any factor of a finite product. Also, note that one object may occur several times in the factors of a product, so the quantification is actually along a certain projection rather than over a certain object. Given any finite product $A_1 \times \cdots \times A_n$ and any i between 1 and n, define l_i to be the obvious isomorphism from $A_1 \times \cdots \times A_n$ to $A_1 \times \cdots \times A_{i-1} \times A_{i+1} \times \cdots \times A_n \times A_i$. For any $r: R \rightarrowtail A_1 \times \cdots x A_n$, define the universal quantification of r over A_i, written $(\forall p_i)r$, to be the pullback of $`t_{Ai}'$ along the transpose of $l_i \circ r$.

13.7 Members of implicates and of universal quantifications

For any sub-objects $q: Q \rightarrowtail A$ and $r: R \rightarrowtail A$ and generalized element $w \in_S A$, we have $w \in Q \cap R$ iff $w \in Q$ and $w \in R$. The situation for material implicates and universal quantifications is more subtle:

THEOREM 13.9 Consider a relation $r: R \rightarrowtail B \times A$. For any $w \in_S B$ we have $w \in (\forall a)R$ iff $w \times 1_A \subseteq R$.

PROOF We have $w \in (\forall a)R$ iff $\overline{\chi_r} \circ w = `t_B' \circ !_S$. Transposing both sides gives $\chi_r \circ (w \times 1_A) = t_{S \times A}$, which is equivalent to $w \times 1_A \subseteq R$. □

COROLLARY For r and w as in the theorem, $w \in (\forall a)R$ iff for every $h: C \longrightarrow S$ and $k: C \longrightarrow A$ we have $\langle w(h), k \rangle \in R$. In other words, for every later stage of definition $h: C \longrightarrow S$, w considered at that stage has R to every element of A defined at that stage.

PROOF The theorem is the case in which $C = S \times A$, $h = p_1$, and $k = p_2$. The general case follows by an easy diagram. Think of $S \times A$, p_1, p_2 as the generic object with an arrow to S and one to A—every such object and two arrows factors uniquely through this one. □

THEOREM 13.10 For w, q, and r as above, define $y: T \rightarrowtail S$ by the following pullback:

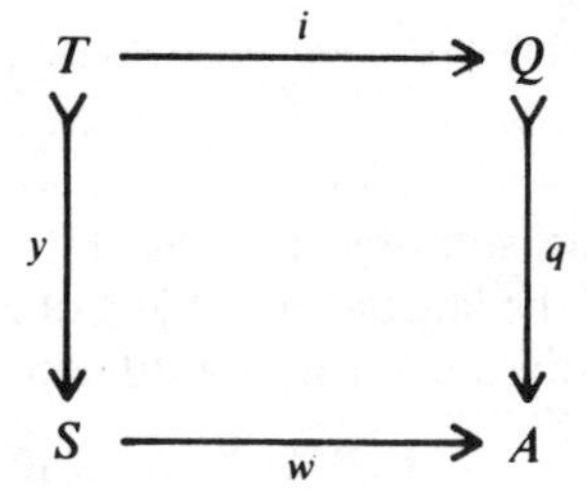

Then $w \in (Q \Rightarrow R)$ iff $w(y) \in R$.

PROOF Note that $w(y) \in Q$. If we assume that $w \in Q \Rightarrow R$ then certainly $w(y) \in Q \Rightarrow R$. Thus $w(y) \in (Q \Rightarrow R) \cap Q$, so the remark following Theorem 13.6 shows that $w(y) \in R$. Conversely, assume that $w(y) \in R$. Since $w(y)$ is the pullback $w^{-1}Q$ this means that $w^{-1}Q \subseteq w^{-1}R$, and so

$$w^{-1}Q \equiv w^{-1}(Q \cap R)$$

This means that $w \circ \chi_q$ classifies the same sub-object as $w \circ \chi_{q \cap r}$ and so these are the same arrow. Therefore w factors through the equalizer of χ_q and $\chi_{q \cap r}$. By Exercise 13.13, that equalizer is $Q \Rightarrow R$. □

Think of y as measuring how much of w is in Q. Then the theorem says that $w \in Q \Rightarrow R$ iff w is in R at least as much as it is in Q.

COROLLARY For w, q, and r as above, $w \in Q \Rightarrow R$ iff: for every $h: C \longrightarrow S$, if $w(h) \in Q$ then $w(h) \in R$.

PROOF This follows by the pullback property of T. Think of $y: T \longrightarrow S$ as the generic arrow to S factoring through Q—every arrow to S factoring through Q factors uniquely through y. □

Exercises

13.1 Show that the singleton category **1** is a topos. A topos in which every object is terminal is called *trivial*. Supposing that a topos **E** has an initial object $\emptyset$ (Theorem 16.12 shows that every topos has an initial object), show that if $\emptyset$ has a global element, **E** is trivial.

For any toposes **E** and **E**$'$ prove that the product category $\mathbf{E} \times \mathbf{E}'$ is a topos. Show that if **E** and **E**$'$ have initial objects $\emptyset$ and $\emptyset'$ then for any objects A and A' of **E** and **E**$'$ respectively the product $\langle \emptyset, A' \rangle \times \langle A, \emptyset \rangle$ is initial in $\mathbf{E} \times \mathbf{E}'$.

13.2 Let $t: 1 \longrightarrow \Omega$ and $t': 1 \longrightarrow \Omega'$ be two sub-object classifiers for one topos. Show that $\Omega \cong \Omega'$ by a unique isomorphism commuting with t and t'.

13.3 Given any $s: S \rightarrowtail A$ and $f: B \longrightarrow A$, show that the classifying arrow of $f^{-1}(s)$ is $\chi_s \circ f$.

13.4 Show, for any sub-object $u: U \rightarrowtail A$ with classifying arrow χ_u, and $v: V \rightarrowtail A$, that we have $V \subseteq U$ iff $\chi_u \circ v = t_V$.

13.5 Let $u: U \rightarrowtail A$ be any sub-object and c any T-element of A with $c \in U$. Show for any $s: S \longrightarrow T$ that $c(s) \in U$. We say that S is a 'later stage of definition' than T and that $c(s)$ is 'c at the later stage S'. Compare with Chapter 18. In these terms, if c is in U it remains in U at all later stages of definition.

13.6 Extend Theorem 13.7 to generalized elements. For any $\langle x, s \rangle: T \longrightarrow A \times \Omega^A$, show that $x \in^A s$ iff $\langle 1_T, x \rangle \in s^*$.

13.7 Prove, for any sub-objects Q and R of one object A, that $[Q \cap (Q \Rightarrow R)] \equiv (Q \cap R)$.

13.8 Show that the intersection of $\leq_1$ and its opposite is the diagonal $\Delta_\Omega: \Omega \rightarrowtail \Omega \times \Omega$. Show that $\leq_1$ is the material implicate of the generic pair of sub-objects.

13.9 For typographical convenience, let PA denote the power object Ω^A. For any object A define an *internal intersection* arrow $\cap : PPA \longrightarrow PA$. [Hint: the transpose $PPA \times A \longrightarrow \Omega$ should take a pair $\langle s, x \rangle$ to *true* iff for every $p \in^{PA} s$ we have $x \in^A p$.]

13.10 For any A, define $\delta_A: A \times A \longrightarrow \Omega$ to be the classifying arrow for $\Delta: A \rightarrowtail A \times A$. Thus for any $\langle h, k \rangle: T \longrightarrow A \times A$, $\delta_A(h, k) = t$ iff $h = k$. The transpose of δ_A is called the *singleton arrow* $\{\ \}_A: A \longrightarrow \Omega^A$. For any $x: T \longrightarrow A$ we write $\{x\}$ to abbreviate $\{\ \}_A \circ x$.

Prove, for any $x \in_T A$ and $y \in_T A$, that $x \in^A \{y\}$ iff $x = y$. [Hint: use the diagram and the fact that $\delta_A \circ \langle y, x \rangle = t$ iff $x = y$.]

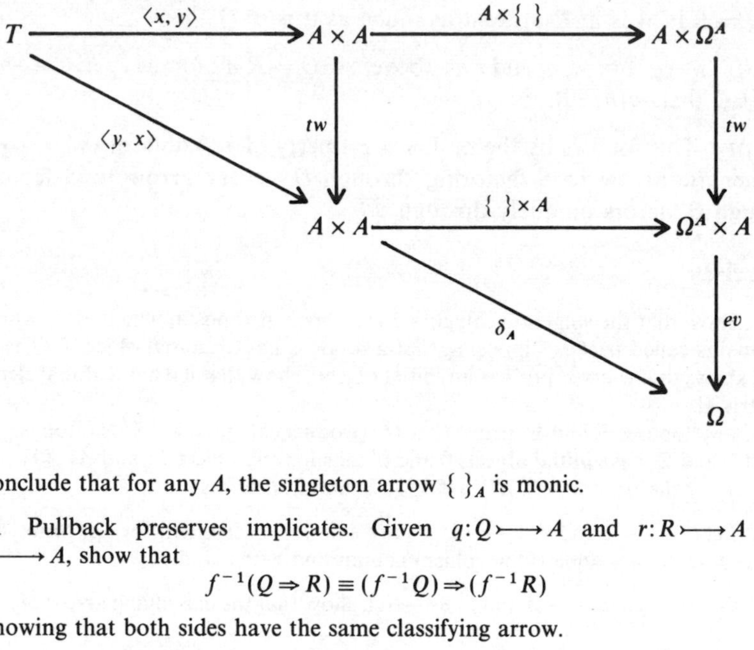

Conclude that for any A, the singleton arrow $\{\ \}_A$ is monic.

13.11 Pullback preserves implicates. Given $q: Q \rightarrowtail A$ and $r: R \rightarrowtail A$ and $f: B \longrightarrow A$, show that

$$f^{-1}(Q \Rightarrow R) \equiv (f^{-1}Q) \Rightarrow (f^{-1}R)$$

by showing that both sides have the same classifying arrow.

13.12 Pullback preserves universal quantification. Given $r: R \rightarrowtail B \times A$ and any $f: C \longrightarrow B$, show that $f^{-1}((\forall a)R) \equiv (\forall a)(f \times A)^{-1}R$. [Hint: consider classifying arrows and transposition.]

13.13 To show that $Q \Rightarrow R$ is the equalizer of χ_q and $\chi_{q \cap r}$, note that these factor as $p_1 \circ \langle \chi_q, \chi_r \rangle$ and $\wedge \circ \langle \chi_q, \chi_r \rangle$ respectively, and prove the following theorem, which holds in any category.

For any arrows as shown:

$$A \xrightarrow{f} B \underset{g}{\overset{h}{\rightrightarrows}} C$$

pulling any equalizer for h and g back along f gives an equalizer for $h \circ f$ and $g \circ f$.

13.14 A *class of generators* for a category **E** is a class **C** of **E** objects such that for every parallel pair of arrows $f, g: A \longrightarrow B$ in **E** either $f = g$ or there is some C in **C** and $h: C \longrightarrow A$ with $f \circ h \neq g \circ h$. Obviously, the class of all objects of **E** is a class of generators, but a given category **E** may also have much smaller classes of generators. The singleton class $\{\mathbf{2}\}$ is a generator for the category of categories.

If **E** is topos show it follows that, for any two sub-objects $u: U \rightarrowtail A$ and $v: V \rightarrowtail A$, $U \subseteq V$ iff for every C in **C** and $x \in_C A$ if $x \in U$ then $x \in V$. The proof uses the sub-object classifier, and the result does not hold in all categories (compare with Theorem 4.3).

13.15 Let **C** be a class of generators for a topos **E**. Given any $u: U \rightarrowtail A$ and $v: V \rightarrowtail A$, let $s: S \rightarrowtail A$ have the following property: for any C in **C** and $x \in_C A$ we have $x \in S$ iff, for every C' in **C** and $h: C' \longrightarrow C$, if $x(h) \in U$ then $x(h) \in V$. Show that $S \equiv U \Rightarrow V$. That is, show that any sub-object that looks like $U \Rightarrow V$ to the objects of **C** actually is $U \Rightarrow V$, up to equivalence of sub-objects.

Given R, a sub-object of $B \times A$, suppose that for any C in **C** and $x \in_C B$, $x \in S$ iff for every $h: C' \longrightarrow B$ and $k: C' \longrightarrow A$ we have $\langle x(h), k \rangle \in R$. Show that $S \equiv (\forall a)R$.

13.16 In any category with pullbacks a *power object* of an object A is an object PA with a *universal relation* from A to PA: a relation $\in^A \rightarrowtail A \times PA$ such that for any relation $R \rightarrowtail A \times B$ there is a unique arrow $f: B \longrightarrow PA$ such that R is a pullback of $\in^A$ along $1_A \times f$. Show that the power objects defined above are power objects in this sense.

Conversely, show that any category with all finite limits and with a power object for each object is a topos. First, show that $P1$ is a sub-object classifier. Then show that for any object B the power object $P(B \times A)$ is an exponential of PB by A. (Any $C \longrightarrow P(B \times A)$ gives a sub-object of $C \times A \times B$, and so an arrow $C \times A \longrightarrow PB$.) Show that any object B is an equalizer of $t \circ !_B$ and the arrow δ_B that classifies the diagonal $B \rightarrowtail B \times B$. Show that the equalizer of the exponential arrows $(t \circ !_B)^A$ and $(\mathrm{d}_B)^A$ is an exponential of B by A.

14

The internal language

For more complex reasoning in a topos **E**, we introduce a notation combining variables with **E** arrows to make up a kind of set-theoretic language with a familiar (although not quite classical) logic. This language is called the *internal language* of **E**. The internal language, and how its terms refer to **E** arrows, is described in Section 14.1. The logic of this language is described in Section 14.2, and various proofs in it are given in Section 14.3.

14.1 The language

The internal language is sorted, with each **E** object A as one sort. We inductively define terms and their sorts:

(LT1) Each **E** object A has a list of *variables over* A, $x_1, x_2, x_3 \ldots$ Every variable over A is a term of sort A.

(LT2) For any arrow $f: A \longrightarrow B$ and term s sort A, fs is a term of sort B. Every arrow $c: 1 \longrightarrow A$ with domain sort 1 is itself a term of sort A: we call it a constant of sort A. We write ! as a constant for the identity arrow on 1.

(LT3) For every term s_1 of sort A and s_2 of sort B, there is a term $\langle s_1, s_2 \rangle$ of sort $A \times B$.

(LT4) For every term s of sort B and variable y of sort A, there is a term $(\lambda y)s$ of sort B^A.

A variable x is *free* unless it is *bound* by a lambda operator (λx). We may write $(\lambda x\,.\,A)$ to indicate the sort of the variable. A term with no free variables is *closed.*

Given a variable x and a term v of the same sort, for any term s we write $s(x/v)$ for the term obtained by substituting v for every free occurrence of x in s. If no free variables of v become bound in $s(x/v)$ we say that v is *free for* x in s.

Suppose that a term s has sort B and all its free variables are in the list $y_1, \ldots, y_k$, where the y's are variables over $A_1, \ldots, A_k$ respectively. Then s refers to an arrow $|s|: A_1 \times \cdots \times A_k \longrightarrow B$, which we call the *interpretation* of s. Intuitively, any assignment of a value to each variable, giving each y_i a value in A_i, determines a value for s. Since the arrow $|s|$ actually depends on the list of variables involved, we should show the list in the notation.

We use $\bar{x}$ to abbreviate a list $x_1, \ldots, x_n$ and then $A_1, \ldots, A_n$ is the list of sorts of the variables in the same order. A variable can only appear once in a list, but an object A will appear as many times as there are variables over A in the list. For a term s of sort B and a list $\bar{x}$ including all the free variables of s, we write $|s|_{\bar{x}}: A_1 \times \cdots \times A_n \longrightarrow B$ for the interpretation of s relative to the list $\bar{x}$. We always assume that our lists of variables include all those that are free in the terms we apply them to. If s has no free variables then $\bar{x}$ can be an empty list of variables, and of course the product of an empty list of sorts is 1.

(I1) For any list $\bar{x}$ and variables x_i in the list, $|x_i|_{\bar{x}}$ is the ith projection

$$A_1 \times \cdots \times A_n \longrightarrow A_i$$

(I2) For any arrow $f: A \longrightarrow B$, if s is a term of sort A then $|fs|_{\bar{x}}$ is

$$A_1 \times \cdots \times A_n \xrightarrow{|s|_{\bar{x}}} A \xrightarrow{f} B$$

For any constant c, $|c|_{\bar{x}}$ is

$$A_1 \times \cdots \times A_n \longrightarrow 1 \xrightarrow{c} B$$

(I3) For any terms s_1 of sort A and s_2 of sort B, $|\langle s_1, s_2 \rangle|_{\bar{x}}$ is the pair arrow to $A \times B$ induced by $|s_1|_{\bar{x}}$ and $|s_2|_{\bar{x}}$.

(I4) For any term s of sort B, if the variable y over A is not in the list $\bar{x}$ then $|(\lambda y)s|_{\bar{x}}$ is the transpose of $|s|_{\bar{x}, y}: \times \cdots \times A_n \times A \longrightarrow B$.

If the bound variable y is in the list $\bar{x}$ that is an irrelevant coincidence. Then we replace y in $(\lambda y)s$ by some variable over A neither in s nor in the list $\bar{x}$.

By LT2, for any term g of sort B^A and s of sort A there is a term $ev(\langle g, s \rangle)$. We will abbreviate this to $g(s)$. Also, we will use set builder notation for lambda abstraction over Ω; that is, rather than write $(\lambda x . A)s$ when s has sort Ω, we write $\{x . A | s\}$. This suits our view of an arrow from A to Ω as a subobject of A. And for a term P of sort Ω^A we write $x \in P$ rather than $ev(\langle P, x \rangle)$. In what follows we will omit angle brackets inside parentheses, writing $f(x, y)$ rather than $f(\langle x, y \rangle)$.

A formula is a term of sort Ω. By LT2, for any formulas φ and ψ there are formulas $\wedge(\varphi, \psi)$ and $\rightarrow(\varphi, \psi)$. We will write these as $\varphi \,\&\, \psi$ and $\varphi \rightarrow \psi$. There are also formulas fa, $\sim\varphi$, and $\varphi \vee \psi$, read as 'false', 'not-φ', and 'φ or ψ', which we describe in this chapter, although the definitions are given in Chapter 15.

For any formula φ and variable y over any object A, we define the formula $(\forall y . A)\varphi$ as an abbreviation for $\forall_A\{y . A | \varphi\}$. So $(\forall y . A)\varphi$ says that $\{y . A | \varphi\}$ is all of A. Its interpretation over a list of variables $\bar{x}$ follows the definition.

It is

$$A_1 \times \cdots \times A_n \xrightarrow{\overline{|\varphi|_{\bar{x},y}}} \Omega^A \xrightarrow{\forall_A} \Omega$$

if the variable y does not occur in the list $\bar{x}$. If y in the list it is first replaced by some new variable. Note that y is bound in $(\forall y\,.\,A)\varphi$. In Chapter 15 we define an existential quantifier $(\exists x\,.\,A)$ for each A.

For any terms a_1 and a_2 of sort A, the arrow $\delta_A : A \times A \longrightarrow \Omega$ gives a term $\delta_A(a_1, a_2)$ which we abbreviate to $a_1 = a_2$. It follows that $|a_1 = a_2|_{\bar{x}}$ is the classifying arrow for the equalizer of $|a_1|_{\bar{x}}$ and $|a_2|_{\bar{x}}$.

Finally, for any monics i: $I \rightarrowtail A$ and $r: R \rightarrowtail A \times B$, we write formulas $I(s)$ and $Rs_1 s_2$ as abbreviations for $\chi_i(s)$ and $\chi_r(s_1, s_2)$, where s and s_1 are any terms of sort A and s_2 is any term of sort B.

14.2 Topos logic

The *extension* of φ over a list of variables $\bar{x}$ is the sub-object of $A_1 \times \cdots \times A_n$ classified by $|\varphi|_{\bar{x}}$. Write $[\bar{x}|\varphi]$ for this extension and think of 'all $\bar{x}$ such that φ'. For example, we have:

$$[\bar{x}|t] \equiv A_i \times \cdots \times A_n$$

$$[\bar{x}|\varphi \,\&\, \psi] \equiv [\bar{x}|\varphi] \cap [\bar{x}|\psi]$$

$$[\bar{x}|\varphi \rightarrow \psi] \equiv [\bar{x}|\varphi] \Rightarrow [\bar{x}|\psi]$$

and $[\bar{x}|\,(\forall y)\varphi]$ is the universal quantification of $[\bar{x}, y|\varphi]$ over the projection corresponding to y.

A formula φ is called *true* if its extension $[\bar{x}|\varphi]$ is all of $A_1 \times \cdots \times A_n$ when $\bar{x}$ lists exactly the variables free in φ.

We say that a formula φ implies ψ if the extension of φ is contained in that of ψ. More generally, for any finite set of formulas Γ, write $[\bar{x}|\Gamma]$ for the intersection of the extensions over $\bar{x}$ of all formulas in Γ. In particular, $[\bar{x}|\ ] = A_1 \times \cdots \times A_n$ for the empty list of formulas. Then Γ *implies* φ iff $[\bar{x}|\Gamma] \subseteq [\bar{x}|\varphi]$ when $\bar{x}$ lists exactly the free variables in Γ and φ.

A *sequent* is an expression $\Gamma : \varphi$, where Γ is a finite (possibly empty) set of formulas and φ is a formula. Think of $\Gamma : \varphi$ as a claim that the formulas in Γ imply φ. The sequent is *true* iff Γ does imply φ. In particular, a sequent $\ :\varphi$ with empty left side is true iff $[\bar{x}|\varphi]$ is all of $A_1 \times \cdots \times A_n$ and so iff φ is true. When we know a sequent $\Gamma : \varphi$ is true we may write

$$\Gamma \vdash \varphi$$

We can give sound rules of inference for these sequents; that is, rules such that applying them to true sequents of **E** always yields true sequents. We call these rules *topos logic*. We will describe these rules and their use in this chapter, and prove that they are sound in Chapter 15.

First are the *structural rules*. From the sequent or sequents above the line we can infer the one below. An asterisk shows that the sequent below follows from an empty set of assumptions:

Trivial sequent: $\dfrac{*}{\varphi:\varphi}$

True and false: $\dfrac{*}{:t} \qquad \dfrac{*}{fa:\varphi}$

Thinning: $\dfrac{\Gamma:\varphi}{\Gamma,\psi:\varphi}$

Substitution: $\dfrac{\Gamma:\varphi}{\Gamma(x/s):\varphi(x/s)}$ for any term s free for x in all the formulas

Cut: $\dfrac{\Gamma,\psi:\varphi \quad \text{and} \quad \Gamma:\psi}{\Gamma:\varphi}$ if every variable free in ψ is free in Γ or in φ

Exercises 14.5, 15.5, and 16.7 explain the restriction on the cut rule. The basic point is that if ψ has free variables over empty objects then the upper sequents are trivially true even if the lower one is false.

There is one reversible *connective rule* for each connective. From the sequent(s) above the line we can infer the one below, and from the one below we can infer the one or either of the ones above:

$$\frac{\Gamma,\varphi:fa}{\overline{\Gamma:\sim\varphi}} \qquad \frac{\Gamma:\varphi \quad \text{and} \quad \Gamma:\psi}{\overline{\Gamma:\varphi\,\&\,\psi}}$$

$$\frac{\Gamma,\varphi:\theta \quad \text{and} \quad \Gamma,\psi:\theta}{\overline{\Gamma,\varphi\vee\psi:\theta}} \qquad \frac{\Gamma,\varphi:\psi}{\overline{\Gamma:\varphi\rightarrow\psi}}$$

and if the variable x is not free in Γ or ψ:

$$\frac{\Gamma:\varphi}{\overline{\Gamma:(\forall x)\varphi}} \qquad \frac{\Gamma,\varphi:\psi}{\overline{\Gamma,(\exists x)\varphi:\psi}}$$

Finally, we have the *axioms of local set theory*:

Singleton: $\vdash (\forall y\,.\,1)y = !$

The equality axioms, for variables q, r, and x all over any one object A,

$$\vdash q = q \quad (q = r), \qquad \varphi(x/r) \vdash \varphi(x/q)$$

where φ is any formula with q and r both free for x in φ:

Biconditional: $(v\rightarrow w)\,\&\,(w\rightarrow v) \vdash v = w \qquad v, w\,.\,\Omega.$

The converse follows from the equality axioms. For formulas φ and ψ we usually write '$\varphi \leftrightarrow \psi$' instead of '$\varphi = \psi$'.

Product: $p_1 u = p_1 u', p_2 u = p_2 u' \vdash u = u' \qquad u, u'.A \times B$

$$\vdash (p_1\langle s, s'\rangle = s \,\&\, p_2\langle s, s'\rangle = s') \qquad s, s'.B$$

Extensionality: $(\forall x.A)f(x) = g(x) \vdash f = g \qquad f, g.B^A$

Comprehension: $\vdash ((\lambda x.A)s)(x) = s \qquad x.A$

for any term s of sort B.

In the case $B = \Omega$ we have an alternative notation, of course. The extensionality and comprehension axioms take the following form:

$$(\forall x.A)(x \in P \leftrightarrow x \in Q) \vdash P = Q \qquad P, Q.\Omega^A$$

$$\vdash x \in \{x.A \mid \varphi\} \leftrightarrow \varphi \qquad x.A$$

for any formula φ.

This is local set theory in that each variable is over one object—there are no global variables.

The structural rules, connective rules, and axioms of local set theory are not really axioms for us. The next chapter proves all of them from the axioms for a topos. But we treat them as axiomatic in the internal language, and in fact they are complete in a certain sense, as discussed at the end of Chapter 15.

14.3 Proofs in topos logic

Typical differences between topos and classical logic are that '$\varphi \vee \sim\varphi$' and '$\sim\sim\varphi \rightarrow \varphi$' are not provable in topos logic, and neither is '$\sim(\varphi \,\&\, \psi) \rightarrow (\sim\varphi \vee \sim\psi)$' or '$\sim(\forall x)\varphi \rightarrow (\exists x)\sim\varphi$', or '$(\forall x)\varphi \rightarrow (\exists x)\varphi$'. In short, it is hard to conclude from a negative, or to a disjunction or an existential statement in topos logic. An assortment of deductions will show how topos logic works. We often omit obvious steps.

Clearly, by the definition of & any list of formulas on the left-hand side of a sequent is equivalent to its conjunction. So we form or separate conjunctions of formulas on the left without comment.

We deduce two forms of the law of non-contradiction:

$$\dfrac{\dfrac{\dfrac{*}{\sim\varphi, \varphi : fa}}{\varphi \,\&\, \sim\varphi : fa}}{: \sim(\varphi \,\&\, \sim\varphi)} \qquad \dfrac{\dfrac{\dfrac{*}{\sim\varphi, \varphi : fa}}{\varphi : \sim\sim\varphi}}{: \varphi \longrightarrow \sim\sim\varphi}$$

Three DeMorgan laws are provable. This deduction:

$$\cfrac{\cfrac{\cfrac{*}{\sim(\varphi \vee \psi), (\varphi \vee \psi), \varphi : fa} \quad \text{and} \quad \cfrac{\cfrac{*}{\varphi : \varphi \vee \psi}}{\sim(\varphi \vee \psi), \varphi : \varphi \vee \psi}}{\sim(\varphi \vee \psi), \varphi : fa}}{\sim(\varphi \vee \psi) : \sim \varphi}$$

shows how to prove

$$\vdash \sim(\varphi \vee \psi) \rightarrow (\sim \varphi \,\&\, \sim \psi)$$

Two more are straightforward applications of $\sim$, &, and $\vee$ rules:

$$\vdash (\sim \varphi \,\&\, \sim \psi) \longrightarrow \sim(\varphi \vee \psi)$$

$$\vdash (\sim \varphi \vee \sim \psi) \longrightarrow \sim(\varphi \,\&\, \psi)$$

As already remarked, the converse of the last is not provable in topos logic.

The distributive laws for conjunction and disjunction are provable. For example:

$$\cfrac{\cfrac{\cfrac{\cfrac{*}{\varphi \,\&\, \theta : (\varphi \,\&\, \theta) \vee (\varphi \,\&\, \psi)}}{\varphi, \theta : (\varphi \,\&\, \theta) \vee (\varphi \,\&\, \psi)} \quad \text{and} \quad \cfrac{\cfrac{*}{\varphi \,\&\, \psi : (\varphi \,\&\, \theta) \vee (\varphi \,\&\, \psi)}}{\varphi, \psi : (\varphi \,\&\, \theta) \vee (\varphi \,\&\, \psi)}}{\varphi, \theta \vee \psi : (\varphi \,\&\, \theta) \vee (\varphi \,\&\, \psi)}}{\varphi \,\&\, (\theta \vee \psi) : (\varphi \,\&\, \theta) \vee (\varphi \,\&\, \psi)}$$

The converse is easier:

$$(\varphi \,\&\, \theta) \vee (\varphi \,\&\, \psi) \quad \vdash \quad \varphi \,\&\, (\theta \vee \psi)$$

and, similarly,

$$\varphi \vee (\theta \,\&\, \psi) \quad \dashv\vdash \quad (\varphi \vee \theta) \,\&\, (\varphi \vee \psi)$$

As to quantifiers and negation, consider the following deduction:

$$\cfrac{\cfrac{\cfrac{\cfrac{*}{(\forall x) \sim \varphi : (\forall x) \sim \varphi}}{(\forall x) \sim \varphi : \sim \varphi}}{(\forall x) \sim \varphi, \varphi : fa}}{(\forall x) \sim \varphi, (\exists x)\varphi : fa}$$

A similar deduction shows that $(\forall x)\varphi$ and $(\exists x) \sim \varphi$ are contradictory. Thus we have

$$\vdash (\forall x) \sim \varphi \longrightarrow \sim (\exists x)\varphi$$

$$\vdash (\exists x)\varphi \longrightarrow \sim (\forall x) \sim \varphi$$

$$\vdash (\exists x) \sim \varphi \longrightarrow \sim (\forall x)\varphi$$

$$\vdash (\forall x)\varphi \longrightarrow \sim (\exists x) \sim \varphi$$

Of all these, only the first has converse provable in topos logic for all formulas φ:

$$\cfrac{\cfrac{\cfrac{*}{\varphi : (\exists x)\varphi}}{\sim (\exists x)\varphi : \sim \varphi}}{\sim (\exists x)\varphi : (\forall x) \sim \varphi}$$

For the step from the first sequent to the second, see Exercise 14.1.

Local set theory proves relativized versions of the usual axioms of set theory; that is, versions with bounded quantifiers. For example, for every object A, the term $\{x.A|fa\}$ is the 'empty set of things in A'. Formally, we show that

$$\vdash (\forall y.A) \sim (y \in \{x.A|fa\})$$

by deducing it from

$$y \in \{x, A|\, fa\} \vdash fa$$

which itself is a case of the comprehension axiom.

Each object A has analogues to the pair set, power set, and union axioms of classical set theory relativized to A:

$$\vdash (\forall x.A)(\forall y.A)(\exists z.\Omega^A)(\forall a.A)(a \in z \leftrightarrow (a = x \vee a = y))$$

$$\vdash (\forall x.\Omega^A)(\exists y.\Omega^{(\Omega^A)})(\forall z.\Omega^A)(z \in y \leftrightarrow (\forall a.A)(a \in z \longrightarrow a \in x))$$

$$\vdash (\forall x.\Omega^{(\Omega^A)})(\exists y.\Omega^A)(\forall a.A)(a \in y \leftrightarrow (\exists z.\Omega^A)(z \in x \,\&\, a \in z))$$

To deduce the first just use the term $\{s.A|s = x \vee s = y\}$. The others are similar.

Exercises

14.1 For any formulas φ and ψ, from the premise $\varphi:\psi$ deduce $\sim \psi : \sim \varphi$.

14.2 Fill out the deductions abbreviated in these inferences, used several times above without comment:

$$\frac{\Gamma:\psi \quad \text{and} \quad \psi:\varphi}{\Gamma:\varphi} \qquad \frac{\Gamma:\varphi}{\Gamma, \varphi \longrightarrow \theta : \theta}$$

where every variable free in ψ is free in either Γ or φ.

14.3 Give deductions in topos logic for each of these:

$$\varphi \vdash \psi \longrightarrow \varphi$$

$$\varphi \longrightarrow {\sim}\varphi \vdash {\sim}\varphi$$

$${\sim}\varphi \longrightarrow \varphi \vdash {\sim}{\sim}\varphi$$

$${\sim}\varphi \vee \psi \vdash \varphi \longrightarrow \psi \qquad \text{(but not the converse)}$$

$${\sim}{\sim}{\sim}\varphi \dashv\vdash {\sim}\varphi$$

$$(\varphi \,\&\, \psi) \longrightarrow \theta \;\dashv\vdash\; \varphi \longrightarrow (\psi \longrightarrow \theta)$$

$$(\varphi \vee \psi) \longrightarrow \theta \;\dashv\vdash\; (\varphi \longrightarrow \theta) \,\&\, (\psi \longrightarrow \theta)$$

$$(\forall x\,.\,A)(\varphi \,\&\, \psi) \;\dashv\vdash\; (\forall x)\varphi \,\&\, (\forall x)\psi$$

$$(\exists x\,.\,A)(\varphi \,\&\, \psi) \vdash (\exists x)\varphi$$

$$(\exists x\,.\,A)(\varphi \vee \psi) \;\dashv\vdash\; (\exists x)\varphi \vee (\exists x)\psi$$

14.4 Assuming that x is not free in φ, give deductions in topos logic for each of these:

$$(\exists x\,.\,A)(\psi \longrightarrow \varphi) \;\vdash\; (\forall x)\psi \longrightarrow \varphi$$

$$(\forall x\,.\,A)(\psi \longrightarrow \varphi) \;\dashv\vdash\; (\exists x)\psi \longrightarrow \varphi$$

14.5 The free variable restriction on the cut rule blocks the attempt to deduce

$$\vdash (\forall x\,.\,A)\varphi \longrightarrow (\exists x\,.\,A)\varphi$$

by cut from $(\forall x)\varphi \vdash \varphi$ and $\varphi \vdash (\exists x)\varphi$. Show that the sequent can be deduced if there is any constant of sort A. Show it can be deduced from the premise $(\exists x\,.\,A)\; x = x$.

In fact, suppose that $x_1 \ldots x_n$ are all the variables free in ψ which are not free in Γ or φ. Then show that from

$$:(\exists x_1\,.\,B_1)\ldots(\exists x_n\,.\,B_n)x_1 = x_1 \,\&\, \ldots \,\&\, x_n = x_n$$

$$\Gamma, \psi : \varphi$$

$$\Gamma : \psi$$

one can deduce

$$\Gamma : \varphi$$

while obeying the free variable restriction.

14.6 Show that in topos logic a contradiction implies anything; that is, deduce

$$\varphi \,\&\, {\sim}\varphi \vdash \psi$$

14.7 From the axioms of equality deduce each of these sequents, for any variables q, r, and s of the same sort:

$$q = r : r = q$$

$$q = r, r = s : q = s$$

14.8 Call the negation rule given above the *topos negation* rule, and call the following the *Boolean negation* rule:

$$\frac{\Gamma, \sim \varphi : fa}{\Gamma : \varphi}$$

Derive the topos rule from cut and the Boolean rule. [Hint: from the Boolean rule derive $\sim \sim \varphi \vdash \varphi$. Use that and cut to derive half the topos rule. From that derive $\varphi \vdash \sim \sim \varphi$ and then the other half of the topos rule.]

Boolean logic is topos logic with the Boolean negation rule. Apart from the free variable restriction on cut, Boolean logic is classical logic.

14.9 Derive each of these rules in Boolean logic:

$$\frac{*}{\sim \sim \varphi : \varphi} \qquad \frac{*}{: \varphi \vee \sim \varphi}$$

Conversely, derive the Boolean negation rule from topos logic (including the topos negation rule) plus either one of these.

14.10 Deduce each of these in Boolean logic:

$$\sim (\varphi \,\&\, \psi) \vdash \sim \varphi \vee \sim \psi$$

$$\sim \psi \longrightarrow \sim \varphi \vdash \varphi \longrightarrow \psi$$

$$\varphi \longrightarrow \psi \vdash \sim \varphi \vee \psi$$

$$(\forall x . A)\varphi \longrightarrow \psi \vdash (\exists x . A)(\varphi \longrightarrow \psi) \qquad \text{if } x \text{ is not free in } \psi$$

and converses to the quantifier negation theorems in Section 14.3.

14.11 Suppose that we extend topos logic by the quantifier negation rule $(\forall x) \leftrightarrow \sim (\exists x) \sim$ for all variables x (or even just for all variables over Ω). Show that this formula is derivable:

$$\vdash (\forall w . \Omega)(w = t \vee w = fa)$$

From that derive the law of excluded middle, $\vdash \varphi \vee \sim \varphi$, and conclude that the extension is Boolean logic. [Hint: show that $\vdash (\varphi = t) \leftrightarrow \varphi$, and show that $\vdash (\varphi = fa) \leftrightarrow \sim \varphi)$ for any formula φ.]

14.12 Show that extending topos logic by the fourth DeMorgan law

$$\sim (\varphi \,\&\, \psi) \vdash \sim \varphi \vee \sim \psi$$

is equivalent to extending it by the law

$$\vdash \sim \varphi \vee \sim \sim \varphi$$

The example of $\mathbf{Set}^2$ among others in Chapter 22 shows this is weaker than excluded middle.

15

Soundness proof for topos logic

In this chapter the truth value false, negation, disjunction, and existential quantifiers are defined, and it is proved that topos logic is sound. A reader who feels no need for the details may omit this chapter.

15.1 Defining *fa*, $\sim$, $\vee$, and $(\exists x)$

We define *fa*, $\sim$, $\vee$, and $(\exists x)$ in terms of &, $\rightarrow$, and $(\forall x)$, using quantification over Ω. In this section we use only the structural rules, not including 'false', and the inference rules for &, $\rightarrow$, and $(\forall x)$.

The formula $(\forall w\,.\,\Omega)w$ implies every formula:

$$\cfrac{\cfrac{*}{(\forall w)w : w}}{(\forall w)w : \varphi}$$

since φ is obviously free for w in w. Thus we can define *fa* to be $(\forall w)w$, and the sequent rule 'false' follows. There is a reversible derived rule of inference:

$$\frac{\Gamma, \varphi : fa}{\Gamma : \varphi \longrightarrow fa}$$

So we can define $\sim\varphi$ to be $\varphi \longrightarrow fa$ and the $\sim$-rule follows.

For any formulas φ and ψ consider the formula

$$(\forall w)([(\varphi \longrightarrow w)\,\&\,(\psi \longrightarrow w)] \longrightarrow w)$$

for w a variable over Ω not free in φ or ψ. Each of φ and ψ implies it:

$$\cfrac{\cfrac{*}{\varphi, (\varphi \longrightarrow w)\,\&\,(\psi \longrightarrow w) : w}}{\varphi : (\forall w)([(\varphi \longrightarrow w)\,\&\,(\psi \longrightarrow w)] \longrightarrow w)}$$

since w is not free in φ or ψ. On the other hand, for every formula θ we have

$$(\forall w)([(\varphi \longrightarrow w)\,\&\,(\psi \longrightarrow w)] \longrightarrow w) \vdash [(\varphi \longrightarrow \theta)\,\&\,(\psi \longrightarrow \theta)] \longrightarrow \theta$$

So we can use cut in the last step to give this deduction:

$$\cfrac{\cfrac{\cfrac{\Gamma, \varphi : \theta \qquad \Gamma, \psi : \theta}{\Gamma : (\varphi \longrightarrow \theta) \,\&\, (\psi \longrightarrow \theta)}}{\Gamma, [(\varphi \longrightarrow \theta) \,\&\, (\psi \longrightarrow \theta)] \longrightarrow \theta : \theta}}{\Gamma, (\forall w)([(\varphi \longrightarrow w) \,\&\, (\psi \longrightarrow w)] \longrightarrow w) : \theta}$$

Altogether, there is a reversible derived rule of inference:

$$\frac{\Gamma, \varphi : \theta \quad \text{and} \quad \Gamma, \psi : \theta}{\overline{\Gamma, (\forall w)([(\varphi \longrightarrow w) \,\&\, (\psi \longrightarrow w)] \longrightarrow w) : \theta}}$$

So we can use the formula to define $\varphi \vee \psi$, and the $\vee$-rule follows (see Exercise 15.6).

The existential quantification of a formula φ at a variable x can be defined to be $(\forall w)((\forall x)(\varphi \longrightarrow w) \longrightarrow w)$, where w is a variable over Ω not free in φ. On one hand:

$$\cfrac{\cfrac{\cfrac{\cfrac{*}{(\forall x)(\varphi \longrightarrow w) : \varphi \longrightarrow w}}{\varphi, (\forall x)(\varphi \longrightarrow w) : w}}{\varphi : (\forall x)(\varphi \longrightarrow w) \longrightarrow w}}{\varphi : (\forall w)((\forall x)(\varphi \longrightarrow w) \longrightarrow w))}$$

since w is not free in φ. On the other hand, if x is not free in a formula θ, then we have

$$(\forall w)((\forall x)(\varphi \longrightarrow w) \longrightarrow w) \vdash (\forall x)(\varphi \longrightarrow \theta) \longrightarrow \theta)$$

So a cut gives the last step in:

$$\cfrac{\cfrac{\cfrac{\cfrac{\Gamma, \varphi : \theta}{\Gamma : \varphi \longrightarrow \theta}}{\Gamma : (\forall x)(\varphi \longrightarrow \theta)}}{\Gamma, (\forall x)(\varphi \longrightarrow \theta) \longrightarrow \theta : \theta}}{\Gamma, (\forall w)((\forall x)(\varphi \longrightarrow w) \longrightarrow w) : \theta}$$

The second step requires that x not be free in Γ. In short, for any Γ and θ with x not free in either there is a reversible derived rule of inference:

$$\frac{\Gamma, \varphi : \theta}{\overline{\Gamma, (\forall w)((\forall x)(\varphi \longrightarrow w) \longrightarrow w) : \theta}}$$

15.2 Soundness

First, variables that are not free in a term have no effect on its interpretation except in determining the domain:

LEMMA 15.1 (the superfluous variable lemma) Let s be any term of sort A. Let $\bar{x}$ be a list of variables over $A_1, \ldots, A_n$, including all those free in s. Let $\bar{y}$ be a list of variables over $B_1, \ldots, B_k$, including all the variables of $\bar{x}$. Then $|s|_{\bar{y}}$ is the arrow

$$B_1 \times \cdots \times B_k \xrightarrow{\text{proj}} A_1 \times \cdots \times A_n \xrightarrow{|s|_{\bar{x}}} A$$

where proj is the obvious projection of $B_1 \times \cdots \times B_k$ on to those factors that are in $A_1 \times \cdots \times A_n$.

PROOF This is immediate if s is a variable and easy for terms given by LT2 or LT3. The result holds for terms $(\lambda x)s$ since the obvious projection of $B_1 \times \cdots \times B_k \times A$ to $A_1 \times \cdots \times A_n \times A$ is just $(\text{proj}) \times A$ and the transpose of any $h \circ (k \times A)$ is $\bar{h} \circ k$. □

COROLLARY For any formula φ and list $\bar{x}$ including all its free variables, and list $\bar{y}$ including all the variables in $\bar{x}$, the extension $[\bar{y}|\varphi]$ is the pullback along proj of $[\bar{x}|\varphi]$. Thus, if $\Gamma{:}\,\varphi$ is true then $[\bar{y}|\Gamma] \subseteq [\bar{y}|\varphi]$ for every list $\bar{y}$ including all variables free in the sequent. □

LEMMA 15.2 (the substitution lemma) Consider a term s of sort B and suppose that $\bar{x}$ contains every variable free in s. For each variable x_i in $\bar{x}$ let c_i be a term free for x_i in s. Let $s(\bar{x}/\bar{c})$ denote the result of substituting each c_i for x_i in s. Suppose that a list $\bar{y}$ of variables over $B_1 \times \cdots \times B_k$ includes all variables free in $s(\bar{x}/\bar{c})$. Then $|s(\bar{x}/\bar{c})|_{\bar{y}}$ is the arrow

$$B_1 \times \cdots \times B_k \xrightarrow{\langle c_1, \ldots, c_n \rangle_{\bar{y}}} A_1 \times \cdots \times A_n \xrightarrow{|s|_{\bar{x}}} B$$

where $\langle c_1, \ldots, c_n \rangle_{\bar{y}}$ is the n-tuple of the arrows $|c_i|_{\bar{y}}$.

PROOF If s is a variable then the theorem merely repeats the definition of $\langle c_1, \ldots, c_n \rangle_{\bar{y}}$. From there, the proof is virtually the same as that of the superfluous variable lemma. □

COROLLARY The extension $[\bar{y}|\varphi(\bar{x}/\bar{c})]$ is the pullback of $[\bar{x}|\varphi]$ along $\langle c_1, \ldots, c_n \rangle_{\bar{y}}$. The same holds for a finite set of formulas, so if $[\bar{x}|\Gamma] \subseteq [\bar{x}|\varphi]$ then

$$[\bar{y}|\Gamma(\bar{x}/\bar{c})] \subseteq [\bar{y}|\varphi(\bar{x}/\bar{c})].$$

The structural rules are sound by simple properties of inclusion and intersection. Certainly we have $[\bar{x}|\varphi] \subseteq [\bar{x}|\varphi]$ as well as $[\bar{x}|\ \ ] \subseteq [\bar{x}|t]$. And if $[\bar{x}|\Gamma] \subseteq [\bar{x}|\varphi]$ then certainly $[\bar{x}|\Gamma, \psi] \subseteq [\bar{x}|\varphi]$. The preceding corollary shows that substitution is sound. The proof in Section 15.1 shows that the rule 'false' is sound if substitution and the $(\forall x)$-rule are.

As to the cut rule, the free variable restriction guarantees that the list $\bar{x}$ of variables free in the sequent $\Gamma{:}\,\varphi$ is also the list of variables free in $\Gamma, \psi{:}\,\varphi$, and includes all variables free in $\Gamma:\psi$. Given $[\bar{x}|\Gamma] \subseteq [\bar{x}|\psi]$ it is easy to see that $[\bar{x}|\Gamma, \psi] \equiv [\bar{x}|\Gamma]$. And in that case $[\bar{x}|\Gamma, \psi] \subseteq [\bar{x}|\varphi]$ itself says that $[\bar{x}|\Gamma\ ] \subseteq [\bar{x}|\varphi]$.

The connective rule for conjunction is clearly sound: we have $[\bar{x}|\Gamma] \subseteq [\bar{x}|\varphi]$ and $[\bar{x}|\Gamma] \subseteq [\bar{x}|\psi]$ iff we have $[\bar{x}|\Gamma] \subseteq [\bar{x}|\varphi] \cap [\bar{x}|\psi]$. The $\rightarrow$-rule is sound in both directions, by Theorem 13.6. The soundness of the $(\forall x)$-rule is by Theorem 13.8 once one notices, by the superfluous variable theorem, that $[\bar{x}, x|\Gamma] \equiv [\bar{x}|\Gamma] \times A$ iff x is a variable over A not occurring free in Γ.

Since the rules for $\sim$, $\vee$, and $(\exists x)$ can be derived from rules that we have proved sound, they are sound.

It remains to show that the axioms of local set theory are true.

LEMMA 15.3 A formula $s_1 = s_2$ is true iff $|s_1|_{\bar{x}} = |s_2|_{\bar{x}}$, where $\bar{x}$ lists exactly the variables free in s_1 and s_2.

PROOF The extension $[\bar{x}|s_1 = s_2]$ is the equalizer of $|s_1|_{\bar{x}}$ and $|s_2|_{\bar{x}}$, and the equalizer is iso iff the arrows are equal. □

Thus every formula $x = x$ is true. Also, $y = !$ is true for any variable y over 1, so by soundness of the $(\forall x)$-rule the formula $(\forall y\,.\,1)y = !$ is true.

LEMMA 15.4 For any formula φ and variables q, r, and x over one object A, the following sequent is true:

$$q = r, \qquad \varphi(x/r) \vdash \varphi(x/q)$$

PROOF Let $\bar{y}$ be a list of variables over $y_1, \ldots, y_k$ such that $\bar{y}, q, r$ is a list of the variables free in the sequent. Then we must show that

$$[\bar{y}, q, r|q = r \,\&\, \varphi(x/r)] \subseteq [\bar{y}, q, r|\varphi(x/q)]$$

The right-hand extension contains all generalized elements $\langle h_1, \ldots, h_k, s, s'\rangle$ of $Y_1 \times \cdots \times Y_k \times A \times A$ such that $\langle h_1, \ldots, h_k, s\rangle$ is in the extension over the list $\bar{y}, q$ of $\varphi(x/q)$. Routine calculation shows that the left-hand extension contains the same generalized elements, but subject to the further condition that $s = s'$. □

Exercise 13.8 proves that the biconditional axiom is true.

LEMMA 15.5 For any objects A and B and variables u and u' over $A \times B$ this sequent is true:

$$p_1 u = p_1 u', \qquad p_2 u = p_2 u' : u = u'$$

PROOF A generalized element $\langle\langle h, k\rangle, \langle m, n\rangle\rangle$ of $(A \times B) \times (A \times B)$ is a member of the extension $[u, u'|p_1 u = p_1 u']$ iff $h = m$. It is a member of $[u, u'|p_2 u = p_2 u']$ iff $k = n$. So it is in both iff it is in $\Delta_{A\times B}: A \times B \rightarrowtail (A \times B) \times (A \times B)$; but that is the extension $[u, u'|u = u']$. □

Each case of the second product axiom is true by an easy application of Lemma 15.3.

LEMMA 15.6 Each case of the extensionality axiom is true.

PROOF Suppose that a generalized element $\langle h, k\rangle$ of $B^A \times B^A$ is a member of $[f, g|(\forall x.A)f(x) = g(x)]$. By Theorem 13.9 $\langle h, k\rangle \times A$ is a member of $[f, g, x|f(x) = g(x)]$. But the latter is the equalizer of $ev \circ (p_1 \times A):(B^A \times B^A) \times A \longrightarrow B$ and $ev \circ (p_2 \times A)$. If $\langle h, k\rangle \times A$ equalizes these two arrows then $ev \circ (h \times A) = ev \circ (k \times A)$ and so, by taking transposes, $h = k$. Thus $\langle h, k\rangle$ is a member of $[f, g|f = g]$. □

Each case of the comprehension axiom follows by Lemma 15.3, including the case $B = \Omega$.

The structural rules, connective rules, and axioms of local set theory are complete in this sense: if in every topos every sequent with the logical form of $\Gamma:\varphi$ is true, then $\Gamma:\varphi$ is deducible from the rules and axioms of topos logic. (See Boileau and Joyal (1981), Lambek and Scott (1986) or Bell (1988*b*) for precise statements and proofs.) But topos logic is not complete for each individual topos. In a given topos **E** it may well be that every sequent with some form $\Gamma:\varphi$ is true while the form is not deducible in topos logic. The Boolean rules are sound in **Set** and many other toposes but not deducible in topos logic, and other intermediate sentential logics are sound and complete for various toposes. The classical result that no axiomatic higher order logic is complete for standard interpretations in sets applies directly to show that there is no complete axiomatization of the internal truths of **Set**, and the same proof holds for many toposes (see further remarks in McLarty 1990*a*). Since we are concerned with particular toposes, rather than with deducibility, we use the soundness results but not the completeness theorem.

Exercises

15.1 Show, for any variables y and y' over B, that the extension $[y, y'|y = y']$ is the diagonal sub-object of $B \times B$. Show, for any $f:A \longrightarrow B$, that the extension $[x.A, y.B|y = fx]$ is the graph of f.

15.2 Define $\sim:\Omega \longrightarrow \Omega$ as the interpretation $|\sim w|_w$. Show that for any formula φ we have $|\sim\varphi|_{\bar{x}} = \sim \circ |\varphi|_{\bar{x}}$. Do the same for a disjunction arrow $\vee:\Omega \times \Omega \longrightarrow \Omega$. Use substitution and

$$fa \quad \dashv\vdash \quad t \longrightarrow fa$$

to show that $\sim$ is the classifying arrow for fa. Some authors use this to define $\sim$.

15.3 Show that every sequent in **E** is true if **E** is a trivial topos.

15.4 Deduce from the substitution lemma that the order of the variables in a list is immaterial. Let $\bar{x}$ list variables over $A_1 \times \cdots \times A_n$ and let $\bar{y}$ list the same variables, possibly in a different order. Write $B_1 \times \cdots \times B_n$ for the objects in the order of $\bar{y}$.

There is a unique isomorphism $I: B_1 \times \cdots \times B_n \longrightarrow A_1 \times \cdots \times A_n$ that commutes with the projections. Show, for any s with all its free variables in $\bar{x}$, that $|s|_{\bar{y}} = |s|_{\bar{x}} \circ I$. Give the corresponding result for extensions.

15.5 The free variable restriction in the cut rule. Let $\bar{x}$ list exactly the variables that are free in Γ and φ. Let $\bar{y}$ list any other variables that are free in ψ, varying over $B_1 \times \cdots \times B_k$. Show that if

$$\Gamma, \psi \vdash \varphi \qquad \text{and} \qquad \Gamma \vdash \psi$$

then $[\bar{x}, \bar{y} | \Gamma] \subseteq [\bar{x}, \bar{y} | \varphi]$.

In general, it does not follow that $[\bar{x} | \Gamma] \subseteq [\bar{x} | \varphi]$. Show that if the product $A_1 \times \cdots \times A_n \times B_1 \times \cdots \times B_k$ is an initial object then $[\bar{x}, \bar{y} | \Gamma] \subseteq [\bar{x}, \bar{y} | \varphi]$ holds whatever the formulas Γ and φ are. Exercise 13.1 shows that a product may be initial while none of the factors is.

Show that we do have $[\bar{x} | \Gamma] \subseteq [\bar{x} | \varphi]$ if the projection

$$A_1 \times \cdots \times A_n \times B_1 \times \cdots \times B_k \longrightarrow A_1 \times \cdots \times A_n$$

is epic. By the description of epics in Chapter 16, the projection is epic iff the first premise in Exercise 14.5 is true.

15.6 Given any formula φ and variable y which does not occur free in φ and is free for x in φ, deduce these in topos logic:

$$(\forall x)\varphi \dashv\vdash (\forall y)(\varphi(x/y)) \qquad (\exists x)\varphi \dashv\vdash (\exists y)(\varphi(x/y))$$

The derivations of the $\vee$-rule and $\exists$-rule above have gaps, as each applies the $\forall$-rule to w, which might be free in Γ or θ. Correct the deductions by showing that w can be replaced by a suitable new variable before that rule is applied.

16

From the internal language to the topos

16.1 Overview

We work with a fixed topos **E**. We show an arrow $f: A \longrightarrow B$ is monic iff it is one-to-one:

$$fx = fx' \vdash x = x' \qquad x, x' . A$$

It is epic iff it is internally onto, meaning that:

$$\vdash (\exists x . A) y = fx \qquad y . B$$

The reader is warned that this sequent does *not* say that for every generalized element $y \in_T B$ there is some $x \in_T A$ with $y = fx$. It says *internally* that every y in B is the image of some x in A, but *externally* this only means that f is epic and not necessarily split epic (see Exercise 18.5).

We show that every arrow $f: A \longrightarrow B$ factors as an epic $q: A \twoheadrightarrow I$ followed by a monic $i: I \rightarrowtail B$. This is called an *image factorization*, or more briefly the *image* of f.

The central result is that every functional relation in **E** defines an arrow. That is, for any R from A to B, if

$$Rxy, Rxy' \vdash y = y' \qquad x . A\ y, y' . B$$

$$\vdash (\exists y . B) Rxy \qquad x . A$$

then there is a unique arrow $f: A \longrightarrow B$ such that:

$$\vdash Rxy \longleftrightarrow y = fx \qquad x . A\ y . B$$

We show that $[x . 1 \mid fa]$ is an initial object. Intuitively, the object of 'all x in 1 such that false is true' is empty. For any objects A and B, we construct a coproduct in the form of a disjoint union of A and B, an object C with monics $A \rightarrowtail C$ and $B \rightarrowtail C$ such that every value in C is either in A or B and none are in both. For any arrows $f: A \longrightarrow T$ and $g: B \longrightarrow T$, we can define a unique arrow $u: C \longrightarrow T$ by the relation '$z = u(w)$ iff either w lies in A and $z = f(w)$, or else w lies in B and $z = g(w)$'.

For every equivalence relation $r: R \rightarrowtail A \times A$ in **E** we construct a quotient, namely the sub-object of Ω^A containing just the equivalence classes for the relation. Then we can take any parallel pair $f, g: A \longrightarrow B$ and construct an equivalence relation on B such that the equivalence classes form a coequalizer for f and g.

16.2 Monics and epics

LEMMA 16.1 Arrows $f, g: A \longrightarrow B$ have $f = g$ iff

$$\vdash fx = gx \qquad x.A$$

PROOF The extension of the formula is the equalizer of f and g, so it is all of A iff $f = g$. □

THEOREM 16.2 An arrow $f: A \longrightarrow B$ is monic iff

$$fx = fx' \vdash x = x' \qquad x.A\, x'.A$$

PROOF The extension $[x, x'.A \mid fx = fx']$ is the equalizer of $f \circ p_1$ and $f \circ p_2$. This is an arrow $\langle h, k\rangle: E \rightarrowtail A \times A$ with h and k a kernel pair for f. And $[x, x'.A \mid x = x']$ is the diagonal $\Delta_A: A \rightarrowtail A \times A$. Exercise 4.9 shows that the kernel pair is contained in the diagonal iff f is monic. □

LEMMA 16.3 An arrow $f: A \longrightarrow B$ is epic iff 1_B is the smallest sub-object of B that f factors through.

PROOF Suppose that f is epic and factors through $s: S \rightarrowtail B$. Then $\chi_s \circ f = t_A$, but also $t_B \circ f = t_A$, so since f is epic $\chi_s = t_B$ and $s \equiv 1_B$. Conversely, suppose that 1_B is the smallest sub-object of B that f factors through. If $h \circ q = k \circ q$ then q factors through the equalizer of h and k, so the equalizer must be all of B. So $h = k$. □

THEOREM 16.4 Every arrow $f: A \longrightarrow B$ has an image factorization as

$$A \overset{q}{\twoheadrightarrow} [y.B \mid (\exists x.A)y = fx] \rightarrowtail B$$

PROOF Consider a sub-object $s: S \rightarrowtail B$. Since $f = p_2 \circ \Gamma_f$ we see that f factors through s iff $\Gamma_f \subseteq p_2^{-1}(s)$. (Draw the pullback square: you may recognize the adjunction $\Sigma_{p2} \dashv p_2^*$.) The graph of f is the extension $[x.A, y.B \mid y = fx]$, so the inclusion holds iff

$$y = fx \vdash S(y) \qquad x.A\, y.B$$

By topos logic that is iff

$$(\exists x.A)y = fx \vdash S(y) \qquad y.B$$

and so iff S contains the extension $[y.B \mid (\exists x.A)y = fx]$. The extension is the smallest sub-object that f factors through. □

COROLLARY An arrow $f: A \longrightarrow B$ is epic iff

$$\vdash (\exists x.A)y = fx \qquad y.B$$

PROOF The formula is true iff its extension is 1_B. □

COROLLARY For any monics $i: I \rightarrowtail A$ and $r: R \rightarrowtail A \times B$, we have

$$\vdash I(x) \longleftrightarrow (\exists w . I) x = i(w) \qquad x . A$$

$$\vdash Rxy \longleftrightarrow (\exists w . R)\langle x, y\rangle = r(w) \qquad x . A\, y . B$$

PROOF The left-hand side of the first formula has extension I, and by Theorem 16.4 so does the right side. The second formula is a case of the first, substituting $\langle x, y\rangle$ for x. □

16.3 Functional relations

Throughout this section we use the following notation for a relation R with h and k its composites with the projections:

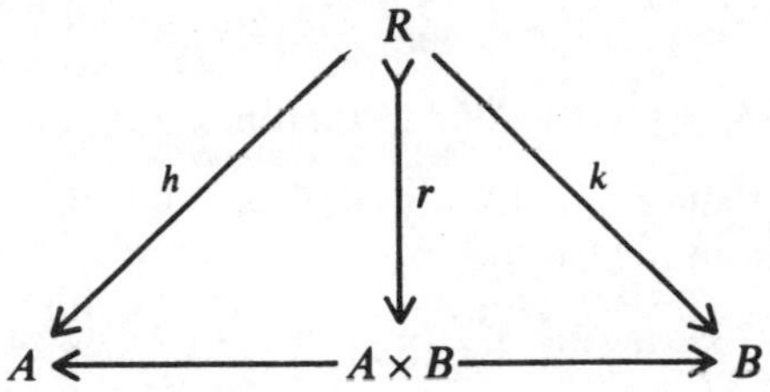

THEOREM 16.5 The arrow h is iso iff R is equivalent to the graph of some $f: A \longrightarrow B$, and in that case $f = k \circ h^{-1}$. (In fact, this holds in any category with binary products.)

PROOF Suppose that R is equivalent to the graph of some f; that is, there exists an iso $i: R \xrightarrow{\sim} A$ such that $\Gamma_f \circ i = r$. Then

$$i = p_1 \circ \Gamma_f \circ i = h$$

so h is iso and $\Gamma_f = r \circ h^{-1}$ and so

$$f = p_2 \circ \Gamma_f = k \circ h^{-1}$$

Conversely, if h is iso then h itself makes R equivalent to the graph of $k \circ h^{-1}$. □

Call R *single-valued* iff

$$Rxy, Rxy' \vdash y = y' \qquad x . A\, y, y' . B$$

THEOREM 16.6 If R is single-valued the arrow h is monic.

PROOF We have

$$\vdash R(h(w), k(w)) \qquad w . R$$

since it says that $\langle h, k\rangle$ factors through r and in fact $\langle h, k\rangle = r$. For any $c\colon T \longrightarrow R$ and $c'\colon T \longrightarrow R$, substitution gives both conjuncts in

$$\vdash R(h(c(s)), k(c(s))) \mathbin{\&} R(h(c'(s)), k(c'(s))) \qquad s.T$$

If we suppose that $h \circ c = h \circ c'$ then we conclude that

$$\vdash R(h(c(s)), k(c(s))) \mathbin{\&} R(h(c(s)), k(c'(s))) \qquad s.T$$

If we suppose that R is single-valued we conclude that

$$\vdash k(c(s)) = k(c'(s))$$

and so $k \circ c = k \circ c'$. But $h \circ c = h \circ c'$ and $k \circ c = k \circ c'$ together imply that $c = c'$. □

Call a relation R *totally defined* iff

$$\vdash (\forall x.A)(\exists y.B)\,Rxy$$

THEOREM 16.7 If R is totally defined then h is epic.

PROOF Suppose that R is totally defined. By the last corollary to Theorem 16.4 this is equivalent to

$$\vdash (\exists y.B)(\exists w.R)\,r(w) = \langle x, y\rangle \qquad x.A$$

But since $h = p_1 \circ r$, composing p_1 with each side gives

$$\vdash (\exists y.B)(\exists w.R)\,h(w) = x$$

Since y does not occur outside the quantifier, this implies that

$$\vdash (\exists w.R)\,h(w) = x.$$

□

Call R *functional* iff it is single-valued and totally defined. We have shown that R is functional iff it is equivalent to the graph of some arrow, and in that case it determines the arrow uniquely:

THEOREM 16.8 A relation R has

$$Rxy, Rxy' \vdash y = y' \qquad x.A\; y, y'.B$$

$$\vdash (\exists y.B)\,Rxy$$

iff there is an arrow $f\colon A \longrightarrow B$ such that

$$\vdash Rxy \longleftrightarrow y = fx$$

and in that case f is uniquely determined by R.

PROOF If f is functional h is monic and epic and thus iso. So R is the graph of a unique arrow, and the third sequent says that R is the graph of f. Conversely, the first two sequents follow from the third by topos logic. □

16.4 Extensions and arrows

We often use Theorem 16.8 to define an arrow from the extension of some formula, $[x|\varphi]$, to another object. Such use depends on:

THEOREM 16.9 Take any formula $\varphi(x)$, the sole free variable of which is x of sort A. Let $d{:}\,D \rightarrowtail A$ abbreviate the extension $[x\,.\,A|\varphi(x)]$. Then

$$\vdash \varphi(d(y)) \qquad y\,.\,D$$

That is, everything in the extension has the property.

PROOF By the substitution lemma, $[y\,.\,D|(d(y))]$ is the pullback along d (as the interpretation of $d(y)$) of d (as the extension of $\varphi(x)$). So it is 1_D. □

On the other hand, to define an arrow to an extension:

THEOREM 16.10 For any formula $\varphi(x)$, the sole free variable of which is x of sort A, an arrow $f{:}\,B \longrightarrow A$ factors through the extension $[x\,.\,A|\varphi(x)]$ iff

$$\vdash \varphi(fy) \qquad y\,.\,B$$

That is, every value of f has φ.

PROOF An arrow $f{:}\,B \longrightarrow A$ factors through any sub-object of A iff the pullback of the sub-object along f is 1_B. By the substitution lemma the extension of the sequent is the pullback along f of $[x\,.\,A|\varphi(x)]$; and the sequent is true iff its extension is 1_B. □

16.5 Initial objects and negation

Abbreviate the extension $[y\,.\,1|fa]$ by $\emptyset \rightarrowtail 1$. So the following is a pullback:

$$\begin{array}{ccc} \emptyset & \rightarrowtail & 1 \\ \downarrow & & \downarrow{\scriptstyle t} \\ 1 & \xrightarrow[fa]{} & \Omega \end{array}$$

In any topos $\emptyset$ is an initial object. From the point of view of logic, the proof turns on the classical fact that any universal quantification over an empty domain is true.

THEOREM 16.11 Every sequent with a free variable x over $\emptyset$ in it is true.

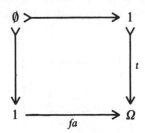

PROOF For x of sort $\emptyset$ we have

$$x = x \vdash fa \qquad x.\emptyset$$

because both $[x.\emptyset | x = x]$ and $[x.\emptyset | fa]$ are $\emptyset$. So if x is free in $\Gamma : \varphi$ there is a deduction:

$$\dfrac{\dfrac{\dfrac{\dfrac{*}{fa:\varphi}}{fa, \Gamma:\varphi} \quad \text{and} \quad \dfrac{*}{x = x : fa}}{x = x, \Gamma : \varphi} \qquad \text{and} \quad \dfrac{*}{: x = x}}{\Gamma : \varphi} \quad x.\emptyset$$

Both cuts satisfy the free variable restriction. □

The free variable restriction keeps us from using similar deductions simply to prove every sequent in **E**.

THEOREM 16.12 The object $\emptyset$ is initial.

PROOF For every object A there is the identity relation $\emptyset \times A \rightarrowtail \emptyset \times A$. The sequents that say that this relation is functional both have free variables over $\emptyset$, so both are true and there is an arrow $\emptyset \longrightarrow A$. And for any arrows $h, k : \emptyset \longrightarrow A$ the sequent that says that $h = k$ also has a free variable over $\emptyset$, so it is true. There is exactly one arrow $\emptyset \longrightarrow A$. □

Since a topos is Cartesian closed, $\emptyset$ is a strict initial object. So every arrow $\emptyset \longrightarrow A$ is monic. It is also the smallest sub-object of A, since for any $p : P \rightarrowtail A$ there is the unique $\emptyset \rightarrowtail P$.

There are several ways to look at the pullback defining $\emptyset$. It says that fa classifies the empty sub-object of 1. In fact, it defines the truth value false as the unique truth value disjoint from true. Exercise 15.2 showed that the following is a pullback:

$$\begin{array}{ccc} 1 & \longrightarrow & 1 \\ {\scriptstyle fa}\downarrow & & \downarrow{\scriptstyle t} \\ \Omega & \xrightarrow[\sim]{} & \Omega \end{array}$$

So $\sim fa = t$, and in fact for any $w : C \longrightarrow \Omega$, $\sim w = t_C$ iff $w = fa \circ !_C$. A truth value with negation true must be false. We also have $\sim t = fa$, since both sides classify $\emptyset \rightarrowtail 1$. But from $\sim w = fa \circ !_C$ we can only conclude that

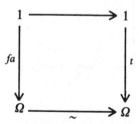

$\sim \sim w = t_C$. We cannot conclude that $w = t_C$. A truth value with negation false has double-negation true but need not itself be true. (See Exercise 16.15 and examples in specific toposes in Chapters 22–24.)

16.6 Coproducts

THEOREM 16.13 Take any two sub-objects $i_1 : A \rightarrowtail C$ and $i_2 : B \rightarrowtail C$ of an object C such that

$$\vdash A(w) \vee B(w) \qquad w.C$$

$$\vdash \sim (A(w) \,\&\, B(w))$$

Then C is a coproduct of A and B, with the injections i_1 and i_2.

PROOF Assume the sequents and take any $f : A \longrightarrow T$ and $g : B \longrightarrow T$. Recalling the last corollary to Theorem 16.5 it is easy to show that the formula on the right below is functional from C to T, so there is a unique $u : C \longrightarrow T$ such that

$$z = u(w) \longleftrightarrow [(\exists x)(w = i_1 x \,\&\, z = fx) \vee (\exists y)(w = i_2 y \,\&\, z = gy)] \qquad w.C\,z.T$$

It follows directly that

$$\vdash fx = u(i_1 x) \qquad x.A$$

$$\vdash gy = u(i_2 y) \qquad y.B$$

so $f = u \circ i_1$ and $g = u \circ i_2$. And for any $v : C \longrightarrow T$ such that $f = v \circ i_1$ and $g = v \circ i_2$ we have

$$\vdash z = u(w) \longleftrightarrow z = v(w) \qquad w.C\,z.T$$

and so $v = u$. □

A coproduct in which the injections are monic and disjoint is often called a *disjoint union.* The two sequents in the theorem say that C is a disjoint union of A and B. (The second sequent says that A and B have intersection $\emptyset$.) The rest of this section shows how to embed any objects A and B into a disjoint union. So every topos has coproducts.

For any object A, let $\emptyset_A : 1 \longrightarrow \Omega^A$ abbreviate the interpretation of the term $\{x.A \mid fa\}$. That is, internally

$$\vdash \emptyset_A = \{x.A \mid fa\}$$

so we have already proved that

$$\vdash \sim (x \in \emptyset_A) \qquad x.A$$

The arrows $\langle\{\ \}_A, \emptyset_B \circ !_A\rangle: A \rightarrowtail \Omega^A \times \Omega^B$ and $\langle \emptyset_A \circ !_B, \{\ \}_B\rangle: B \rightarrowtail \Omega^A \times \Omega^B$ are monic since $\{\ \}_A$ and $\{\ \}_B$ are already monic. By Theorem 16.10 both factor through the extension

$$[Q . \Omega^A \times \Omega^B | (\exists x . A)(Q = \langle\{x\}, \emptyset_B\rangle) \vee (\exists y . B)(Q = \langle \emptyset_A, \{y\}\rangle)]$$

where $\{x\}$ abbreviates $\{\ \}_A(x)$. So if $c: C \rightarrowtail \Omega^A \times \Omega^B$ abbreviates the extension then there are monics i_1 and i_2 that commute with those monics:

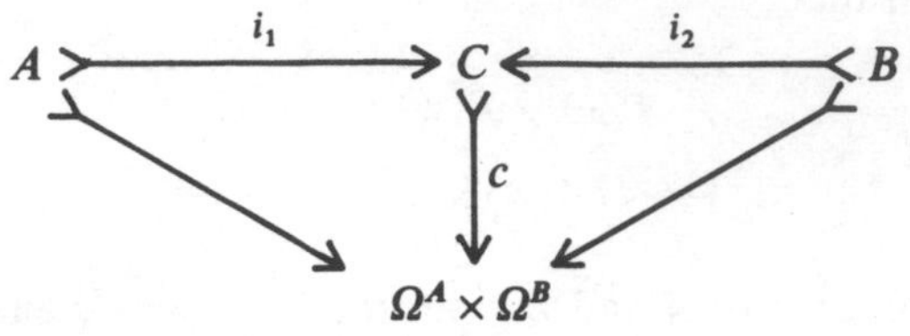

LEMMA 16.14 We have

$$\vdash (\exists x . A)(w = i_1 x) \vee (\exists y . B)(w = i_2 y) \qquad w . C$$

$$\vdash \sim [(\exists x . A)(w = i_1 x) \,\&\, (\exists y . B)(w = i_2 y)]$$

PROOF Since c is monic the first sequent is equivalent to

$$\vdash (\exists x . A)(c(w) = c(i_1 x)) \vee (\exists y . B)(c(w) = c(i_2 y))$$

and by the definition of i_1 and i_2 that is equivalent to

$$\vdash (\exists x . A)(c(w) = \langle\{x\}, \emptyset_B\rangle) \vee (\exists y . B)(c(w) = \langle \emptyset_A, \{y\}\rangle)$$

which is true by Theorem 16.9.

As to the second, it is easy to deduce that

$$i_1 x = i_2 y \vdash \{x\} = \emptyset_A \qquad x . A\, y . B$$

and so since no singleton is empty (see Exercise 16.9)

$$\vdash \sim (i_1 x = i_2 y)$$

and the second sequent in the lemma follows by topos logic. □

THEOREM 16.15 Any pair of objects A and B has a coproduct with the inclusions i_1 and i_2 that is monic and disjoint.

PROOF Lemma 16.14 showed that A and B have a pair of monics meeting the conditions in Theorem 16.13. □

Since coproducts are unique up to isomorphism, the inclusions to any coproduct (in a topos) are monic and disjoint.

We could use the logical construction of a coproduct and the substitution lemma to show that coproducts are stable under pullback. That is, pulling i_1

and i_2 back along any arrow $g: D \longrightarrow C$ gives a coproduct diagram D, $g^{-1}(i_1)$, $g^{-1}(i_2)$. But, instead, we prove stability of coproducts from the fundamental theorem of toposes in the next chapter.

16.7 Equivalence relations

We show that every equivalence relation $r: R \rightarrowtail B \times B$ in a topos is a kernel pair, namely for the epic projecting B on to the object of equivalence classes for R, and every epic is a co-equalizer for its kernel pair.

LEMMA 16.16 Given a relation $\langle h, k\rangle: R \rightarrowtail B \times B$, h and k are a kernel pair for $c: B \longrightarrow C$ iff

$$\vdash cy = cy' \longleftrightarrow Ryy' \qquad y, y'.B$$

PROOF The extension $[y, y'.B|cy = cy']$ is the pullback of c along c, i.e. the kernel pair as a sub-object of $B \times B$. □

LEMMA 16.17 For any equivalence relation $r: R \rightarrowtail B \times B$, there is an epic $q: B \twoheadrightarrow Q$ with h and k a kernel pair for Q.

PROOF Let $I: B \longrightarrow \Omega^B$ abbreviate the interpretation $|\{w.B|Ryw\}|_y$. We easily deduce that

$$\vdash \{w|Ryw\} = \{w|Ry'w\} \longleftrightarrow Ryy'$$

since R is an equivalence relation. Let $q: B \twoheadrightarrow Q$ and $i: Q \rightarrowtail C$ be an epic–monic factorization of I. Thus, $I = i \circ q$, and since i is monic we can deduce that

$$\vdash qy = qy' \longleftrightarrow Ryy'$$

which says that R is the kernel pair of q. □

By definition, Q is (up to equivalence) the extension

$$[P.\Omega^B|(\exists y)(P = \{w.B|Ryw\})]$$

and q maps each value in B to its equivalence class.

LEMMA 16.18 Any epic is a coequalizer for its kernel pair.

PROOF Let $q: B \longrightarrow Q$ be an epic with kernel pair $\langle h, k\rangle: R \rightarrowtail B \times B$ and take any $c: B \longrightarrow C$ such that $c \circ h = c \circ k$. Define a relation V from Q to C by

$$Vwz \longleftrightarrow (\exists y.B)(w = qy \;\&\; z = cy) \qquad w.Q\, z.C$$

Then V is single-valued, as follows from the sequent

$$qy = qy' \vdash cy = cy' \qquad y, y'.B$$

which in turn is true since the extension of the left-hand side is the kernel pair of q, and c coequalizes the kernel pair.

Since q is epic we can deduce that V is totally defined:

$$\dfrac{\dfrac{\dfrac{*}{:(\exists y.B)(w=qy\,\&\,cy=cy)}\quad w.Q}{:(\exists y.B)\,V(w,cy)}}{:(\exists z.C)\,Vwz}$$

So there is a unique $u\colon Q\longrightarrow C$ with

$$\vdash Vwz \longleftrightarrow z=u(w) \qquad w.Q\,z.C$$

From the definition of V,

$$\vdash cy=u(qy) \qquad y.B$$

Conversely, for any $v\colon Q\longrightarrow C$ with

$$\vdash cy=v(qy) \qquad y.B$$

the definition of V implies

$$\vdash Vwz \Leftrightarrow z=v(w) \qquad w.Q\,z.C$$

and since V uniquely defines u, $u=v$. □

16.8 Coequalizers

As $\Omega^{B\times B}$ is the object of relations from B to B, so the object of equivalence relations on R is the extension

$$\begin{aligned}[r.\Omega^{B\times B}|(\forall y,y',y''.B)(&\langle y,y\rangle\in r\,\&\\ &(\langle y,y'\rangle\in r\longrightarrow\langle y',y\rangle\in r)\,\&\\ &((\langle y,y'\rangle\in r\,\&\,\langle y',y''\rangle\in r)\longrightarrow\langle y,y''\rangle\in r))]\end{aligned}$$

that is, all reflexive, symmetric, and transitive relations r. Abbreviate the extension as $\mathrm{EqRel}\rightarrowtail\Omega^{B\times B}$.

For any parallel pair of arrows $f, g\colon A\longrightarrow B$ we say that a relation r on B *respects* f and g if, for all x in A, the values of x under f and g are r-related. We want to express this internally, so we define a sub-object $R_{fg}\rightarrowtail\Omega^{B\times B}$ as the extension

$$[r.\Omega^{B\times B}|(\forall x.A)\langle fx,gx\rangle\in r]$$

To construct a coequalizer for f and g we define a relation on B, holding for just those pairs $\langle y,y'\rangle$ such that y is r-related to y' for every equivalence relation r that respects f and g. Specifically, let $R\rightarrowtail B\times B$ be the extension

$$[y,y'.B|(\forall r.\Omega^{B\times B})((\mathrm{EqRel}(r)\,\&\,R_{fg}(r)\longrightarrow\langle y,y'\rangle\in r)]$$

Clearly R is an equivalence relation on B. As usual, we write h and k for the projections from R to B.

LEMMA 16.19 For any arrow $c: B \longrightarrow C$ we have $c \circ f = c \circ g$ iff $c \circ h = c \circ k$.

PROOF The kernel pair of c is an equivalence relation on B (see Chapter 5). On one hand, we have $c \circ f = c \circ g$ iff the kernel pair of c respects f and g. On the other hand, by construction of R, that kernel pair respects f and g iff R is contained in it, and so iff $c \circ h = c \circ k$. □

The projection $q: B \twoheadrightarrow Q$ on to equivalence classes for R is a coequalizer for h and k, so Exercise 16.11 proves that it is a coequalizer for f and g. Every parallel pair in **E** has a coequalizer.

We could follow through the logical construction and use the substitution lemma to show that the pullback of a coequalizer is a coequalizer, and thus the pullback of any epic is epic. But, instead, we will prove these from the fundamental theorem of toposes.

Exercises

16.1 Show that an object A is subterminal iff

$$\vdash x = x' \qquad x, x'.A$$

[Hint: show that the relation R defined by

$$\vdash Rxy \longleftrightarrow y = ! \qquad x.A\,y.1$$

is functional and gives a monic arrow.] Show that A is terminal iff we also have $\vdash (\exists x.A)x = x$.

16.2 Show that P, f, g is a product diagram for A and B iff

$$fp = fp', \quad gp = gp' \vdash p = p' \qquad p, p'.P$$

$$\vdash (\exists p.P)(fp = x \,\&\, gp = y) \qquad x.A\,y.B$$

16.3 Show that E, e is an equalizer for $f, g: A \longrightarrow B$ iff

$$e(w) = e(w') \vdash w = w' \qquad w, w'.E$$

$$\vdash f(e(w)) = g(e(w)) \qquad w.E$$

$$fx = gx \vdash (\exists w.E)x = e(w) \qquad x.A$$

16.4 Show that $h: P \longrightarrow A$ and $k: P \longrightarrow B$ are a pullback for $f: A \longrightarrow C$ and $g: B \longrightarrow C$ iff

$$hw = hw', \quad kw = kw' \vdash w = w' \qquad w, w'.P$$

$$\vdash f(hw) = g(kw) \qquad w.P$$

$$fx = gy \vdash (\exists w.P)(x = hw \,\&\, y = kw) \qquad x.A\,y.B$$

Compare with Lemma 16.16.

16.5 In any category, an *image factorization* for an arrow $f: A \longrightarrow B$ consists of an epic $q: A \twoheadrightarrow I$ and a monic $i: I \rightarrowtail B$ with $i \circ q = f$. Show that if the category has pullbacks, and all epic–monic arrows are iso, then images are unique up to isomorphism. That is, given q, i as above and another $q': A \twoheadrightarrow I'$ and $i': I' \rightarrowtail B$ with $i' \circ q' = f$, show there is an iso $h: I \xrightarrow{\sim} I'$ with $i = i' \circ h$ and $q' = h \circ q$. Show that either one of those equations uniquely determines h. [Hint: show that the intersection $i \cap i'$ is equivalent as a sub-object of B to both i and i'. Note that if $p_1 \circ u = q$ then p_1 must be epic since q is.]

16.6 Prove the converse to Theorem 16.6, if h is monic then R is single-valued. Prove the converse to Theorem 16.7.

16.7 Give a quick proof directly from the fact that $\emptyset$ is initial, that every sequent with a free variable over $\emptyset$ is true. [Hint: $\emptyset$ has only one sub-object, up to equivalence.] Show that if we did not have the free variable restriction on the cut rule we could derive every sequent from a true premise.

16.8 Show that an object A is initial iff $\vdash \sim (x = x)$ for x a variable over A.

16.9 Using $\{y\}$ to abbreviate $\{\ \}_A(y)$, show that

$$\vdash x \in \{y\} \longleftrightarrow x = y \qquad x, y . A$$

by showing that both sides of the biconditional have the same interpretation. Using topos logic conclude that $\vdash x \in \{x\}$. Then with $\vdash \sim (x \in \emptyset_A)$ conclude that $\vdash \sim (\{x\} = \emptyset_A)$.

16.10 Exercise 6.10 has already shown that if there is a coproduct $A + B$ then $\Omega^{A+B} \cong \Omega^A \times \Omega^B$. Show that under this isomorphism the singleton arrow $\{\ \}_{A+B}$ corresponds to the unique $u: A + B \longrightarrow \Omega^A \times \Omega^B$ with $u \circ i_A = \langle \{\ \}_A, \emptyset_B \circ !_A \rangle$ and $u \circ i_B = \langle \emptyset_A \circ !_{B'} \{\ \}_{B'} \rangle$. Show that the sub-object of $\Omega^A \times \Omega^B$ which we took as the coproduct of A and B is (up to this isomorphism) the image of $\{\ \}_{A+B}$.

16.11 Take any parallel pairs $f, g: A \longrightarrow B$ and $f', g': A' \longrightarrow B$ (in any category). Show directly from the definition of coequalizer that if, for every $c: B \longrightarrow C$, we have $c \circ f = c \circ g$ iff $c \circ f' = c \circ g'$, then any coequalizer for f and g is also a coequalizer for f' and g'.

16.12 Show that the sequent $\vdash fa$ is true in **E** iff E is trivial. [Hint: show that the sequent is true iff $\emptyset \cong 1$.]

16.13 Show that if **E** has an object A with an epic $r: A \twoheadrightarrow \Omega^A$ then **E** is trivial. (The converse is easy.) That is, prove that

$$ry = \{x . A \mid \sim (x \in rx)\} \ \vdash \ \sim (y \in ry) \ \& \sim \sim (y \in ry) \qquad y . A$$

for any $r: A \longrightarrow \Omega^A$ in any topos, and from the assumption that r is epic conclude that $\vdash fa$.

16.14 Show that if **E** has an object A with a monic $i: \Omega^A \rightarrowtail A$ then **E** is trivial. Use the term $\{x . A \mid \sim (\exists P \cdot \Omega^A)(iP = x \ \& \ x \in P\}$.

16.15 This exercise is for those familiar with topology. For every topological space S there is a topos of sheaves on S. These are described in most texts on topos theory (for details, see Tennison 1975). In this topos the truth values, global elements of Ω, correspond to open subsets of S. The truth value t is the whole of S. Conjunction is defined by intersection of the open subsets. Show that the negation of a subset, the largest open subset disjoint from it, is the interior of the complement. Show that the truth value fa is the empty set. Show that an open subset has negation fa iff it is dense. Every open subset of the real line R with countable complement is dense.

In such a topos, disjunction of truth values is union. Show, for any open subset w, that the union of w and $\sim w$ is all of S iff w is also closed.

17

The fundamental theorem

The fundamental theorem of toposes says that every slice of a topos is also a topos and, for every arrow $f: A \longrightarrow B$ of **E**, the functor $f^*: \mathbf{E}/B \longrightarrow \mathbf{E}/A$ has a right adjoint Π_f and a left adjoint Σ_f. This makes indexed families of objects and arrows in a topos easy to handle, and it proves a great deal about the structure of a topos. The last two sections of this chapter discuss aspects of Boolean logic in toposes, and show that in toposes the axiom of choice implies the law of excluded middle. Unless otherwise noted, objects and arrows are understood to be those of a fixed topos **E**.

17.1 Partial arrow classifiers

A *partial arrow* from A to B consists of a monic $i: I \rightarrowtail A$ and an arrow $f: I \longrightarrow B$. We call i the *domain of definition* of the partial arrow. Partial arrows i, f and i', f', both from A to B, are *equivalent* iff there is an isomorphism $h: I \longrightarrow I'$ such that $i = i' \circ h$ and $f = f' \circ h$. That is, the domains of definition are equivalent as sub-objects of A, and the arrows f and f' agree up to that isomorphism. The *graph* of a partial arrow i, f is the induced arrow to the product, $\langle i, f \rangle: I \longrightarrow A \times B$. It is monic since i is.

A *partial arrow classifier* for an object B is a monic $\eta: B \rightarrowtail \tilde{B}$ such that for every partial arrow i, f from any object A to B there is a unique arrow h that makes the following a pullback:

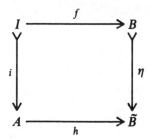

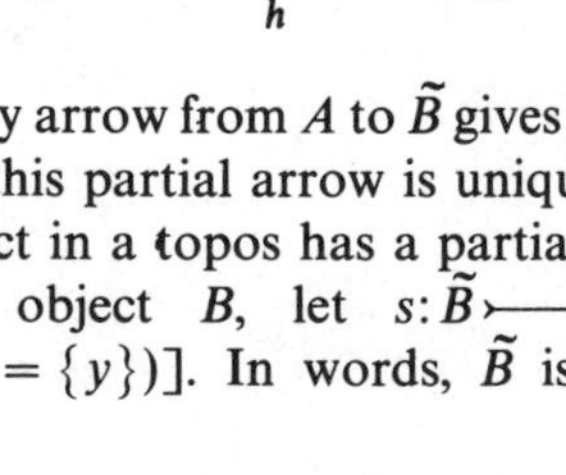

The converse follows: every arrow from A to $\tilde{B}$ gives a partial arrow from A to B, by pullback of η, and this partial arrow is unique up to equivalence. We will show that every object in a topos has a partial arrow classifier.

Specifically, for any object B, let $s: \tilde{B} \rightarrowtail \Omega^B$ be the extension $[P.\Omega^B | (\forall y.B)(y \in P \rightarrow P = \{y\})]$. In words, $\tilde{B}$ is the object of 'at most singleton' subsets of B. If the logic of the topos $\mathbf{E}$ is Boolean then every 'at most singleton' subset is 'singleton or empty', but this is not so in general.

The singleton arrow $\{\ \}_B: B \rightarrowtail \Omega^B$ factors through $\tilde{B}$ since every singleton is 'at most singleton'. A formal proof could use Theorem 16.10 and

$$\vdash y \in \{x\} \to \{x\} = \{y\} \qquad x, y.B$$

So there is a unique $\eta: B \rightarrowtail \tilde{B}$ with $s \circ \eta = \{\ \}_B$, and η is monic.

THEOREM 17.1 The monic $\eta: B \rightarrowtail \tilde{B}$ is a partial arrow classifier.

PROOF Take any partial arrow i, f from A to B. It suffices to show there is a unique arrow $k: A \longrightarrow \Omega^B$ such that k factors through s, and i, f form a pullback for k and $\{\ \}_B$. Then there is an arrow $h: A \longrightarrow \tilde{B}$ with $k = s \circ h$. Since s is monic, h is unique and i, f form a pullback for h and η.

So we must prove that there is a unique k such that:

$$\vdash y \in kx \to kx = \{y\} \qquad x.A\ y.B$$

$$\vdash k(i(w)) = \{fw\} \qquad w.I$$

$$\vdash kx = \{y\} \to (\exists w.I)(x = iw \,\&\, y = fw)$$

The first sequent for a pullback in Exercise 16.4 is trivial in this case as i is monic. These are equivalent to:

$$\vdash y \in kx \longleftrightarrow (\exists w.I)(x = iw \,\&\, y = fw)$$

and so

$$\vdash kx = \{y.B \mid (\exists w.I)(x = iw \,\&\, y = fw)\}$$

But that means that k is uniquely determined as the interpretation $|\{y.B \mid (\exists w.I)(x = iw \,\&\, y = fw)\}|_x$. □

We define a functor $\tilde{\_}: \mathbf{E}/A \longrightarrow \mathbf{E}/A$ so that the arrows η make a natural transformation $\eta: 1_{\mathbf{E}/A} \longrightarrow \tilde{\_}$. For any arrow $f: A \longrightarrow B$ we have a partial arrow η, f from $\tilde{A}$ to B. Let $\tilde{f}$ be the unique arrow that makes the following a pullback:

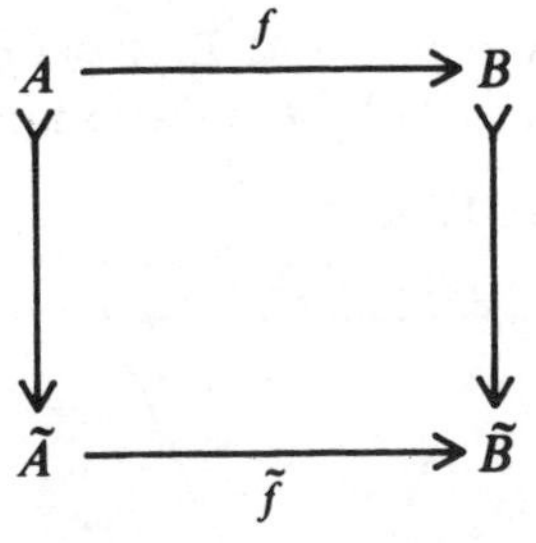

Verification that this is a functor is left to the reader.

17.2 Local Cartesian closedness

We can show that each slice $\mathbf{E}/A$ is Cartesian closed. Take any $g: C \longrightarrow A$ and consider its twisted graph as a partial arrow from $A \times C$ to A:

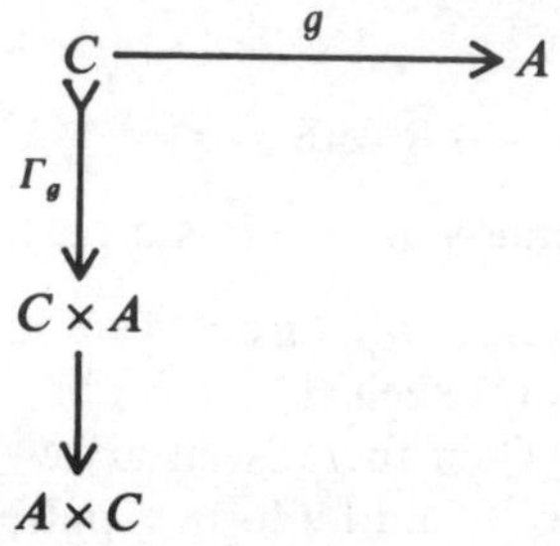

Intuitively, the partial arrow is defined for pairs $\langle x, y \rangle$, where y is in the fibre of g over x, that is $gy = x$, and then it takes $\langle x, y \rangle$ to the base point x. The transposed classifying arrow $\bar{h}: A \longrightarrow \tilde{A}^C$ intuitively takes each x in A to a partial arrow from C to A, namely the one defined on the fibre of g over x and taking it all to x.

Now take any $k: D \longrightarrow A$ and define k^g by the pullback

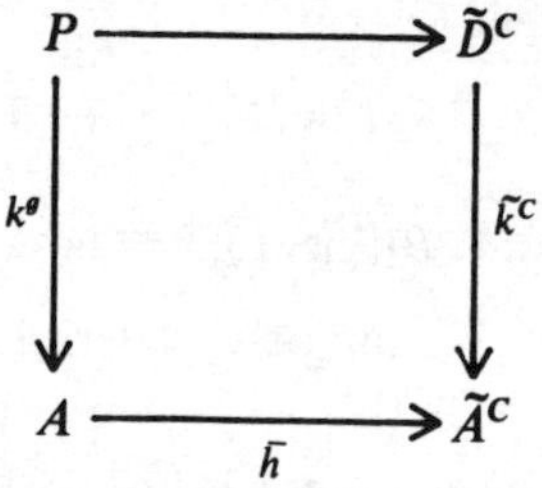

Then P indexes certain partial arrows from C to D; namely, all those which, for some x in A, are defined on the fibre of g over x and take values in the fibre of k over x. The arrow k^g takes each such partial arrow to the base point x it is defined over.

THEOREM 17.2 For every object g of $\mathbf{E}/A$ the operation $\_^g$ extends to a functor right adjoint to $\_ \times g$. So $\mathbf{E}/A$ is Cartesian closed.

PROOF Take any arrow of $\mathbf{E}/A$:

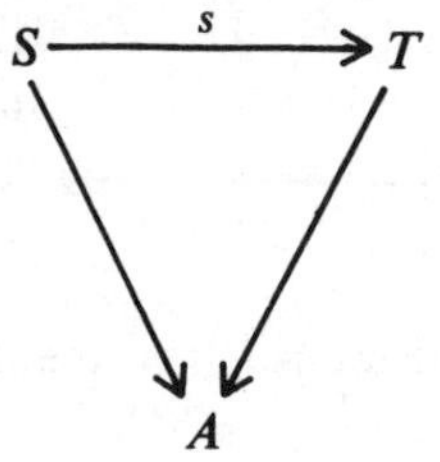

To define s^g first apply $\tilde{\_}$ to the whole triangle, exponentiate by C, and then pull the resulting triangle back along $\bar{h}$. The pullback theorems show that this is a functor. To show that $\_^g$ is right adjoint to $\_\times g$ consider any $\mathbf{E}/A$ objects $f\colon T\longrightarrow A$ and $k\colon D\longrightarrow A$ and series of arrows:

$$
\begin{array}{ll}
f \xrightarrow{m} k^g & \text{in } \mathbf{E}/A \\
\hline
T \xrightarrow{m} P & \text{with } k^g \circ m = f \ \text{ in } \ \mathbf{E} \\
\hline
T \xrightarrow{n} \tilde{D}^C & \text{with } \tilde{k}^C \circ n = \bar{h} \circ f \\
\hline
T\times C \xrightarrow{\bar{n}} \tilde{D} & \text{with } \tilde{k} \circ \bar{n} = h \circ (f\times C) \\
\hline
T\times_A C \xrightarrow{r} D & \text{with } k \circ r = g \circ p_2 \\
\hline
f\times g \xrightarrow{r} k & \text{in } \mathbf{E}/A
\end{array}
$$

To explain the fourth step, note that every square here is a pullback:

$$
\begin{array}{ccccc}
T\times_A C & \xrightarrow{p_2} & C & \xrightarrow{g} & A \\
{\scriptstyle e}\downarrow & & \downarrow & & \downarrow \\
T\times C & \xrightarrow[f\times C]{} & A\times C & \xrightarrow[h]{} & \tilde{A}
\end{array}
$$

where e is the monic from $T\times_A C$ to $T\times C$ described in Theorem 4.5 and the middle arrow is the twisted graph of g. The right-hand square is a pullback by definition, the left-hand square can be checked using pullback theorems, and so the outer is also.

For any arrow $\bar{n}\colon T\times C\longrightarrow\tilde{D}$ with $\tilde{k}\circ\bar{n} = h\circ(f\times C)$ there is a unique r with $k\circ r = g\circ p_2$, namely the arrow that makes the outer square here a pullback:

$$
\begin{array}{ccccc}
T\times_A C & \xrightarrow{r} & D & \xrightarrow{k} & A \\
{\scriptstyle e}\downarrow & & \downarrow{\scriptstyle \eta_D} & & \downarrow{\scriptstyle \eta_A} \\
T\times A & \xrightarrow{\bar{n}} & \tilde{D} & \xrightarrow{\tilde{k}} & \tilde{A}
\end{array}
$$

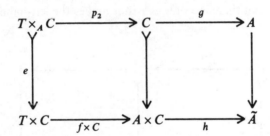

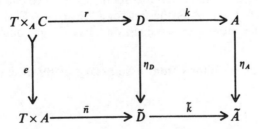

Conversely, any $r: T\times_A C \longrightarrow D$ gives a partial arrow e, r and so determines a unique $\bar{n}$ that makes the left-hand square a pullback. But then $\tilde{k}\circ\bar{n}$ classifies the partial arrow $e, k\circ r$. If $k\circ r = g\circ p_2$ then the first pullback in this proof shows that $h\circ(f\times C)$ classifies the same partial arrow and so $\tilde{k}\circ\bar{n} = h\circ(f\times C)$. □

17.3 The fundamental theorem

Every slice of a category with all finite limits has all finite limits, and Theorem 17.2 shows that every slice of a topos is Cartesian closed. To complete the proof that every slice $\mathbf{E}/A$ is a topos we must show that $\mathbf{E}/A$ has a sub-object classifier. In fact, we prove that the slice functor A^* preserves the sub-object classifier:

THEOREM 17.3 For every object A, $A^*t: A^*1 \longrightarrow A^*\Omega$ is a sub-object classifier in $\mathbf{E}/A$.

PROOF By Exercise 11.1 the monics to an $\mathbf{E}/A$ object $f: B \longrightarrow A$ are precisely the arrows $h: f\circ h \longrightarrow f$ for $h: C \rightarrowtail B$ monic in $\mathbf{E}$. We must show that for each such h there is a unique $u: f \longrightarrow A^*\Omega$ that makes the requisite pullback. But a square in $\mathbf{E}/A$ is a pullback iff Σ_A of it is a pullback in $\mathbf{E}$. So we must show that there is a unique v that makes the left-hand square here a pullback:

$$\begin{array}{ccccc}
C & \xrightarrow{\langle f\circ h, !\rangle} & A\times 1 & \xrightarrow{p_2} & 1 \\
{\scriptstyle h}\downarrow & & \downarrow{\scriptstyle A\times t} & & \downarrow{\scriptstyle t} \\
B & \xrightarrow[\langle f, v\rangle]{} & A\times\Omega & \xrightarrow[p_2]{} & \Omega
\end{array}$$

Composing with the projection arrows shows $\chi_h: B \longrightarrow \Omega$ is this unique v. □

THEOREM 17.4 (the fundamental theorem) Every slice of a topos is a topos. For every arrow $f: A \longrightarrow B$, the functor $f^*: \mathbf{E}/B \longrightarrow \mathbf{E}/A$ preserves exponentials and the sub-object classifier, and has a right adjoint Π_f and left adjoint Σ_f.

PROOF The proof follows from a summary of the results above and in Chapter 11. □

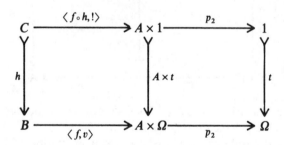

A functor $\mathbf{F}:\mathbf{E}\longrightarrow\mathbf{E}'$ between toposes is a *logical functor* if it preserves limits, exponentials, and the sub-object classifier. So the pullback functors f^* between slices are logical morphisms. Preserving the sub-object classifier means that $\mathbf{F}t:\mathbf{F}1\longrightarrow\mathbf{F}\Omega$ is a sub-object classifier in $\mathbf{E}'$, but not that $\mathbf{F}t$ is necessarily the selected one in $\mathbf{E}'$. In fact, $\mathbf{F}1$ might not be the selected terminator in $\mathbf{E}'$. Preserving exponentials means that for any objects A and B in $\mathbf{E}$ the object and arrow $\mathbf{F}(B^A)$, $\mathbf{F}(ev_{A,B})$ form an exponential of $\mathbf{F}B$ by $\mathbf{F}A$ in $\mathbf{E}'$. So $\mathbf{F}(B^A)$ is iso to the selected exponential $\mathbf{F}B^{\mathbf{F}A}$ in $\mathbf{E}'$ but again the two need not be equal.

A logical functor also preserves implicates, universal quantifications, all finite colimits, and anything else that can be constructed from limits, exponentials, and Ω (see Exercise 17.8).

More important than logical functors are *geometric morphisms*, which are actually adjunctions. A geometric morphism from $\mathbf{E}$ to $\mathbf{E}'$ is any adjoint pair of functors $f^* \dashv f_*$, where $f_*:\mathbf{E}\longrightarrow\mathbf{E}'$ and f^* preserves all finite limits. Note that f^* necessarily preserves all colimits, and f_* all limits, because they are adjoints. We may compress the notation to $f^* \dashv f_*:\mathbf{E}\longrightarrow\mathbf{E}'$, and call f^* the *inverse image functor* and f_* the *direct image functor* of the morphism. If f^* has a left adjoint it is usually written $f_! \dashv f^*$, and the morphism $f^* \dashv f_*$ is called *essential.* Here f^* is not necessarily pullback along an arrow f.

So, for every arrow $f:A\longrightarrow B$ of $\mathbf{E}$ there is an essential geometric morphism $f^* \dashv \Pi_f:\mathbf{E}/A\longrightarrow\mathbf{E}/B$, with further left adjoint Σ_f. Inverse image functors of geometric morphisms do not generally preserve exponentials or sub-object classifiers, but pullback functors do.

Geometric morphisms are given this name because they arise from geometric functions. Every topological space S has a topos of sheaves $\mathrm{Shv}(S)$, and every continuous map between spaces $f:S\longrightarrow T$ gives a geometric morphism $f^* \dashv f_*:\mathrm{Shv}(S)\longrightarrow\mathrm{Shv}(T)$. Similar facts apply to various other notions of spaces (see Tennison 1975).

One of the main uses of logical functors is in giving geometric morphisms. This uses the fact that if a logical functor has either a right or left adjoint it has both (see Barr and Wells 1985, p. 180) and thus is the inverse image functor of an essential geometric morphism.

17.4 Stability

For any $f:A\longrightarrow B$ the pullback functor $f^*\,\mathbf{E}/B\longrightarrow\mathbf{E}/A$ preserves all colimits since it is a left adjoint. But there is another important way to look at pullbacks of colimits:

THEOREM 17.5 Suppose that $q:B\longrightarrow C$ is a coequalizer for f and g, and consider any $r:C'\longrightarrow C$:

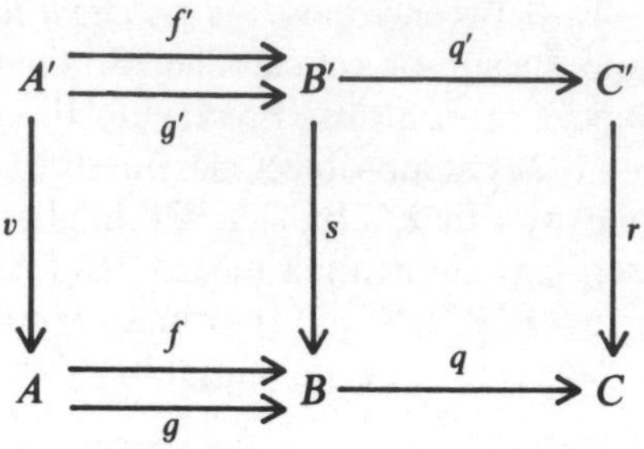

Let s be the pullback of r along q, and v the pullback of r along $q \circ f$. Since $q \circ f = q \circ g$, there is some f' that makes v, f', s, f a pullback, and also some g' that makes v, g', s, g one. Then q' is a coequalizer for f' and g'.

Proof If q is a coequalizer for f and g, the following is a coequalizer diagram in $\mathbf{E}/C$:

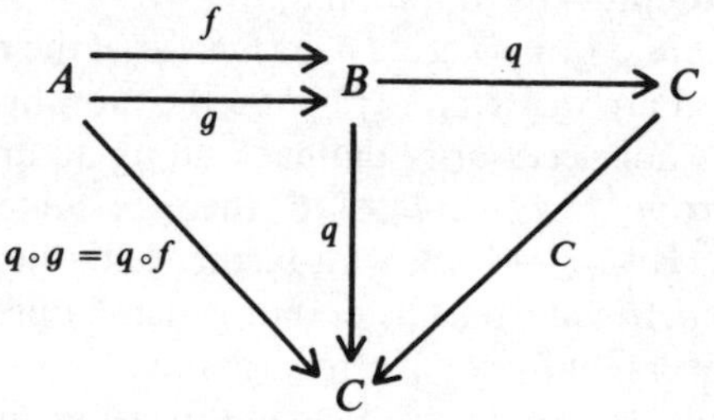

But f^* and $\Sigma_{C'}$ both preserve coequalizers, and $\Sigma_{C'} \circ f^*$ of this is the upper fork in the theorem. □

Suppose that the lower row here is a coproduct diagram, and that both squares are pullbacks:

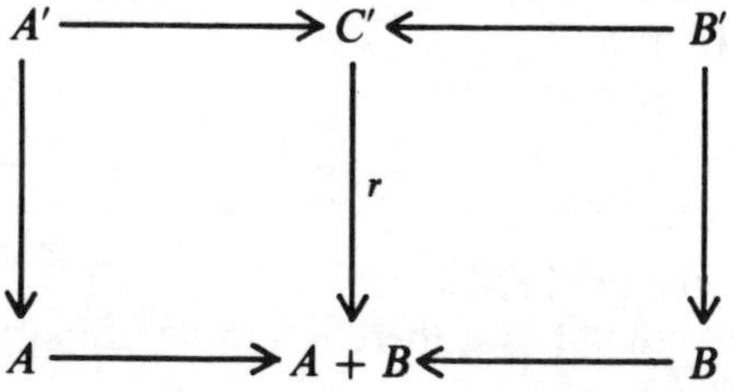

Then a similar proof shows that the upper row is also a coproduct, $C' \cong A' + B'$. An even simpler proof shows that pulling $\emptyset \longrightarrow C$ back along any $r: C' \longrightarrow C$ gives $\emptyset \longrightarrow C'$.

We say that coequalizers, coproducts, and $\emptyset$ are *stable under pullback*. It follows that all finite colimits are stable. The reader may give the precise statement and proof.

Since every epic in a topos is a coequalizer, it follows that the pullback of an epic is epic. There is also a dual result that the pushout of a monic is monic:

THEOREM 17.6 If $i: I \rightarrowtail A$ is monic and the following square is a pushout:

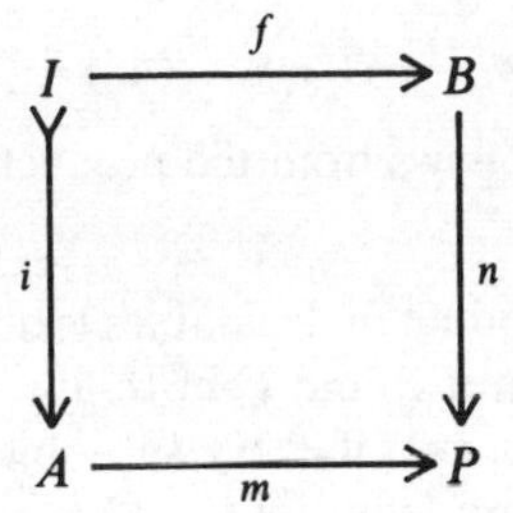

then n is monic and the square is also a pullback.

PROOF Consider the arrow $h: A \longrightarrow \tilde{B}$ that classifies i, f and the corresponding pullback square. Since P is a pushout there is an arrow $u: P \longrightarrow \tilde{B}$ with $u \circ m = h$ and $u \circ n = \eta$. Thus i, f is a pullback for m and n. And since η is monic, n is too. □

17.5 Complements and Boolean toposes

Define the *union* $i \cup j$ of two monics $i: I \rightarrowtail A$ and $j: J \rightarrowtail A$ to be the extension $[x . A | I(x) \vee J(x)]$.

A *complement* to a sub-object $i: I \rightarrowtail A$ is a sub-object $j: J \rightarrowtail A$ such that $i \cap j \equiv \emptyset$ and $i \cup j \equiv 1_A$. In general, a sub-object in a topos does not have a complement; but if it does then the complement is the negation (Exercise 17.10). Generally, $i \cup \sim i$ falls short of 1_A, although its negation $\sim (i \cup \sim i)$ is $\emptyset_A$, as is easily proved by a De Morgan law (compare with Exercise 16.15).

Note that if $i: I \rightarrowtail A$ and $j: J \rightarrowtail A$ are complementary then their union as sub-objects of A is a disjoint union; so A is the coproduct of I and J, with i and j as the injections. Conversely, the two summands A and B of a coproduct $A + B$ are always complementary sub-objects of $A + B$.

Complements to diagonals have a special role. Call an object B *decidable* if the diagonal $\Delta: B \rightarrowtail B \times B$ has a complement. In other words, B is decidable iff

$$\vdash (y = y') \vee \sim (y = y') \qquad y, y' . B$$

For example, in Chapter 19 it is shown that if a topos has a natural number object, that is a model of Peano's axioms for arithmetic, then that object is decidable. Certainly 1 is decidable.

THEOREM 17.7 Every sub-object of a decidable is decidable, and any product of two decidables is decidable. Thus every finite limit of decidables is decidable.

PROOF Take any decidable objects B and C and monic $i: I \rightarrowtail B$. Decidability of I follows from

$$\vdash w = w' \longleftrightarrow i(w) = i(w') \qquad w, w'.I$$

and decidability of $B \times C$ follows from the product axioms in local set theory. □

It is easy to see that a coproduct of decidables is decidable. An exponential B^C need not be decidable even if B and C both are.

A topos is said to be *Boolean* if every sub-object in it has a complement. There are several equivalent definitions:

THEOREM 17.8 The following conditions on a topos **E** are equivalent:

(i) Every sub-object in **E** has a complement.
(ii) The arrow $\binom{t}{fa}: 1 + 1 \longrightarrow \Omega$ is an isomorphism.
(iii) The arrow $\sim\sim : \Omega \longrightarrow \Omega$ is 1_Ω.
(iv) Boolean logic is sound in **E**.
(v) Ω is decidable.

PROOF Assume (i). Then the complement of t must be its negation fa, and in that case $\Omega \cong 1 + 1$, so (ii) holds. In any case, $\sim\sim \binom{t}{fa}$ equals $\binom{t}{fa}$, so if we assume that $\binom{t}{fa}$ is iso, or even just epic, then $\sim\sim = 1_\Omega$. If we assume (iii), then for every formula φ in **E**,

$$\vdash \sim\sim\varphi \longrightarrow \varphi$$

and so Boolean logic is sound. But Boolean logic makes every object decidable. Finally, if Ω is decidable and $i: I \rightarrowtail A$ is any sub-object, then $[x.A \mid \sim I(x)]$ is a complement to i. □

17.6 The axiom of choice

The axiom of choice says that given a family of non-empty objects there is a way of selecting one value from each family. A B-indexed family of objects is an arrow $f: A \longrightarrow B$, and we understand 'non-empty' to mean that each fibre is occupied. That is:

$$\vdash (\forall y.B)(\exists x.A) fx = y$$

but that just says f is epic. By a 'selection of values' we understand an arrow $s: B \longrightarrow A$ that takes each value in B to one in its fibre; that is, $f \circ s = 1_B$. So, to be precise, the *axiom of choice* for a topos **E** says that every epic in **E** is split epic. This is also a traditional form of the axiom in set theory: every onto function has a right inverse.

Diaconescu (1975) discovered that the axiom of choice implies excluded middle. The rest of this section gives a proof and then a more intuitive discussion.

THEOREM 17.9 If the axiom of choice holds for **E**, then **E** is Boolean.

PROOF Suppose that choice does hold. To construct a complement for a arbitrary monic $h: Y \rightarrowtail X$, consider the pushout

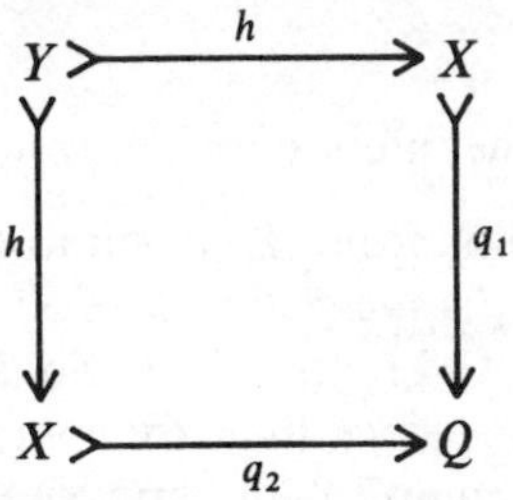

By Theorem 17.6 the arrows q_1 and q_2 are monic and Y is their intersection. In other words, Q consists of two copies of X overlapping precisely on Y. As a pullback is an equalizer of a product, so a pushout is a coequalizer of a coproduct. Thus the arrows q_1 and q_2 induce an epic arrow $q: X + X \twoheadrightarrow Q$, and by choice this has a splitting $s: Q \rightarrowtail X + X$.

Consider the following pullbacks, for $m = 1$ or 2 and $n = 1$ or 2:

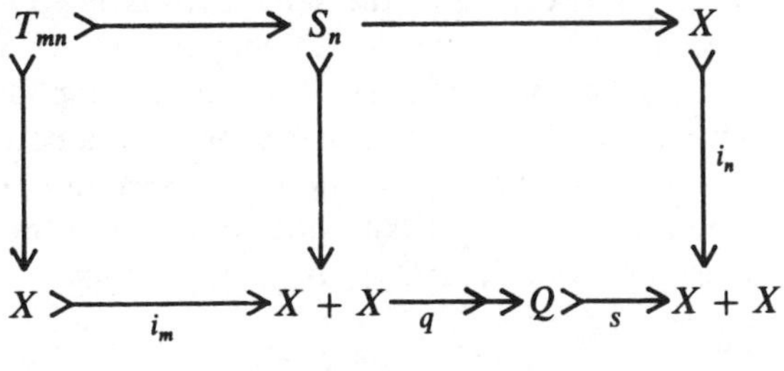

The two copies of X are complementary sub-objects of $X + X$, so by stability of co-products S_1 and S_2 are complementary sub-objects of $X + X$. And so T_{11} and T_{12} are complements in X, as are T_{21} and T_{22}. It follows by distributivity that $T_{11} \cap T_{22}$ and $T_{12} \cup T_{21}$ are complements. The proof is finished by showing that $T_{12} \cup T_{21} \equiv Y$, which we do in two steps.

Step 1 $T_{12} \cup T_{21}$ is contained in Y. For this it suffices to show that each of T_{12} and T_{21} is contained in Y, and by symmetry it suffices to prove that one of them is. Take T_{12}. Composing the two routes around the pullback defining T_{12} with q shows that the following square commutes:

$$\begin{array}{ccc} T_{12} & \rightarrowtail & X \\ \downarrow & & \downarrow q_2 \\ X & \xrightarrow[q_2]{} & Q \end{array}$$

Thus T_{12} is contained in the intersection of q_1 and q_2, which is Y.

Step 2 $T_{11} \cap T_{22}$ is disjoint from Y. It suffices to show that $T_{11} \cap Y$ is disjoint from T_{22}. Now T_{11} is the pullback of S_1 along i_1, so $T_{11} \cap Y$ is the pullback of S_1 along $i_1 \circ h$. And $i_1 \circ h = i_2 \circ h$. So $T_{11} \cap Y$ is also $T_{21} \cap Y$ and T_{21} itself is disjoint from T_{22} since they are complements.

Since $T_{12} \cup T_{21}$ has a complement, any sub-object containing it and disjoint from its complement must be it. So $T_{12} \cup T_{21} \cong Y$. □

Intuitively, $q: X + X \twoheadrightarrow Q$ takes two copies of X and pastes them together, overlapping precisely on Y. Then $s: Q \rightarrowtail X + X$ takes each value in Q to one of the values in $X + X$ that it came from. The sub-object T_{12} contains those values in X which, when mapped into $X + X$ by i_1 and then back to $X + X$ by $s \circ q$, land in the image of i_2. That implies that the value is mapped into the region of overlap in Q, so it is in Y. Thus T_{12} is contained in Y, as proved in step 1 above, and by the same reasoning so is T_{21}. On the other hand, T_{11} contains those values which, mapped into $X + X$ by i_1 and back to $X + X$ by $s \circ q$, land in the image of i_1. The corresponding route for T_{22} lands in the image of i_2. Now a value in Y cannot be in both T_{11} and T_{22}, since $q \circ i_1$ agrees with $q \circ i_2$ for any value in Y; and s cannot take that single result into both the image of i_1 and the image of i_2, since those images are disjoint. So Y is disjoint from $T_{11} \cap T_{22}$, as in step 2 above.

Exercises

17.1 Show that $t: 1 \rightarrowtail \Omega$ is a partial arrow classifier for 1.

17.2 Show that a relation from A to B is the graph of a partial arrow iff it is single-valued, and that partial arrows are equivalent iff their graphs are equivalent sub-objects. Therefore partial arrows are just a way of looking at single-valued relations.

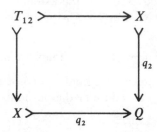

17.3 Using Boolean logic prove that

$$(\forall y.B)(y \in P \longrightarrow P = \{y\}) \vdash P = \emptyset \vee (\exists y.B)P = \{y\} \qquad P.\Omega^B$$

So in a Boolean topos an 'at most singleton' is either a singleton or empty.

17.4 Give a quick proof that if **E** is Boolean then $i_1: B \rightarrowtail B + 1$ is a partial arrow classifier for B. [Hint: if **E** is Boolean every domain of definition has a complement.]

17.5 Show that the composite of logical functors is logical, so that toposes and logical functors form a category **LOG**. Show that the topos **1** is terminal in **LOG** and that a topos **E** has a global element in **LOG** iff **E** is trivial. Show, for any toposes **E** and **E**′, that the topos $\mathbf{E} \times \mathbf{E}'$ is a product in **LOG**; that is, the projection functors are logical, and the functor induced by logical functors to **E** and **E**′ is logical.

17.6 Given geometric morphisms $f^* \dashv f_*: \mathbf{E} \longrightarrow \mathbf{E}'$ and $g^* \dashv g_*: \mathbf{E}' \longrightarrow \mathbf{E}''$, show that $f^* \circ g^* \dashv g_* \circ f_*$ is geometric, and so toposes and geometric morphisms form a category **GEOM**.

17.7 For any topos **E** the functor $!: \mathbf{E} \longrightarrow \mathbf{1}$ preserves finite limits. Show that ! has as many right adjoints as there are terminal objects in **E**. Thus there are as many geometric morphisms from **1** to **E** as there are terminal objects in **E**; but all have ! as inverse image functor, and all the direct image functors are naturally isomorphic. In short, **1** is an initial object in **GEOM** 'up to natural isomorphism of functors'. Show that the topos $\mathbf{E} \times \mathbf{E}'$ is a *co-product* for **E** and **E**′ in **GEOM** up to the same kind of natural isomorphism, with the projection functors to **E** and **E**′ as the inverse image parts of the geometric morphism injections.

The best approach to **GEOM** uses 2-categories, which we will not consider here (see geometric morphisms in Johnstone 1977).

17.8 Let $\mathbf{F}: \mathbf{E} \longrightarrow \mathbf{E}'$ be a logical functor. Show that if f has transpose $\bar{f}: C \longrightarrow B^A$ then $\mathbf{F}f$ has transpose $\mathbf{F}\bar{f}$ (up to the isomorphism between $\mathbf{F}(B^A)$ and $\mathbf{F}B^{\mathbf{F}A}$). For any monic i in **E**, show that $\mathbf{F}(i)$ has classifying arrow $\mathbf{F}(\chi_i)$. Use the constructions of implicates and universal quantifications in Chapter 13, and of finite co-limits in Chapter 16, to show that **F** preserves them. [Hint: you could use the result in the next exercise.]

17.9 A logical functor $\mathbf{F}: \mathbf{E} \longrightarrow \mathbf{E}'$ interprets the internal language of **E** in **E**′. For any formula φ of **E**, let $\mathbf{F}\varphi$ be the **E**′ formula obtained by replacing every variable x_i of sort A in φ by the first variable y_j of sort $\mathbf{F}A$ that has not already been used, and every arrow f by $\mathbf{F}f$. Show that $\mathbf{F}\varphi$ is a formula of $L\mathbf{E}'$. Show that **F** preserves extensions, as follows. If φ has free variables $x_1 \ldots x_n$ of sorts $A_1 \ldots A_n$ then there is an isomorphism h that makes the following commute:

$$\begin{array}{ccc}
\mathbf{F}[x_1 \ldots x_n | \varphi] & \xrightarrow{\;h\;} & [y_1 \ldots y_n | \mathbf{F}\varphi] \\
\downarrow & & \downarrow \\
\mathbf{F}(A_1 \times \cdots \times A_n) & \xrightarrow{\;\sim\;} & \mathbf{F}A_1 \times \cdots \times \mathbf{F}A_n
\end{array}$$

Conclude that if φ is true in $L\mathbf{E}$ then $\mathbf{F}\varphi$ is true in $L\mathbf{E}'$.

17.10 Show that if $i: I \rightarrowtail A$ has a complement $j: J \rightarrowtail A$ then $J \equiv {\sim} I$. [Hint: ${\sim} I$ is the coproduct of its pullbacks along i and j.] Then conclude that $I \equiv {\sim}{\sim} I$.

17.11 Use stability of epics to show that $f: A \longrightarrow B$ is epic iff: for every T-element $c: T \longrightarrow B$ there is an epic $s: S \twoheadrightarrow T$ such that $c \circ s$ factors through f. For one direction form the pullback of f and c, for the other consider the generic element 1_B.

17.12 For any object A define the inclusion relation on Ω^A by

$$P \subseteq Q \longleftrightarrow (\forall x . A)(x \in P \longrightarrow x \in Q) \qquad P, Q . \Omega^A$$

Prove that Ω^A is a poset ordered by inclusion. Use $\{x . A \mid P(x) \,\&\, Q(x)\}$ to prove that Ω^A has binary intersections, and that these are meets for the poset; that is, show that

$$\vdash (\exists M . \Omega^A)(\forall N . \Omega^A) N \subseteq M \longleftrightarrow (N \subseteq P \,\&\, N \subseteq Q) \qquad P, Q . \Omega^A$$

Use disjunction and the material conditional to show that Ω^A has joins and implicates, and thus is a Heyting algebra. Use

$$\{x . A \mid (\forall P . \Omega^A)(P \in F \longrightarrow x \in P)\}$$

to show that

$$\vdash (\exists M . \Omega^A)(\forall N . \Omega^A) N \subseteq M \longleftrightarrow (\forall P . \Omega^A)(P \in F \longrightarrow N \subseteq P)$$

where F varies over the power object of the power object of A. In words, any family F of sub-objects of A has an intersection, a greatest lower bound for F in the order on Ω^A. Show that the union is a least upper bound for F.

A *Boolean algebra* can be defined as a Heyting algebra in which every member has a complement; so **E** is Boolean iff every Ω^A is a Boolean algebra, and thus iff Ω is.

A Heyting algebra in which every family has a least upper bound is called *complete*. Thus Ω^A is a complete Heyting algebra. Complete Heyting algebras and their relation to toposes are discussed in Fourman and Scott (1979), and from another point of view in Joyal and Tierney (1984) under the name *locales*.

17.13 For any monic $i: I \rightarrowtail A$ write $[i]$ for the selected pullback of t along χ_i. Then $[i] = [j]$ iff $i \equiv j$, and the selected monics $[i]$ form a poset ordered by inclusion, and not just a preorder. Call the poset $\mathrm{Sub}(A)$, as in Exercise 8.13. Conclude from Exercise 17.12 that $\mathrm{Sub}(A)$ is a Heyting algebra, with $[1_A]$ as supremum, $[\emptyset]$ as infimum, and $\cap$, $\cup$, and $\Rightarrow$ giving meets, joins, and implicates.

For every arrow $f: A \longrightarrow B$, show that $f^{-1}: \mathrm{Sub}(B) \longrightarrow \mathrm{Sub}(A)$ preserves all the Heyting algebra operations.

Exercise 17.12 does not prove that every family of sub-objects in $\mathrm{Sub}(A)$, defined by any means whatsoever, has a g.l.b. or a l.u.b. It generally is not so. Any *internal* family of sub-objects of A has a l.u.b. and a g.l.b.; that is, any family defined by a sub-object in **E** of Ω^A. But, in general, not every collection of sub-objects definable in some way is definable internally.

18

External semantics

We have interpreted formulas in a topos **E** by assigning each an extension. This is called the *internal semantics*. The *external semantics*, also called Kripke–Joyal semantics, describes which generalized elements satisfy each formula. The idea is that the extension collects the instances of a formula internally to **E**, while the class of generalized elements satisfying it is an external collection. A generalized element satisfies a formula iff it is a member of the formula's extension, and this external viewpoint is often handy for more complicated formulas.

18.1 Satisfaction

As before, $\bar{x}$ is a list of variables $x_1, \ldots, x_n$ over objects $A_1, \ldots, A_n$. Let $\bar{x}$ include all the variables that are free in φ. Then $c: T \longrightarrow A_1 \times \cdots \times A_n$ *satisfies* φ with respect to $\bar{x}$ iff c is a member of $[\bar{x}|\varphi]$. The following theorem gives conditions which are sometimes used for an inductive definition of satisfaction:

THEOREM 18.1 Let $\bar{x}$ include all the variables that are free in φ and ψ, while θ may have a further free variable y of sort B. Then, for any $c: T \longrightarrow A_1 \times \cdots \times A_n$, we have:

(i) $c \in [\bar{x}|\varphi \,\&\, \psi]$ iff $c \in [\bar{x}|\varphi]$ and $c \in [\bar{x}|\psi]$;
(ii) $c \in [\bar{x}|fa]$ iff T is initial;
(iii) $c \in [\bar{x}|\varphi \longrightarrow \psi]$ iff, for every $s: S \longrightarrow T$, if $c \circ s \in [\bar{x}|\varphi]$ then $c \circ s \in [\bar{x}|\psi]$;
(iv) $c \in [\bar{x}|\varphi \vee \psi]$ iff there exist S_1 and S_2 and an epic $s: S_1 + S_2 \twoheadrightarrow T$ such that $c \circ s \circ i_1 \in [\bar{x}|\varphi]$ and $c \circ s \circ i_2 \in [\bar{x}|\psi]$;
(v) $c \in [\bar{x}|(\forall y . B)\theta]$ iff for every $s: S \longrightarrow T$ and $r: S \longrightarrow B$ we have $\langle c \circ s, r\rangle \in [\bar{x}, y|\theta]$ (see Exercise 18.4);
(vi) $c \in [\bar{x}|(\exists y . B)\theta]$ iff there is some epic $s: S \twoheadrightarrow T$ and an $r: S \longrightarrow B$ such that $\langle c \circ s, r\rangle \in [\bar{x}, y|\theta]$ (see Exercise 18.4).

PROOF Clauses (i), (iii), and (v) were proved in discussing intersections, universal quantifications, and implicates. Clause (ii) is trivial, since $\emptyset$ is a strict initial object. For clause (iv), suppose that c is in the extension, so that c factors through some $h: T \longrightarrow [\bar{x}|\varphi \vee \psi]$. Let S_1 be the preimage $h^{-1}[\bar{x}|\varphi]$

and let S_2 be $h^{-1}[\bar{x}|\psi]$. The obvious arrow from $[\bar{x}|\varphi]+[\bar{x}|\psi]$ to $[\bar{x}|\varphi \vee \psi]$ is epic, so by stability of coproducts and epics the induced arrow from $S_1 + S_2$ to T is epic. Conversely, suppose that there are S_1 and S_2, as in the clause. Then s factors through the preimage $c^{-1}[\bar{x}|\varphi \vee \psi]$. Then since s is epic the monic of the preimage is also epic, and so it is iso, which proves that c factors through $[\bar{x}|\varphi \vee \psi]$. Similarly, stability of epics proves clause (vi). □

Given a T-element $c: T \longrightarrow A$, we call $s: S \longrightarrow T$ a *change of stage* and say that $c \circ s: S \longrightarrow A$ is c at the *later stage of definition* S. The terminology is slightly ambiguous, since what matters is the change of stage arrow s and not just the domain S. Clause (iii) says that c satisfies $\varphi \rightarrow \psi$ iff, at any stage of definition at which c satisfies φ, c satisfies ψ. Think of this clause as saying 'to whatever extent c satisfies φ it also satisfies ψ'. Clause (v) says that c satisfies $(\mathrm{A}y\,.\,B)\theta$ iff at every later stage of definition S and for every S-element r of B, c and r satisfy θ.

We say that an epic arrow $s: S \twoheadrightarrow T$ *covers* T. Then clause (iv) says that c satisfies a disjunction iff T is covered by some later stage $S_1 + S_2$, itself a coproduct such that c at stage S_1 satisfies one disjunct while c at stage S_2 satisfies the other. Clause (vi) says that c satisfies $(\exists y\,.\,B)\theta$ iff there is a cover S of T and an S-element r of B such that c and r satisfy θ.

The idea of a later stage of definition appears in Kripke's semantics (although not the term 'stage of definition', which seems to be derived from algebraic geometry, where a given polynomial has solutions 'defined over' various fields). Joyal first saw how to generalize to topos semantics, and in particular he noticed the role of coverings. Lawvere (1976) discusses all this. Bell (1988*b*) describes Kripke models in topos terms.

A closed existential formula $(\exists y\,.\,B)\varphi$ is true iff 1_1 satisfies it, and so iff there is some object S with $S \longrightarrow 1$ epic and some $r: S \longrightarrow B$ that satisfies φ. There need not be a global element $c: 1 \longrightarrow B$ that satisfies φ. But the corollary below shows that unique existence does imply a global instance. We introduce unique existential quantifiers $(\exists!x\,.\,A)$ as abbreviations in the usual way: for any formula φ, $(\exists!x\,.\,A)\varphi$ abbreviates

$$(\exists x\,.\,A)\varphi \,\&\, (\forall x, x'\,.\,A)[(\varphi \,\&\, \varphi(x/x')) \longrightarrow x = x']$$

where x' is any variable over A that is free for x in φ.

We can neatly express the theorem on functional relations using the unique existential quantifier (compare this with Theorem 16.8):

THEOREM 18.2 Suppose, for some relation $R \rightarrowtail A \times B$, that we have

$$\vdash (\exists!y\,.\,B)Rxy \qquad x\,.\,A$$

Then

$$\vdash (\exists!g\,.\,B^A)(\forall x\,.\,A)(\forall y\,.\,B)(Rxy \longleftrightarrow g(x) = y)$$

Proof $(\exists! y\,.\,B)Rxy$ says that R is totally defined and single-valued, so R is the graph of an actual arrow, which corresponds to a unique global element $g^\circ\colon 1 \longrightarrow B^A$ that makes this formula true:

$$\vdash (\forall x\,.\,A)(\forall y\,.\,B)(Rxy \longleftrightarrow g^\circ(x) = y)$$

While this proves internal existence, the uniqueness of the global instance is not enough to prove internal unique existence (i.e. just as there may be internal existence without global existence, so there may be many internal instances while there is only one global instance). But from

$$(\forall x\,.\,A)(\forall y\,.\,B)(Rxy \longleftrightarrow g(x) = y \,\&\, Rxy \longleftrightarrow g'(x) = y)$$

we easily deduce that

$$g = g'$$

and this proves internal uniqueness. □

Corollary If a closed formula $(\exists! x\,.\,A)\varphi$ is true then a unique global element of A satisfies φ. In fact, the closed formula is true iff $[x\,.\,A\,|\,\varphi]$ itself is a global element of A.

Proof For any object T define a relation $R \rightarrowtail T \times A$ by the formula $Rzx \longleftrightarrow \varphi$; that is, everything in T is related to everything with the property φ. Then $(\exists! x\,.\,A)\varphi$ implies that this is the unique functional relation from T to A with all values in $[x\,|\,\varphi]$. By Theorem 18.2 there is a unique arrow from T to $[x\,|\,\varphi]$. So $[x\,|\,\varphi]$ is terminal. □

18.2 Generic elements

By topos logic a closed universal formula $(\forall y\,.\,B)\varphi$ is true iff the formula φ with y free is true, and that is iff the identity arrow 1_B satisfies φ. This gives a rapid technique for proving universal formulas when we reconstrue it in terms of the generic element of B as a global element of B^*B in the slice topos $\mathbf{E}/B$; that is, as $\Delta_B\colon B \longrightarrow B \times B$. The next theorem shows that a closed formula $(\forall y\,.\,B)\varphi$ is true iff Δ_B satisfies $B^*\varphi$ in the slice topos $\mathbf{E}/B$. ($B^*\varphi$ is a case of the notation introduced in Exercise 17.9.)

We often speak of 'the formula φ in $\mathbf{E}/B$', meaning the formula $B^*\varphi$. We may speak of 'the extension of φ in $\mathbf{E}/B$' to mean the extension of $B^*\varphi$ which, after all, is just the functor B^* applied to the extension of φ in the base topos $\mathbf{E}$.

Theorem 18.3 A formula φ, the only free variable of which is y over B, is true iff the generic element Δ_B satisfies it in $\mathbf{E}/B$.

Proof Any $c\colon T \longrightarrow B$ satisfies φ iff the $\mathbf{E}/T$ arrow $\langle 1_T, c\rangle\colon T^*1 \longrightarrow T^*B$ satisfies φ in $\mathbf{E}/T$, since each of these triangles commutes iff the other does:

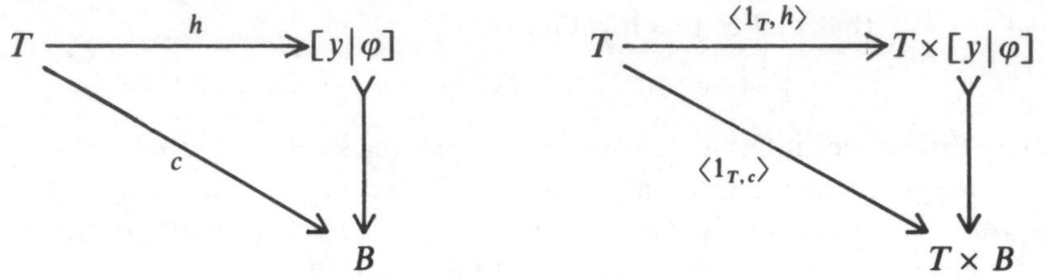

To prove the theorem, let $c = 1_B$. □

Theorem 18.3 is typically used when we know that in every topos every formula of the form of φ is satisfied by all global elements (of the relevant object). Then Δ_B satisfies φ in $\mathbf{E}/B$, and so φ and $(\forall y\,.\,B)\varphi$ are true in $\mathbf{E}$. Exercise 19.8 is one example. Here we use this method to extend Theorem 18.2, after a technical lemma.

LEMMA 18.4 Take any closed formula θ and subterminator $U \rightarrowtail 1$. Then θ is true in $\mathbf{E}/U$ iff U is included in $[\_|\theta] \rightarrowtail 1$.

PROOF The extension of θ in $\mathbf{E}/U$ is the pullback of its extension in $\mathbf{E}$ along $U \rightarrowtail 1$. But the pullback of any monic $V \rightarrowtail 1$ along $U \rightarrowtail 1$ is all of U iff $U \subseteq V$. □

COROLLARY Any closed formula θ is true in the slice over its own extension. A closed conditional $\sigma \longrightarrow \theta$ is true iff θ is true in the slice over the extension of σ, since each claim is equivalent to $[\_|\sigma] \subseteq [\_|\theta]$. □

In any topos $\mathbf{E}$, for any objects A and B of $\mathbf{E}$, consider the following formula:

$$(\forall x\,.\,A)(\exists! y\,.\,B)\,\langle x, y\rangle \in r \longrightarrow (\exists! g\,.\,B^A)((\forall x)(\forall y)\,\langle x, y\rangle \in r \longleftrightarrow y = g(x))$$

with r a variable over $\Omega^{A\times B}$. Abbreviate it as $\varphi(r) \longrightarrow \psi(r)$ to emphasize the free variable.

LEMMA 18.5 Every global element $r^\circ\colon 1 \longrightarrow \Omega^{A\times B}$ satisfies $\varphi(r) \longrightarrow \psi(r)$. Equivalently, substituting the constant r° for the variable r gives a true closed formula $\varphi(r^\circ) \longrightarrow \psi(r^\circ)$.

PROOF Consider the slice of $\mathbf{E}$ over the extension of $\varphi(r^\circ)$. In this slice $\varphi(r^\circ)$ is true, so r° classifies a functional relation from A to B in the slice (that is, the pullback of r° into the slice classifies a functional relation from the pullback of A to the pullback of B) so $\psi(r^\circ)$ is true in the slice. Thus the conditional $\varphi(r^\circ) \longrightarrow \psi(r^\circ)$ is true in $\mathbf{E}$. □

THEOREM 18.6 The formula $\varphi(r) \longrightarrow \psi(r)$ is true.

PROOF Consider the slice of **E** over $\Omega^{A\times B}$. The pullback of $\varphi(r) \longrightarrow \psi(r)$ to this slice is another formula of the same form, so applying the last lemma in the slice topos shows that the generic element $\Delta_{(\Omega^{A\times B})}$ satisfies it. Therefore Theorem 18.3 shows that $\varphi(r) \longrightarrow \psi(r)$ is true in **E**. □

The advantage of Theorem 18.6 over Theorem 18.2 is that Theorem 18.6 allows us to substitute any term for r, and not only closed terms. Therefore it applies to all T-elements of $\Omega^{A\times B}$ rather than just global elements; so we can obtain T-elements of B^A, for any object T of **E**.

Exercises

18.1 Let $\bar{x}$ list exactly the variables that are free in φ. Show that φ is true iff the identity arrow on $A_1 \times \cdots \times A_n$ satisfies it, and thus iff every generalized element of $A_1 \times \cdots \times A_n$ satisfies it w.r.t. $\bar{x}$.

18.2 Suppose that φ has only one free variable, x of sort A. Show that any $c: T \longrightarrow A$ satisfies φ iff the closed formula $\varphi(x/c)$ is true, taking c as a constant. [Hint: calculate $[\_|\varphi(x/c)]$ using the substitution lemma.]

18.3 Let $\bar{x}$ include all the variables free in φ. Show that if $c: T \longrightarrow A_1 \times \cdots \times A_n$ satisfies φ w.r.t. $\bar{x}$ then so does $c \circ r$ for any $r: S \longrightarrow T$. Prove that the extension $[\bar{x}|\varphi] \rightarrowtail A_1 \times \cdots \times A_n$ satisfies φ, and that any $c: T \longrightarrow A_1 \times \cdots \times A_n$ that satisfies φ is $[\bar{x}|\varphi]$ at the later stage T.

18.4 This is a technical detail. If the list of variables $\bar{x}$ is empty (so that θ must have no free variables except y), then clauses (v) and (vi) of Theorem 18.1 must be understood as follows:
(v) $c \in [\_|(\forall y.B)\theta]$ iff for every $s: S \longrightarrow T$ and $r: S \longrightarrow B$ we have $\langle c \circ s, r\rangle\, [\_, y|\theta]$;
(vi) $c \in [\_|(\exists y.B)\theta]$ iff there is some epic $s: S \twoheadrightarrow T$ and an $r: S \longrightarrow B$ such that $\langle c \circ s, r\rangle \in [\_, y|\theta]$;
where '_' is interpreted in the way a variable of sort 1 would be. If θ has no free variables, except possibly y, show that there is a pullback as shown below, so that h is iso:

$$\begin{array}{ccc} [\_, y|\theta] & \xrightarrow{\quad h \quad} & [y|\theta] \\ \downarrow & & \downarrow \\ 1 \times B & \xrightarrow{\quad \sim \quad} & B \end{array}$$

18.5 Show that $(\exists y.B)(y = y)$ is true iff $B \longrightarrow 1$ is epic.

18.6 This exercise is about replacing many variables by one. Suppose that all of the free variables of φ are among $x_1, \ldots, x_n$, and let z be a variable over $A_1 \times \cdots \times A_n$ not in φ. Let $\varphi(\bar{x}/z)$ be obtained by substituting $p_i(z)$ for x_i for every i, where p_i is the projection to A_i. Show that $s: T \longrightarrow A_1 \times \cdots \times A_n$ satisfies φ w.r.t. the list $x_1, \ldots, x_n$ iff s satisfies $\varphi(\bar{x}/z)$.

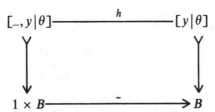

19

Natural number objects

In this chapter, unless otherwise stated, we deal with a topos **E** that has a natural number object N, 0, s, defined below. Not every topos has a natural number object.

19.1 Definition

A *natural number object* is an object N plus a global element $0: 1 \longrightarrow N$ and an arrow $s: N \longrightarrow N$, called the *successor arrow*, with the following property: for any object A and global element $q: 1 \longrightarrow A$ and arrow $f: A \longrightarrow A$, there exists a unique $u: N \longrightarrow A$ such that $u \circ 0 = q$ and $u \circ s = f \circ u$:

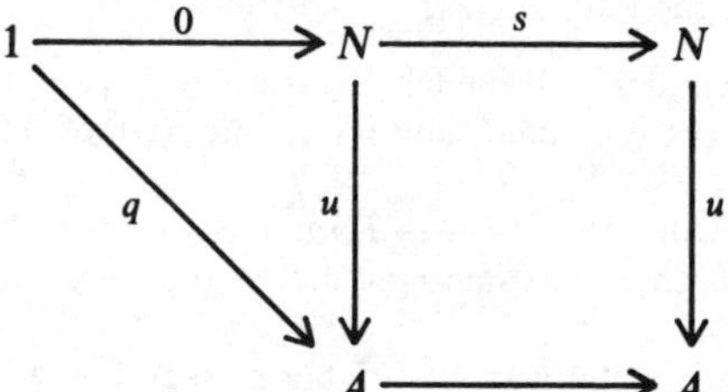

Call q and f *recursion data* for u. Internally, u is defined by

$$\vdash u(0) = q \quad \& \quad u(sy) = f(u(y)) \qquad y.N$$

It follows that indexed recursion data define indexed families of arrows from N:

THEOREM 19.1 Any $q: P \longrightarrow A$ and $f: A \times P \longrightarrow A$ determine a unique $u: N \times P \longrightarrow A$ such that

$$\vdash u(0, p) = q(p) \quad \& \quad u(sy, p) = f(u(y, p), p) \qquad p.T\, y.N$$

PROOF Consider recursion data 'q': $1 \longrightarrow A^P$ and $k: A^P \longrightarrow A^P$, where k is the transpose of $f \circ \langle ev, p_2 \rangle: A^P \times P \longrightarrow A$. These give a unique $v: N \longrightarrow A^P$ and the transpose of v is the desired u. □

19.2 Peano's axioms

A natural number object N, 0, s supports proof by induction; that is, the only sub-object of N including 0 and closed under successor is the whole of N. To say that a monic $i: I \rightarrowtail A$ is *closed under* $f: A \longrightarrow A$ means that these equivalent conditions are satisfied:

1. For any T and $x \in_T A$, if $x \in i$ then $fx \in i$.
2. There is some $h: I \longrightarrow I$ with $f \circ i = i \circ h$.
3. $i \subseteq f^{-1}(i)$ as sub-objects of A.
4. We have $\vdash \; I(y) \longrightarrow I(fy) \qquad y.A$.

The first three are equivalent in any category with pullbacks, and external semantics shows that the fourth is equivalent to the first.

THEOREM 19.2 If $i: I \rightarrowtail N$ has $0 \in i$ and is closed under s, then $i \cong 1_N$.

PROOF Assume that there are some q and h that make the lower triangle and square here commute:

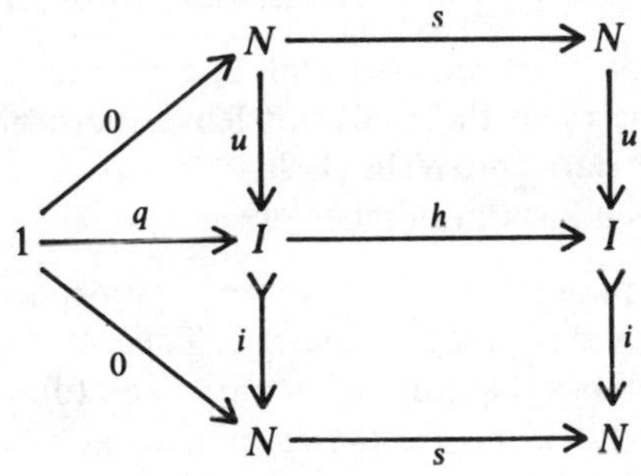

Define u by recursion data q, h. Then $i \circ u$ is defined by recursion data 0, s, which means that $i \circ u = 1_N$. So i is split epic, and thus iso. □

THEOREM 19.3 Suppose that arrows $f, g: N \longrightarrow A$ have $f(0) = g(0)$ and, for every T and $y \in_T N$, if $f(y) = g(y)$ then $f(sy) = g(sy)$. Then $f = g$.

PROOF By Theorem 19.2, the equalizer of f and g is an isomorphism. □

THEOREM 19.4 (Freyd) The following is a coproduct diagram:

$$1 \xrightarrow{\;0\;} N \xleftarrow{\;s\;} N$$

PROOF For any object A and arrows $q: 1 \longrightarrow A$ and $f: N \longrightarrow A$, the recursion data (see Exercise 19.4)

$$u(0) = q \quad \text{and} \quad u(sy) = fy$$

define the unique $u: N \longrightarrow A$ to show that the diagram is a coproduct. □

Thus s is monic and N is the disjoint union of 0 and s. Altogether, the following Peano axioms are true in **E**:

$$\sim (0 = sy) \qquad y.N$$

$$sx = sy \to x = y \qquad x, y.N$$

$$[(0 \in I) \& (\forall x.N)(x \in I \to sx \in I)] \to (\forall x.N)(x \in I) \qquad I.\Omega^N$$

Actually, Theorem 19.2 only shows that the induction axiom is true for every global sub-object of N; that is, for every constant of sort Ω^N in place of the variable I (but see Exercise 19.8).

The next theorem says that everything in N is connected to everything else by way of s.

THEOREM 19.5 (Freyd) The arrow $!_N: N \longrightarrow 1$ is a coequalizer for s and 1_N.

PROOF Suppose that an arrow $f: N \longrightarrow A$ has $f = f \circ s$. Then $f = f \circ 0 \circ !_N$, since both sides correspond to recursion data $f \circ 0$ and 1_A, so f factors through $!_N$. Since $!_N$ is epic, split by 0, any factorization through $!_N$ is unique. □

Freyd showed that Theorem 19.4 and 19.5 characterize natural number objects. We state the main theorems but leave the proofs to Freyd (1972), Johnstone (1977), or Barr and Wells (1990). They hold in any topos, without assuming that it has a natural number object.

THEOREM 19.6 Suppose that there is an isomorphism $f: 1 + M \longrightarrow M$ such that 1 is a coequalizer for $f \circ i_2$ and 1_M. Then $M, f \circ i_1, f \circ i_2$ is a natural number object. It follows that any functor between toposes that preserves 1, coproducts, and coequalizers also preserves natural number objects.

THEOREM 19.7 Suppose that an object M has an isomorphism $f: 1 + M \longrightarrow M$. Let $q: M \longrightarrow Q$ be the co-equalizer of $f \circ i_2$ and 1_M. Then the pullback of $q \circ f \circ i_1$ along q is a natural number object.

19.3 Arithmetic

Define $+ : N \times N \longrightarrow N$ by the N-indexed recursion data 1_N and $s \circ p_1: N \times N \longrightarrow N$. Internally, $+$ is uniquely determined by these equations, abbreviating $+(x, y)$ by $x + y$:

$$\vdash 0 + y = y \quad \& \quad (sx) + y = s(x + y) \qquad x, y.N$$

THEOREM 19.8 These equations also hold:

$$x + 0 = x \qquad x, y, z.N$$

$$x + sy = s(x + y)$$

$$x + y = y + x$$

$$x + (y + z) = (x + y) + z$$

Proof For the first use this induction:

Basis: $0 + 0 = 0$
Induction: $x + 0 = x$ implies $(sx) + 0 = s(x + 0) = sx$.

Formally, using the recursion equation for $+$, we prove that

$$\vdash 0 + 0 = 0 \quad \& \quad (x + 0 = x \rightarrow (sx) + 0 = sx) \qquad x.N$$

Then using the internal Peano axiom of induction we deduce:

$$\vdash (\forall x.N)\, x + 0 = x$$

All thc usual inductive proofs are sound (see recursive functions in Kleene 1952, Mendelson 1987, or other texts). □

Define multiplication, $.: N \times N \longrightarrow N$, by N-indexed recursion data $0 \circ !_N$ and $+ : N \times N \longrightarrow N$. In other words, with the obvious abbreviation $x.y$,

$$\vdash 0.y = 0 \quad \& \quad (sx).y = (x.y) + y \qquad x, y.N$$

The usual inductions show that:

$$x.0 = 0 \qquad x, y, z.N$$

$$x.s0 = x$$

$$x.y = y.x$$

$$x.(y + z) = (x.y) + (x.z)$$

and so on. So we have elementary arithmetic. In fact, all primitive recursive functions can be defined and the usual results on them follow by the usual inductions. Lambek and Scott (1986) give further results on exactly what functions exist in toposes with natural number objects.

19.4 Order

Define a relation $\leq$ on N by

$$x \leq y \longleftrightarrow (\exists z.N) x + z = y \qquad x, y.N$$

Theorem 19.9 The relation $\leq$ is reflexive and transitive. In other words, it is a weak order relation.

Proof Prove that $x \leq x$ from $x + 0 = x$. The formula

$$(x \leq y \,\&\, y \leq z) \rightarrow x \leq z \qquad x, y, z.N$$

follows from associativity of $+$. □

Arithmetic proves formulas such as

$$x \le sx \qquad x, y, z \,.\, N$$

$$sx \le sy \longleftrightarrow x \le y$$

$$x \le y \longrightarrow x + z \le y + z$$

$$x \le y \longrightarrow x \,.\, z \le y \,.\, z$$

THEOREM 19.10 We have:

(1) $x \le y \longleftrightarrow (x = y \vee sx \le y) \qquad x, y \,.\, N;$
(2) $x \le y \vee y \le x;$
(3) $\sim (sx \le x).$

PROOF Theorem 14.4, or an easy induction, shows that

$$z = 0 \vee (\exists w)(z = sw) \qquad z \,.\, N$$

Thus $x \le y$ is equivalent to

$$x = y \vee (\exists z \,.\, N)(x + sz = y)$$

and so to the right-hand side of formula (1).

Formula (2) follows by induction on y. For the basis step, $(x \le 0 \vee 0 \le x)$ follows from $0 \le x$. For the induction, prove that

$$(x \le y \vee y \le x) \;\vdash\; (x \le sy \vee sy \le x)$$

by cases, that is by the $\vee$-rule. Clearly, $x \le y$ implies $x \le sy$. On the other hand,

$$y \le x \vdash (y = x \vee sy \le x)) \vdash (x \le sy \vee sy \le x)$$

Formula (3) is proved by induction, with basis step

$$s0 \le 0 \vdash (\exists z \,.\, N) s0 + z = 0 \vdash (\exists z \,.\, N) sz = 0 \vdash fa$$

For the induction step, given $sx \le 0 \vdash fa$ we can deduce $ssx \le 0 \vdash fa$, since we have $sx \le ssx$. □

THEOREM 19.11 N is decidable.

PROOF Each step here can be justified using Theorem 19.10:

$$\dfrac{\dfrac{\dfrac{\dfrac{*}{sx \le y, \quad x = y \;\vdash\; sx \le x}}{sx \le y \;\vdash\; \sim (x = y)}}{x \le y \;\vdash\; x = y \vee \sim (x = y)}}{\vdash x = y \vee \sim (x = y)} \qquad x, y \,.\, N$$

The last step uses the $\vee$-rule and Theorem 19.10, formula (2). □

We also define a strong order $<$ on N by:

$$x < y \longleftrightarrow sx \leq y \qquad x.N\ y.N$$

Its properties are fairly immediate from Theorem 19.10 (see Exercise 19.9).

The internal logic of a natural number object includes much of the law of excluded middle, because of decidability, even if the general form of that law is not valid in the topos.

19.5 Rational and real numbers

The classicality of N extends part way into arithmetized analysis. For example, we construct an object of positive rational numbers Q^+. Define an equivalence relation Eq on $N \times N$ by

$$\vdash \langle x, y\rangle \mathrm{Eq} \langle z, w\rangle \longleftrightarrow x.sw = z.sy \qquad x, y, z, w.N$$

and let $q: N \times N \twoheadrightarrow Q^+$ be its quotient. The idea is that $q(x, y)$ is the rational x/sy, using s to preclude denominator zero. By definition we have $q(x, y) = q(z, w)$ iff $x.sw = z.sy$, the desired definition of equality of rationals.

One can show that $Q^+ \times Q^+$ is the quotient of $\mathrm{Eq} \times \mathrm{Eq}$ and define an arrow $+_{\mathrm{Eq}}$ from $(N \times N) \times (N \times N)$ to Q^+ by

$$+_{\mathrm{Eq}}(\langle x, y\rangle, \langle z, w\rangle) = \langle (x.sw) + (z.sy), (y.w) + y + w\rangle$$

This is the usual addition rule for fractions, allowing for the use of s in denominators, and $+_{\mathrm{Eq}}$ co-equalizes the two projections from the relation $\mathrm{Eq} + \mathrm{Eq}$ and so induces an arrow from the quotient, $+: Q^+ \times Q^+ \longrightarrow Q^+$. The multiplication of rationals can be defined in the same way and, since internally every rational is $q(x, y)$ for some x and y, arithmetic of rationals can be performed in the usual way using fractions.

A quotient of a decidable object need not be decidable, but Q^+ is because Eq is decidable on $N \times N$. That is, internally, for any r and r' there are x, y, z, w in N with $q(x, y) = r$ and $q(z, w) = r'$. By decidability of N we have

$$(x.sw = z.sy) \vee \sim(x.sw = z.sy)$$

which means that either $\langle x, y\rangle \mathrm{Eq} \langle z, w\rangle$ or not and so either $r = r'$ or not. So all the above theorems on order and arithmetic in N transfer easily to Q^+. To obtain all the rationals, Q, we take an object of integers Z (see Exercise 19.15) and carry out the construction with $Z \times N$ in place of $N \times N$ (so the denominator of a fraction is a non-zero natural number, while the numerator is any integer).

The real numbers are trickier. In most toposes, constructing the reals by Cauchy sequences and by Dedekind cuts does not give isomorphic results.

Clearly, Q^N is an object of sequences of rationals, and we can define the Cauchy sequences in the internal language and the desired equivalence relation. The usual definitions work, but Johnstone (1977) gives versions that are more useful in topos logic. The result is called the object of Cauchy reals, or, R_c. A Dedekind cut is a suitable pair of sub-objects of Q, so the Dedekind reals R_d are a sub-object of $\Omega^Q \times \Omega^Q$ (details are given in Johnstone (1977) and Bell (1988*b*), among other works). In general R_d is not isomorphic to R_c.

There is a canonical inclusion $R_c \rightarrowtail R_d$ in any topos, since topos logic proves that every Cauchy sequence gives a Dedekind cut; but in general R_d is much larger and richer than R_c. For a topological space S, the natural numbers in $\mathrm{Shv}(S)$ are the locally constant functions from S to the actual natural numbers, the rationals are the locally constant functions from S to the actual rationals, and the Cauchy reals in $\mathrm{Shv}(S)$ are the locally constant functions to the actual real numbers $\mathbb{R}$. The Dedekind reals are all continuous functions from S to $\mathbb{R}$.

19.6 Finite cardinals

In addition to the global element $0: 1 \longrightarrow N$, there are $s0$, $ss0$, and so on. These are the *standard* natural numbers, but they need not be the only global elements of N. For example, the generic element of N as a global element of N^*N in $\mathbf{E}/N$ is not standard unless $\mathbf{E}$ is trivial (see Exercises 19.10 and 19.11).

For any global element $p: 1 \longrightarrow N$, define the *cardinal* of p to be the extension $[x.N|x < p]$, a sub-object of N. Since we can show that

$$\vdash \sim(x < 0) \qquad x.N$$

$$\vdash x < sp \longleftrightarrow (x < p \vee x = p)$$

it follows that $[0] \cong \emptyset$ and for every p, $[sp] \cong [p] + 1$. In particular, the cardinal of the nth standard natural number is a coproduct of n copies of 1.

Slice functors preserve natural number objects (see Exercise 19.3 or 19.14), so N^*N, N^*0, N^*s is a natural number object in $\mathbf{E}/N$. Since N^* is logical it preserves $<$, meaning that $N^* <$ is the strong order relation on N^*N in $\mathbf{E}/N$. Similarly, pullback along any $p: 1 \longrightarrow N$ preserves all these, but more is true. Since N^* is the pullback along $!_N$, and $!_N \circ p$ is the identity arrow for 1, for every object A we have $p^*N^*A \cong A$.

Let $\mathbf{n}$ be the generic element $\Delta_N: N \rightarrowtail N \times N$ as a global element of N^*N. Then the cardinal $[\mathbf{n}]$ in $\mathbf{E}/N$ is useful. From the point of view of $\mathbf{E}$, it is an arrow to N, which we write $[\mathbf{n}] \longrightarrow N$. It is the generic cardinal:

THEOREM 19.12 For any global element $p: 1 \longrightarrow N$, the following is a pullback:

$$\begin{array}{ccc} [p] & \rightarrowtail & [\mathbf{n}] \\ \downarrow & & \downarrow \\ 1 & \xrightarrow[p]{} & N \end{array}$$

PROOF The pullback of the generic element $\mathbf{n}$ along p is p, and the pullback functor p^* preserves every construction used in defining cardinals, so it takes $[\mathbf{n}]$ in $\mathbf{E}/N$ to $[p]$ in $\mathbf{E}$. □

In other words, $[\mathbf{n}] \longrightarrow N$ is an N-indexed family of cardinals, where the fibre over any value in N is the cardinal of that value.

For any object A, consider the exponential $(N^*A)^{[\mathbf{n}]}$ in $\mathbf{E}/N$. From the viewpoint of $\mathbf{E}$ this is an arrow $(N^*A)^{[\mathbf{n}]} \longrightarrow N$. Its pullback along any $p: 1 \longrightarrow N$ is, up to isomorphism, $A^{[p]}$, since p^* preserves exponentials.

Think of $A^{[p]}$ as containing all p-tuples, or p-length lists, of values in A. Since $[0] \cong \emptyset$, we have $A^{[0]} \cong 1$, which intuitively says that there is just one empty list. And $[sp] \cong [p] + 1$ implies that $A^{[sp]} \cong A^{[p]} \times A$. A $(p + 1)$-length list consists of a p-length list plus one more value. Notice, however, that this does *not* require p to be a standard natural number, and so $A^{[p]}$ need not actually be a product of copies of A (compare with Exercise 19.13).

The arrow $(N^*A)^{[\mathbf{n}]} \longrightarrow N$ combines all the $A^{[p]}$ into one family indexed over N, and thus gives a meaning internal to $\mathbf{E}$ to the phrase 'all finite length lists of values in A'. For applications of this construction to inductive definitions in toposes, and algebraic theories in toposes, and much more, see Johnstone (1977) and references therein.

Exercises

19.1 Natural number objects make sense in any category with a terminator. Show that natural number objects are unique up to isomorphism.

19.2 For any category $\mathbf{A}$ with terminator define a category $N\mathbf{A}$ as follows. An object of $N\mathbf{A}$ is an object A of $\mathbf{A}$ and arrows $q: 1 \longrightarrow A$ and $f: A \longrightarrow A$. An $N\mathbf{A}$ arrow from A, q, f to A', q', f' is an $\mathbf{A}$ arrow $h: A \longrightarrow A'$ such that $h \circ q = q'$ and $h \circ f = f' \circ h$. Show that a natural number object for $\mathbf{A}$ is an initial object for $N\mathbf{A}$.

19.3 Let N, 0, s be a natural number object in $\mathbf{E}$, and let $\mathbf{F}: \mathbf{E} \longrightarrow \mathbf{E}'$ be any functor preserving 1 and with a right adjoint. Show that $\mathbf{F}N$, $\mathbf{F}0$, $\mathbf{F}s$ is a natural number object in $\mathbf{E}'$. This includes slice functors A^* from any Cartesian closed category and the inverse image functors of geometric morphisms.

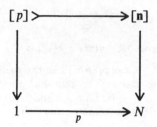

19.4 Show that, for any $q: 1 \longrightarrow A$ and $h: N \times A \longrightarrow A$, there is a unique $u: N \longrightarrow A$ such that

$$\vdash u(0) = q \quad \& \quad u(sx) = h(x, u(x)) \qquad x . N$$

[Hint: use recursion data $\langle 0, q \rangle$, $\langle s \circ p_1, h \rangle$.] Use this to define the factorial function. Combine the reasoning here and in Theorem 19.1 to show we can obtain every primitive recursive function $N \longrightarrow N$. Lambek and Scott (1986) or any treatment of recursive functions define primitive recursive functions.

19.5 Use the predecessor function $\mathrm{pr}: N \longrightarrow N$, defined by $\mathrm{pr}(0) = 0$ and $\mathrm{pr}(sx) = x$, to show that s is split monic.

19.6 Show that Theorems 19.1–19.5, and all the exercises up to this point, hold in any Cartesian closed category with a natural number object. Use Theorem 19.3 to prove the equations in Section 19.3 for generalized elements of N in any such category.

19.7 Prove that a terminator in any preorder is a natural number object. Use a Cartesian closed preorder to show that 0 and s need not be disjoint, and may actually be the same arrow in a Cartesian closed category.

19.8 Show that the induction axiom with variable I over Ω^N is true in any topos **E** with natural number object. [Hint: $(\Omega^N)^* N$ is a natural number object in $\mathbf{E}/(\Omega^N)$, so the axiom's pullback is the induction axiom for $\mathbf{E}/(\Omega^N)$. Use the generic element of Ω^N.]

19.9 Prove that $<$ is transitive and that:

$$\sim (x < x)$$

$$x \leq y \longleftrightarrow [x = y \vee x < y]$$

$$x < y \vee x = y \vee y < x$$

19.10 Show that if $0 = s0$ then **E** is trivial. [Hint: one method is to show that if $0 = s0$ then $[0] = [s0]$.] Conversely, show that every object in a trivial topos is a natural number object. Show that if **E** is not trivial then not all global elements of the natural number object in $\mathbf{E} \times \mathbf{E}$ are standard (even if all in **E** are). We can describe a global element $\langle p, q \rangle : \langle 1, 1 \rangle \longrightarrow \langle N, N \rangle$ in $\mathbf{E} \times \mathbf{E}$, with $p \neq q$ in **E**, as a natural number in $\mathbf{E} \times \mathbf{E}$ with 'internal variation'.

There are toposes built along the lines of non-standard models of type theory, as in non-standard analysis (see Robinson 1974). These toposes have 'infinite' non-standard natural numbers. But internal variation, like the simple example above, is the more typical source of non-standard natural numbers.

19.11 Show that the generic natural number $\mathbf{n}$ for **E**, viewed as a global element of the natural number object in $\mathbf{E}/N$, is neither the zero in $\mathbf{E}/N$ nor the successor of any other global element. [Hint: Exercise 19.3 shows that zero and the successor are preserved by pullback.]

Explain how, nevertheless, '$\mathbf{n} = 0 \vee (\exists y . N) n = sy$' is true in $L(\mathbf{E}/N)$. [Hint: interpret $0: 1 \longrightarrow N$ and $s: N \rightarrowtail N$ as sub-objects of the terminator in $\mathbf{E}/N$, the coproduct of which is the terminator. Show that they are the extensions of '$\mathbf{n} = 0$' and '$(\exists y . N)\mathbf{n} = sy$' respectively in $\mathbf{E}/N$.] Again, $\mathbf{n}$ has internal variation in $\mathbf{E}/N$.

19.12 (Bénabou's original definition of $[p]$) For any $p: 1 \longrightarrow N$ prove that $[p]$ is isomorphic to the pullback of p along $+ : N \times N \longrightarrow N$. [Hint: prove internally that for every $x < p$ there exists exactly one y with $x + y = p$.]

19.13 Let p be an actual natural number (i.e. a natural number from our external point of view rather than a global element of N in **E**). Let p° be the global element $s \circ \ldots \circ s0$ of N in **E**, with s repeated p times. Show that $[p^\circ]$ is the coproduct of p copies of 1, and so for any object A, $A^{[p^\circ]}$ is the product of p copies of A.

19.14 Let **E** be any topos, without supposing that it has a natural number object. Show that if an object N with $0: 1 \longrightarrow N$ and $s: N \longrightarrow N$ satisfies the Peano axioms in **E**, then N, 0, s is a natural number object. [Hint: show that the usual argument from the Peano axioms to recursive definition of functions is sound in **E**.] Conclude that every logical functor preserves natural number objects.

19.15 Define an equivalence relation R on $N \times N$ by

$$\vdash \langle x, y \rangle R \langle z, w \rangle \longleftrightarrow x + w = y + z$$

and define $p: N \times N \twoheadrightarrow Z$ as the quotient, so that intuitively $p(x, y)$ is the integer $x - y$ (compare with Beth 1968, p. 95ff.). Johnstone (1977) constructs the integers as a pushout, but our Chapter 24 uses the fact that Z is a quotient of $N \times N$.

20

Categories in a topos

In this chapter small categories and functors in a topos **E** are defined. Functors are then defined from a small category **A** to **E**, and the category $\mathbf{E}^{\mathbf{A}}$ of such functors, and it is proved that $\mathbf{E}^{\mathbf{A}}$ is a topos. Examples of functor categories on the topos **Set** are given in Chapter 22.

Section 20.1 really only applies naive constructions given in Part II to small categories in **E**. It generalizes the results in Chapter 12 on small categories in a universe of sets.

20.1 Small categories

A *small category* in a topos **E** is a pair of objects A_0 and A_1 and arrows $D: A_1 \longrightarrow A_0$, $C: A_1 \longrightarrow A_0$, $i: A_0 \longrightarrow A_1$, and $M: A_2 \longrightarrow A_1$, where A_2 is the following pullback:

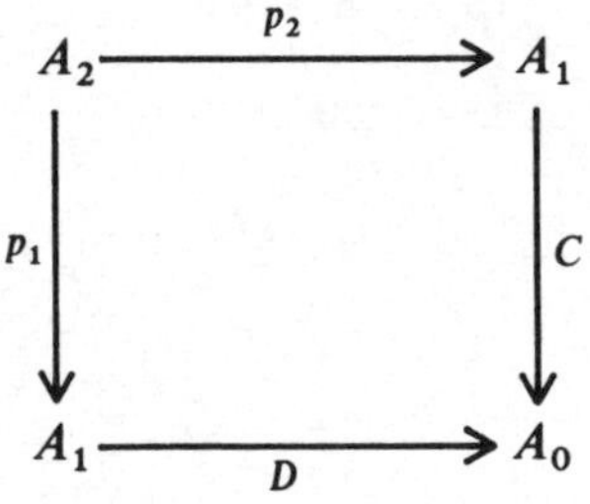

satisfying these equations:

$$D \circ M = D \circ p_2 \quad \text{and} \quad C \circ M = C \circ p_1$$

$$D \circ i = A_0 \quad \text{and} \quad C \circ i = A_0$$

$$M \circ \langle i \circ C, A_1 \rangle = A_1 \quad \text{and} \quad M \circ \langle A_1, i \circ D \rangle = A_1$$

$$M \circ \langle M \times A_1 \rangle = M \circ \langle A_1 \times M \rangle$$

Note that $\langle i \circ C, A_1 \rangle : A_1 \longrightarrow A_2$, among others, is an arrow to the pullback and not to the product $A_1 \times A_1$. But the equation $D \circ i = A_0$ implies the pullback condition $D \circ i \circ C = C \circ A_1$, so in fact $\langle i \circ C, A_1 \rangle$ is well defined to A_2.

The equations look more familiar internally, with variables over objects and arrows of **A**; and writing i_o in place of $i(o)$, and $f \circ g$ in place of $M \circ \langle f, g \rangle$.

$$Df = Cg \vdash D(f \circ g) = Dg \quad \& \quad C(f \circ g) = Cf \quad f, g, h . A_1$$

$$\vdash D(i_o) = o \quad \& \quad C(i_o) = o \quad o . A_0$$

$$\vdash i_{Cf} \circ f = f \quad \& \quad g \circ i_{Dg} = g$$

$$Df = Cg,\ Dg = Ch \vdash (f \circ g) \circ h = f \circ (g \circ h)$$

(Exercise 20.21 discusses '$f \circ g$' as an abuse of notation.)

In a small category $\mathbf{A} = \langle A_0, A_1, D, C, i, M \rangle$, we call A_0 the *object of objects*, and A_1 the *object of arrows*. D, C, i, and M are the *domain arrow*, *codomain arrow*, *identities arrow*, and *composition arrow*.

A *functor*, $\mathbf{F}: \mathbf{A} \longrightarrow \mathbf{B}$, from $\mathbf{A} = \langle A_0, A_1, D, C, i, M \rangle$ to $\mathbf{B} = \langle B_0, B_1, D', C', i', M' \rangle$ is a pair of arrows $F_0: A_0 \longrightarrow B_0$ and $F_1: A_1 \longrightarrow B_1$ such that

$$\vdash D'(F_1 h) = F_1(Dh) \quad \& \quad C'(F_1 h) = F_1(Ch) \quad g, h . A_1$$

$$\vdash F_1(i_o) = i'_{F0(o)} \quad o . A_0$$

$$Dh = Cg \vdash F_1(h \circ g) = (F_1 h) \circ' (F_1 g)$$

Functors compose, with $\mathbf{F} \circ \mathbf{G}$ consisting of $F_0 \circ G_0$ and $F_1 \circ G_1$. Every small category **A** has an identity functor consisting of the identity arrows for A_0 and A_1.

Small categories and functors in **E** themselves form a (non-small) category $\mathrm{Cat}_{\mathbf{E}}$. The constructions in Part II, of products, coproducts, and equalizers of categories, apply in any topos (see Exercise 20.1), so $\mathrm{Cat}_{\mathbf{E}}$ has all of these. Given functors $\mathbf{F}, \mathbf{G}: \mathbf{A} \longrightarrow \mathbf{B}$, the reader can define natural transformations from **F** to **G** and show that there is a small category $\mathbf{B}^{\mathbf{A}}$, an exponential of **B** by **A** in $\mathrm{Cat}_{\mathbf{E}}$ (see Exercise 20.5). Coequalizers are more complicated. In fact, $\mathrm{Cat}_{\mathbf{E}}$ has coequalizers iff **E** has a natural number object. The results on adjunctions in Chapter 10 all hold for $\mathrm{Cat}_{\mathbf{E}}$, but of course we do not have the axiom of choice in every topos **E**.

20.2 E-valued functors

In addition to functors between small categories, we can define functors from a small category **A** to the base topos **E**. These are called **E**-*valued functors on* **A**, or *diagrams* on **A**. The idea, naturally, is that a diagram S on **A** assigns to each object o of **A** an **E** object $S(o)$, and to each arrow $o \longrightarrow o'$ of **A** an **E** arrow $S(o) \longrightarrow S(o')$, preserving identity arrows and composition. But (unless **E** is a topos of sets) the object of objects A_0 is not determined by its global elements alone, and so we cannot formalize this definition simply by assigning an **E** object to each global element of A_0. We need an A_0-indexed family of objects, that is an arrow $s: S \longrightarrow A_0$, and a suitable action of A_1 on that family.

Precisely, define a diagram on **A** to consist of a *structure arrow* $s: S \longrightarrow A_0$ plus an *action* $\mathrm{ac}: A_1 \times_{A0} S \longrightarrow S$, where $A_1 \times_{A0} S$ is the pullback of s along D. In other words, it is the extension $[\langle f, x\rangle . A_1 \times S \mid Df = s(x)]$. The arrows s and ac are required to meet the following conditions:

$$Df = s(x) \vdash s(\mathrm{ac}(f, x)) = Cf \quad f, g . A_1 \; x . S$$

$$s(x) = o \vdash ac(i_o, x) = x$$

$$Dg = s(x), Df = Cg \vdash ac(f \circ g, x) = \mathrm{ac}(f, ac(g, x))$$

We may refer to the diagram $\langle s, \mathrm{ac}\rangle$ as S for short, naming it by the domain of its structure arrow.

Intuitively, the first equation says that when x lies in the fibre over the domain of f, then f acts on x to give a value in the fibre over the codomain of f. The second says that when x lies in the fibre over an object o then the identity arrow on o does not change x. The last says that if f and g are composable and x lies in the fibre over the domain of g, then letting g act on x and letting f act on the result is the same as letting $f \circ g$ act on x. Think of the fibres over objects as the values of the functor on those objects, and the actions of the arrows as the values of the functor for those arrows. This is further spelled out in the next two paragraphs.

Let s, ac be a diagram on **A**. Then each generalized object of **A**, that is each $o: T \longrightarrow A_1$, gives as object the fibre of s over o:

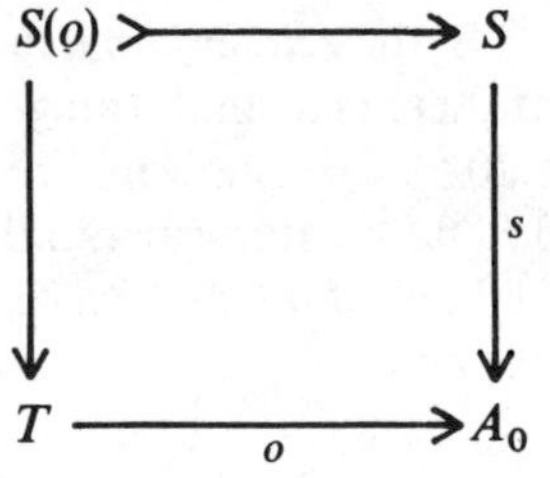

If o is a global object, that is $T = 1$, then we regard $S(o)$ as an **E** object. Otherwise we regard $S(o) \longrightarrow T$ as an object in $\mathbf{E}/T$. In Chapter 11 the subscript notation S_o was used for fibres, as is customary for families of sets. Here we shift to the notation $S(o)$, which is customary for values of functors.

Suppose that a global arrow of **A**, $f: 1 \longrightarrow A_1$, has domain and co-domain $f: o \longrightarrow o'$ in **A**. Then the relation

$$\mathrm{ac}(f, x) = x' \qquad x, x' . S$$

is single-valued from x to x', defined for x in $S(o)$, and takes values x' in $S(o')$.

So the **E** arrow $S(f): S(o) \longrightarrow S(o')$ is uniquely defined by

$$\vdash S(f)(x) = \mathrm{ac}_s(f, x) \qquad x . S(o)$$

The second equation on s and ac says that $S(i_o)$ is the identity on $S(o)$. The third says that S preserves composition. This is extended to generalized **A** arrows in Exercises 20.8–20.10.

Take a pair of diagrams on **A**, one with structure arrow $s: S \longrightarrow A_0$ and action ac_s; and one with $w: W \longrightarrow A_0$ and ac_w. Define a *natural transformation* $v: S \longrightarrow W$ as an $\mathbf{E}/A_0$ arrow, $v: s \longrightarrow w$, which commutes with the actions as follows:

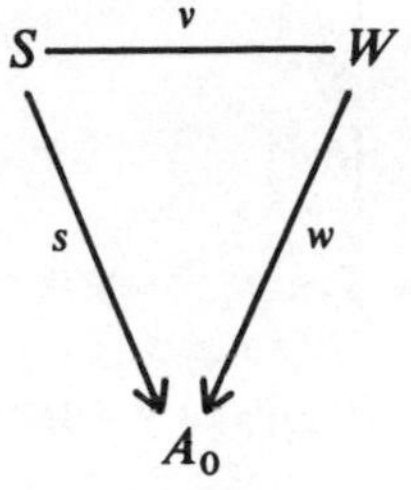

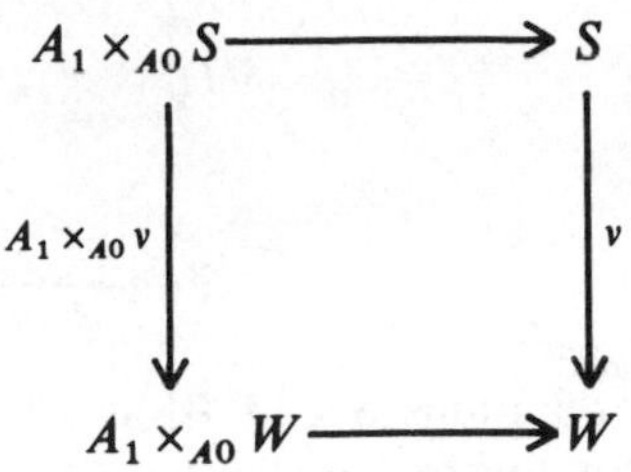

Internally this says that:

$$s(x) = o \vdash w(v(x)) = o \qquad o . A_0\, x . S$$

$$Df = s(x) \vdash \mathrm{ac}_w(f, v(x)) = v(\mathrm{ac}_s(f, x) \qquad f . A_1$$

So pulling the triangle back along any global **A** object $o: 1 \longrightarrow A_A$ gives an **E** arrow $v_o: S(o) \longrightarrow W(o)$, and for any global **A** arrow $f: o \longrightarrow o'$ the naturality square commutes in **E**:

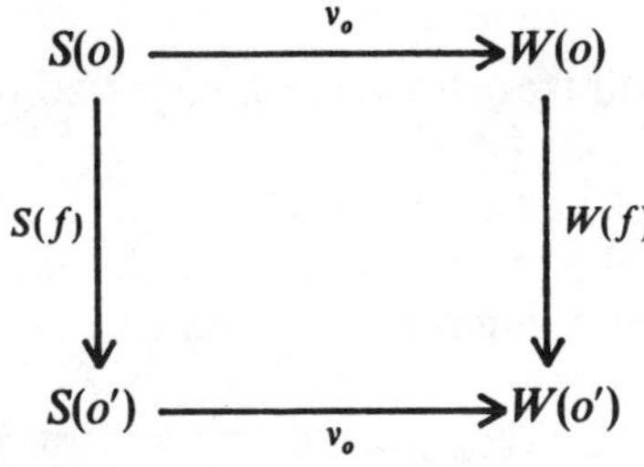

But, again, the real point is that such conditions hold for all generalized objects and arrows (see Exercise 20.10).

Natural transformations $v: S \longrightarrow W$ and $\mu: W \longrightarrow Y$ compose simply by composing the arrows to give $\mu \circ v: S \longrightarrow Y$. And for any diagram S the identity $1_S: S \longrightarrow S$ is natural. Thus diagrams on **A** and natural transformations form a category called the diagram category on **A**, or simply $\mathbf{E}^{\mathbf{A}}$.

Limits in $\mathbf{E}^{\mathbf{A}}$ (as well as colimits) are calculated *pointwise*, meaning that you obtain a given limit diagram by taking the limit in each fibre. A quick description of the fibres will motivate the precise constructions. The terminator in $\mathbf{E}^{\mathbf{A}}$ is called $\Delta 1$, or the *constant diagram* with fibre 1. The fibre over each $\mathbf{A}$ object is 1, and so for each $\mathbf{A}$ arrow f the arrow $S(f)$ has to be the identity $1 \longrightarrow 1$. This is clearly functorial; it preserves identity arrows and composition. And for any diagram S there is a unique natural transformation $!: S \longrightarrow \Delta 1$ with the only components it could have:

$$\begin{array}{ccc} S(o) & \longrightarrow & 1 \\ {\scriptstyle S(f)}\downarrow & & \downarrow \\ S(o') & \longrightarrow & 1 \end{array}$$

Given diagrams S and W as above, define a diagram $S \times W$ with fibre $S(o) \times W(o)$ for any $\mathbf{A}$ object o, and arrow

$$S(o) \times W(o) \xrightarrow{S(f) \times W(f)} S(o') \times W(o')$$

for each $\mathbf{A}$ arrow $f: o \longrightarrow o'$. This is clearly functorial, and the projections $p_1: S(o) \times W(o) \longrightarrow S(o)$ are components of a natural transformation $p_1: S \times W \longrightarrow S$. For any diagram T and natural transformations $\nu: T \longrightarrow S$ and $\mu: T \longrightarrow S$, the components

$$T(o) \xrightarrow{\langle \nu_o, \mu_o \rangle} S(o) \times W(o)$$

give the unique natural transformation required to show that $S \times W$ is a product of S and W in $\mathbf{E}^{\mathbf{A}}$.

For diagrams S and W and natural transformations ν and μ, both from S to W, define a diagram E the fibres of which are the equalizers shown, and the arrows of which are those induced by the equalizers:

$$\begin{array}{ccccc} E(o) & \rightarrowtail & S(o) & \rightrightarrows & W(o) \\ {\scriptstyle E(f)}\downarrow & & \downarrow{\scriptstyle S(f)} & & \downarrow{\scriptstyle W(f)} \\ E(o') & \rightarrowtail & S(o') & \rightrightarrows & W(o') \end{array}$$

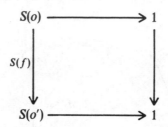

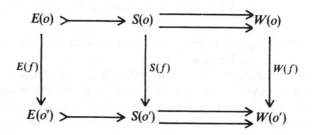

By their definition, the inclusions $E(o) \rightarrowtail S(o)$ are components of a natural transformation $E \rightarrowtail S$, easily seen to be an equalizer for ν and μ in $\mathbf{E}^{\mathbf{A}}$.

Thus $\mathbf{E}^{\mathbf{A}}$ has all finite limits. In fact, like all toposes it has limits over all diagrams that can be defined within it, including infinite ones, of course, if **E** has any infinite objects to begin with (see Exercise 20.18).

It remains to actually construct diagrams with the fibres described. Each diagram must not only have fibres in **E**, as described for global objects and arrows of the diagrams, but it must also have them in $\mathbf{E}/T$ for all objects and arrows of **A** defined over any **E** object T. So it must have the appropriate fibre in $\mathbf{E}/A_0$ over the generic element of A_0, the identity arrow. And, conversely, since all the constructions involved are preserved by pullbacks, if we get this fibre right all the others will follow. Of course, for any diagram S the fibre over the identity arrow on A_0 is simply the structure arrow $s: S \longrightarrow A_0$ itself. So we calculate each limit (or colimit) of structure arrows in $\mathbf{E}/A_0$ and the corresponding actions to obtain limits (resp. colimits) in $\mathbf{E}^{\mathbf{A}}$.

Therefore the terminator $\Delta 1$ must have the terminator in $\mathbf{E}/A_0$ as structure arrow; that is, the identity $A_0 \longrightarrow A_0$. The action is $C: A_1 \longrightarrow A_0$. It can easily be seen that is the only action that this structure arrow can have, and that this diagram is terminal in $\mathbf{E}^{\mathbf{A}}$. For the product $S \times W$, form the product in $\mathbf{E}/A_0$, that is the pullback $S \times_{A0} W$ in **E**, with action $\mathrm{ac}_s \times_{A0} \mathrm{ac}_w$. This is the unique action with projections ac_s and ac_w. For the equalizer of ν and μ, form their equalizer as $\mathbf{E}/A_0$ arrows, which is also their equalizer as **E** arrows, by Theorem 11.1. The action is uniquely defined by the requirement that the insertion of the equalizer be natural, so that it can be an equalizer of diagrams, i.e. by the requirement that the following square commutes:

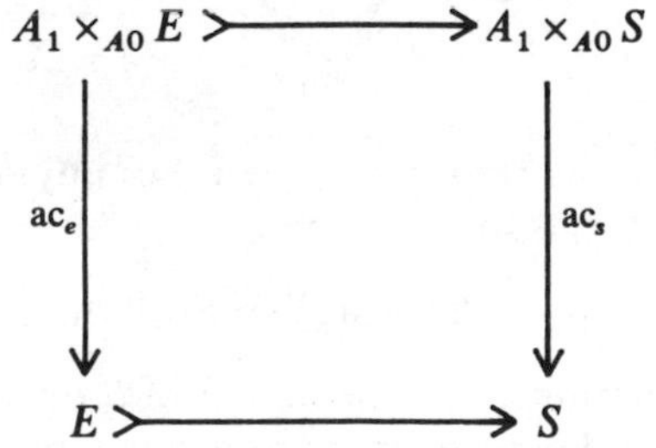

20.3 The Yoneda lemma

For each generalized **A** object $o: T \longrightarrow A_0$ we define a diagram R_o, called the *representable functor* on o. For any **A** object o', $R_o(o')$ consists of all **A** arrows from o to o'. For any **A** arrow $f: o' \longrightarrow o''$, $R_o(f)$ is the function that takes each **A** arrow $g: o \longrightarrow o'$ and composes with f to give $f \circ g: o \longrightarrow o''$. Clearly,

$R_o(f)$ is well defined from $R_0(o')$ to $R_o(o'')$, and R_o preserves identity arrows and composition.

For any $o: T \longrightarrow A_0$, let the object R_o be $A_1 \times_{A0} T$, the pullback of o along $D: A_1 \longrightarrow A_0$. Therefore R_o consists of all **A** arrows with domain o or, more strictly, with domain $o(x)$, where x is a parameter in T. Let the structure arrow be:

$$A_1 \times_{A0} T \xrightarrow{p_1} A_1 \xrightarrow{C} A_0$$

Therefore the fibre over any **A** object o' consists of all pairs $\langle g, x\rangle$ with g an **A** arrow and x in T such that $Dg = o(x)$ and $Cg = o'$. Let the action be:

$$A_1 \times_{A0} A_1 \times_{A0} T \xrightarrow{M \times_{A0} T} A_1 \times_{A0} T$$

Therefore the action applies to triples $\langle f, g, x\rangle$ with $Df = Cg$ and $Dg = o(x)$, and it composes the arrows leaving x alone to give $\langle f \circ g, x\rangle$.

The Yoneda lemma says that for any diagram S and T-indexed object $o: T \longrightarrow A_0$, the T-elements of the fibre $S(o)$ correspond exactly to the natural transformations from R_o to S. Of course, a T-element of the fibre is the same thing as an arrow from o to s over A_0:

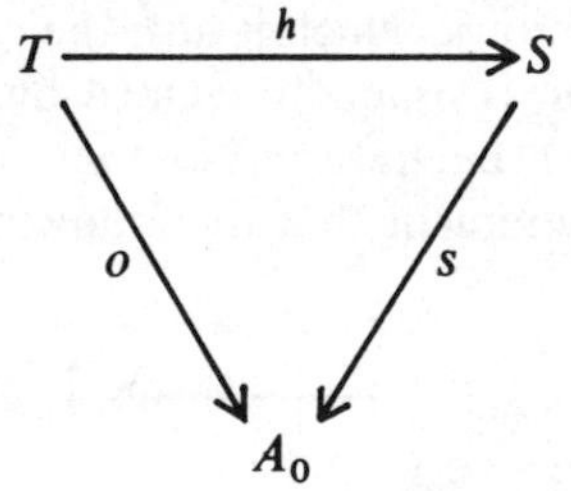

The idea is simply this. Take any $v: R_o \longrightarrow S$. For any **A** arrow g with $Dg = o$ we have $g = g \circ i_o$ in **A** and so

$$v(g) = v(R_o(i_o)(g)) = S(g)(v(i_o))$$

So v is completely determined by the action of S once we know $v(i_o)$. And, conversely, we can use those equations to define a natural transformation v starting with any value in $S(o)$ for $v(i_o)$.

We state the Yoneda lemma as an adjunction. There is a functor $\mathbf{U}: \mathbf{E}^{\mathbf{A}} \longrightarrow \mathbf{E}/A_0$ which 'forgets' the action of a diagram. That is, $\mathbf{U}(S)$ is the structure arrow of S, an object of $\mathbf{E}/A_0$. For any natural transformation $\varphi: S \longrightarrow W$, $\mathbf{U}(\varphi)$ is just φ as an $\mathbf{E}/A_0$ arrow. The theorem shows that $\mathbf{U}$ has a left adjoint which assigns to each $\mathbf{E}/A_0$ object $o: T \longrightarrow A_0$ the representable functor R_o.

THEOREM 20.1 (the Yoneda lemma) Every $o: T \longrightarrow A_0$ has an $\mathbf{E}/A_0$ arrow $h: o \longrightarrow R_o$ universal from o to $\mathbf{U}$.

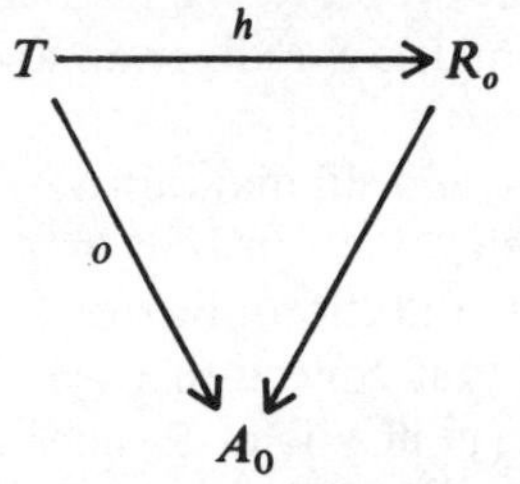

PROOF Let $h: T \longrightarrow A_1 \times_{A0} T$ be the arrow induced by $i \circ o: T \longrightarrow A_1$ and $1_T: T \longrightarrow T$. So h takes any value x in T to the pair $\langle i_{o(x)}, x \rangle$. It is easy to see that pair is in $A_1 \times_{A0} T$ and that the structure arrow takes that pair to $o(x)$, so h is well defined as an arrow over A_0. Let S be any diagram on $\mathbf{A}$ and take any $y: o \longrightarrow S$ in $\mathbf{E}/A_1$. We must show that there is a unique $v: R_o \longrightarrow S$ with $v \circ h = y$. But to say that $v \circ h = y$ is the same as saying that

$$\vdash v(i_{o(x)}, x) = y(x) \quad x . T$$

and we also have

$$Dg = o(x) \vdash \langle g, x \rangle = \mathrm{ac}_{Ro}(g, (i_{ox}, x)) \qquad g . A_1 \; x . T$$

and so naturality of v says that

$$Dg = o(x) \vdash v(g, x) = \mathrm{ac}_s(g, v(i_{o(x), x})) = \mathrm{ac}_s(g, y(x))$$

and that fully defines $v: A_1 \times_{A0} T \longrightarrow S$ in terms of ac_s. □

20.4 $\mathbf{E}^{\mathbf{A}}$ is a topos

We begin proving that $\mathbf{E}^{\mathbf{A}}$ has a sub-object classifier by describing the monics in $\mathbf{E}^{\mathbf{A}}$.

LEMMA 20.2 A natural transformation $v: S \longrightarrow W$ is monic in $\mathbf{E}^{\mathbf{A}}$ iff it is monic in $\mathbf{E}/A_0$, and so iff it is monic in $\mathbf{E}$.

PROOF Obviously, if v is monic in $\mathbf{E}/A_0$ it is monic in $\mathbf{E}^{\mathbf{A}}$. The converse holds since $\mathbf{U}$ is a right adjoint and any right adjoint preserves monics. The second claim is a general fact about slice categories. □

LEMMA 20.3 A subdiagram $v: S \rightarrowtail W$ consists of a sub-object $v: S \rightarrowtail W$ of the $\mathbf{E}$ object W with the following property. Suppose that y in W has $w(y) = o$ and y is in the sub-object S; then for every $\mathbf{A}$ arrow $f: o \longrightarrow o'$ we have $\mathrm{ac}_w(f, y)$ also in S.

PROOF If S is a subdiagram, to say that y is in it means (expressing it internally) that for some x in S we have $v(x) = y$. But then by naturality of v, $\mathrm{ac}_w(f, y)$ equals $v(\mathrm{ac}_s(f, x))$ and so it is in the sub-object. Conversely, given an **E** monic $S \rightarrowtail W$ with that property, the restriction of ac_w to S is clearly the one and only action that makes S a diagram and v natural. □

Consider any diagram W and subdiagram $S \rightarrowtail W$. We can do more than just say whether or not each value y in W is in S. We can measure the 'extent to which y is in S' by the collection of arrows that carry y into S; that is, $\{g . A_1 | \mathrm{ac}_w(g, y) \in S\}$. Call that collection $\chi_s(y)$. For example, $\chi_s(y)$ contains all arrows with domain $w(y)$ iff y is in S, since $\mathrm{ac}_w(i_{w(y)}, y) = y$. The larger $\chi_s(y)$ is, the closer we think of y as being to S.

If $w(y) = o$ then for any $h: o \longrightarrow o'$ and $y' = \mathrm{ac}_w(h, y)$ we have

$$\chi_s(y') = \{f . A_1 | f \circ h \in \chi_s(y)\}$$

since we have

$$\mathrm{ac}_w(f \circ h, y) = \mathrm{ac}_w(f, \mathrm{ac}_w(h, y))$$

Define a *cosieve* on an object o to be a collection of arrows with domain o such that if an arrow g is in the collection and $Df = Cg$ then $f \circ g$ is also in it (compare with sieves, in Section 21.4). The maximal cosieve on o contains all arrows with domain o, and a cosieve on o is maximal iff it includes i_o. Note that $\chi_s(y)$ is always a cosieve. Our 'measures of the extent' to which a value is in a given subdiagram are cosieves. The sub-object classifier for $\mathbf{E}^{\mathbf{A}}$ will be the diagram of cosieves.

THEOREM 20.4 $\mathbf{E}^{\mathbf{A}}$ has a sub-object classifier.

PROOF Define the object $\Omega_{\mathbf{A}}$ as the object of all cosieves (this can be defined internally as a sub-object of the power object $\Omega^{(A_1)}$ in **E**). Define the structure arrow $\gamma: \Omega_{\mathbf{A}} \longrightarrow A_0$ as taking each cosieve to the object that it is a cosieve on. For any arrow $h: o \longrightarrow o'$ and cosieve C on o, define the action this way:

$$\mathrm{ac}_\gamma(h, C) = \{f . A_1 | f \circ h \in C\}$$

This clearly preserves identity arrows and composition. Note that if C is maximal on o then $\mathrm{ac}_\gamma(h, C)$ is maximal on o', and so there is a natural transformation $t: \Delta 1 \longrightarrow \Omega_{\mathbf{A}}$ given by the **E** arrow $A_0 \longrightarrow \Omega_{\mathbf{A}}$ which takes each o in A_0 to the maximal cosieve on o.

Take any subdiagram $S \rightarrowtail W$. The natural transformation $\chi_s: W \longrightarrow \Omega_{\mathbf{A}}$ that takes each y in W to $\chi_s(y)$ is the only one which takes y to the maximal cosieve on $w(y)$ iff y is in S, the only one to make this square a pullback in $\mathbf{E}^{\mathbf{A}}$:

$$\begin{array}{ccc} S & \longrightarrow & \Delta 1 \\ \downarrow & & \downarrow{\scriptstyle t} \\ W & \underset{\chi_s}{\longrightarrow} & \Omega_{\mathbf{A}} \end{array}$$

So $\Omega_{\mathbf{A}}$ is a sub-object classifier. □

Given diagrams S and W on $\mathbf{A}$, define the exponential diagram W^S by saying, for any object o of $\mathbf{A}$, that $W^S(o)$ is the collection of all natural transformations from $R_o \times S$ to W (see Exercise 20.16). For any $\mathbf{A}$ arrow $f\colon o \longrightarrow o'$ there is a natural transformation $f^*\colon R_{o'} \longrightarrow R_o$ defined by $f^*(h) = h \circ f$. So we can define $W^S(f)$ as taking each $v\colon R_o \times S \longrightarrow W$ to

$$R_{o'} \times S \xrightarrow{f^* \times S} R_o \times S \xrightarrow{v} W$$

The reader can verify that W^S is a well-defined diagram on $\mathbf{A}$. Do not confuse the diagram W^S with the exponential of objects W^S in $\mathbf{E}$!

THEOREM 20.5 The diagram W^S is an exponential of W by S in $\mathbf{E}^{\mathbf{A}}$.

PROOF To define the natural transformation $ev\colon W^S \times S \longrightarrow W$, for each $v\colon R_o \times S \longrightarrow W$ and each x in S with $s(x) = o$, let $ev(v, x)$ be $v(i_o, x)$. Take any natural transformation $\varphi\colon V \times S \longrightarrow W$. For each y in V with $v(y) = o$ define a natural transformation $\bar{\varphi}(y)\colon R_o \times S \longrightarrow W$ by saying for each $h\colon o \longrightarrow o'$ and x' in $S(o')$ we have

$$(\bar{\varphi}(y))\,(h, x') = \varphi((Vh)(y), x')$$

Define $\bar{\varphi}\colon V \longrightarrow W^S$ as taking each y to $\bar{\varphi}(y)$. Routine though copious details verify that each of these is natural, and that $\bar{\varphi}$ is the transpose of φ. □

Exercises

20.1 Prove that $\mathrm{Cat}_{\mathbf{E}}$ has a terminator with object of objects and object of arrows both 1. Given small categories $\mathbf{A}$ and $\mathbf{B}$, show that there is a product $\mathbf{A} \times \mathbf{B}$ with object of objects $A_0 \times B_0$ and object of arrows $A_1 \times B_1$. Given functors $\mathbf{F}$ and $\mathbf{G}$, both from $\mathbf{A}$ to $\mathbf{B}$, construct an equalizer for $\mathbf{F}$ and $\mathbf{G}$ using the equalizer of F_0 and G_0, and also the equalizer of F_1 and G_1 in $\mathbf{E}$. Do the same for an initial category, using $\emptyset$, and for coproducts in $\mathrm{Cat}_{\mathbf{E}}$, using coproducts in $\mathbf{E}$.

20.2 Call a small category $\mathbf{A}$ a preorder if $\langle D, C\rangle\colon A_1 \longrightarrow A_0 \times A_0$ is monic, indiscrete if $\langle D, C\rangle$ is iso, and discrete if i is iso. Explain these in terms of our previous definitions of preorders, and indiscrete and discrete categories. What is the dual of $\mathbf{A}$?

20.3 Define an underlying object functor $\mathbf{U}:\mathrm{Cat}_{\mathbf{E}} \longrightarrow \mathbf{E}$ that takes $\mathbf{A}$ to A_0 and $\mathbf{F}$ to F_0. Show that the four adjoint functors of Exercises 12.1–12.8 exist over any topos $\mathbf{E}$. [Hint: Comp$(\mathbf{A})$ is the coequalizer of the domain and codomain arrows of $\mathbf{A}$; that is, functors from $\mathbf{A}$ to Disc(S) correspond to arrows $f: A_0 \longrightarrow S$ with $f \circ D = f \circ C$.] Show there is no further adjoint on either end if $\mathbf{E}$ is not trivial (i.e. if $1 + 1$ is not iso to 1 in $\mathbf{E}$ and thus—comparing with Exercise 20.6 below—the finite category $\mathbf{2}$ in $\mathbf{E}$ is not isomorphic to the category unfortunately also called $\mathbf{E}$ in Exercise 10.6).

20.4 Define a natural transformation between small functors $\mathbf{F}, \mathbf{G}: \mathbf{A} \longrightarrow \mathbf{B}$ as an arrow $v: A_0 \longrightarrow B_1$ that meets certain conditions.

20.5 For any small categories $\mathbf{A}$ and $\mathbf{B}$ in $\mathbf{E}$, construct a functor category $\mathbf{B}^{\mathbf{A}}$ and show that it is an exponential of $\mathbf{B}$ by $\mathbf{A}$ in $\mathrm{Cat}_{\mathbf{E}}$. [Hint: for the object of objects $(\mathbf{B}^{\mathbf{A}})_0$ use the extension

$$[f.\, B_1^{(A_1)} \mid f \text{ preserves identity arrows and composition}]$$

State that condition internally. Define the object of arrows as the sub-object of $(\mathbf{B}^{\mathbf{A}})_0 \times (\mathbf{B}^{\mathbf{A}})_0 \times (B_1^{(A_0)})$ that contains all $\langle f, g, v\rangle$ with v natural from f to g.]

20.6 Show that every finite category appears as a small category in $\mathbf{E}$ with finite coproducts of 1 for object of objects and object of arrows, and that the functors between them all appear in $\mathbf{E}$.

20.7 Given a natural number object N, construct a category $\mathbf{N}$ with object of objects 1 and object of arrows N and $+: N \times N \longrightarrow N$ as composition. Show that any functor $\mathbf{F}:\mathbf{N} \longrightarrow \mathbf{A}$ is fully determined by $\mathbf{F}(1)$, since $\mathbf{F}(sn) = \mathbf{F}(n) \circ \mathbf{F}(1)$. Show that the functor $q: \mathbf{2} \longrightarrow \mathbf{N}$ that takes the non-identity arrow of $\mathbf{2}$ to the natural number 1 is a coequalizer in $\mathrm{Cat}_{\mathbf{E}}$ for the two functors from $\mathbf{1}$ to $\mathbf{2}$.

Conversely, those functors have a coequalizer only if $\mathbf{E}$ has a natural number object. This could be proved by reasoning internally, but it is perhaps better to see Johnstone (1977, Theorem 6.41).

20.8 Given any small category $\mathbf{A}$ in $\mathbf{E}$ and any $\mathbf{E}$ object T, show that applying the functor $T^*: \mathbf{E} \longrightarrow \mathbf{E}/T$ to each object and arrow in $\langle A_0, A_1, D, C, i, M\rangle$ gives a small category in $\mathbf{E}/T$. (This just means that T^* preserves the pullback A_2 and the equations defining a category.) Call this small category $T^*\mathbf{A}$. Show that T^* also preserves small functors, so that there is a (non-small) functor $T^*: \mathrm{Cat}_{\mathbf{E}} \longrightarrow \mathrm{Cat}_{\mathbf{E}/T}$.

20.9 Given a diagram s, ac on $\mathbf{A}$ and any $\mathbf{E}$ object T, show that T^*s, T^*ac is a diagram on $T^*\mathbf{A}$ in $\mathbf{E}/\mathrm{T}$. (As in Exercise 20.8, the verification is trivial.) Show that there is a (non-small) functor T^* from $\mathbf{E}^{\mathbf{A}}$ to $(\mathbf{E}/T)^{T^*\mathbf{A}}$.

20.10 A T-indexed $\mathbf{A}$ object $o: T \longrightarrow A_0$ can be regarded as a global object of $T^*\mathbf{A}$, and similarly for T-indexed arrows. Show that the fibre of T^*S over o as global object of $T^*\mathbf{A}$ is the same as the fibre $S(o)$ defined in Section 20.2 as a pullback in $\mathbf{E}$.

Suppose that an $\mathbf{A}$ arrow $f: T \longrightarrow A_1$ has $Df = o$ and $Cf = o'$, for o and o' T-indexed objects of $\mathbf{A}$. Transfer all of this to the slice $\mathbf{E}/T$ to conclude that there is an $\mathbf{E}/T$ arrow $S(f): S(o) \longrightarrow S(o')$ such that S preserves identity arrows and composition for T-elements of A_0 and A_1. Also conclude that natural transformations give naturality squares in $\mathbf{E}/T$ similar to those in $\mathbf{E}$ described in the text of the chapter for global arrows and objects.

20.11 Generalize Exercise 20.8 by showing that any geometric morphism $f^* \dashv f_*: \mathbf{E} \longrightarrow \mathbf{E}'$ gives a functor $f^*: \mathbf{Cat_E} \longrightarrow \mathbf{Cat_E}$, and a functor $f_*: \mathbf{Cat_{E'}} \longrightarrow \mathbf{Cat_E}$ which are also adjoint $f^* \dashv f_*$.

20.12 For diagrams S and W, define a diagram with structure arrow $\binom{s}{w}: S + W \longrightarrow A_0$ and action $ac_s + ac_w$. Show that the inclusions $S \rightarrowtail S + W$ and $W \rightarrowtail S + W$ and $S + W$ are a coproduct for S and W in $\mathbf{E^A}$. Show that the fibres of $S + W$ over $\mathbf{A}$ objects (including fibres in $\mathbf{E}/T$ for fibres of objects defined over any T) are coproducts of the fibres of S and W.

20.13 Given S and W and natural transformations ν and μ, both from S to W, define a coequalizer for ν and μ. As usual, the basic step is to form the coequalizer of ν and μ as $\mathbf{E}/A_0$ arrows. [Hint: recall Exercise 11.4 on coequalizers in a slice.]

20.14 Sum up the results of finite limits and colimits in $\mathbf{E^A}$ as saying that the forgetful functor $\mathbf{U}: \mathbf{E^A} \longrightarrow \mathbf{E}/A_0$ creates them ('creating' is defined in Exercise 11.5).

20.15 Show that a cosieve on o amounts to a subdiagram of R_o (recall Lemma 20.3). Thus the Yoneda lemma explains why the fibres of the sub-object classifier for $\mathbf{E^A}$ contain the cosieves.

Use the Yoneda lemma to explain why the fibre over o of the exponential diagram W^S contains the natural transformations from $R_o \times S$ to W.

20.16 Show that a natural transformation from $R_o \times S$ to W can be defined internally as a partial function from $A_1 \times S$ to W and so, treating the partial function as a single-valued relation, as a member of the power object of $A_1 \times S \times W$. Thus the object of all natural transformations from $R_o \times S$ to W, for all $\mathbf{A}$ objects o, can be defined internally as a sub-object of the power object of $A_1 \times S \times W$.

20.17 Show that for every $\mathbf{E}$ object B there is an $\mathbf{A}$ diagram with structure arrow $p_1: A_0 \times B \longrightarrow A_0$ and action $C \times B: A_1 \times B \longrightarrow A_0 \times B$. Call this diagram ΔB, or the *constant diagram* with fibre B. Show that its fibre over any global object of $\mathbf{A}$ is B and that the action of any global arrow is 1_B. Similarly, show that the fibre over any T-indexed object is T^*B and that the action of any T-indexed arrow is the identity.

For any $\mathbf{E}$ arrow $f: B \longrightarrow B'$, show that $A_0 \times f: A_0 \times B \longrightarrow A_0 \times B'$ gives a natural transformation between constant diagrams. Conclude that there is a functor $\Delta: \mathbf{E} \longrightarrow \mathbf{E^A}$ that takes each object to its constant diagram. Show that Δ preserves all finite limits.

20.18 For any diagram S, define $\mathrm{Colim}(S)$ by the following coequalizer:

$$A_1 \times_{A0} S \underset{p_2}{\overset{ac}{\rightrightarrows}} S \twoheadrightarrow \mathrm{Colim}(S)$$

Show for any $\mathbf{E}$ object B that a natural transformation $\nu: S \longrightarrow B$ gives an $\mathbf{E}$ arrow $\mathrm{Colim}(S) \longrightarrow B$ and vice versa, so that Colim is left adjoint to Δ. Use A_0-indexed and A_1-indexed products to internalize the construction of limits in Section 4.6 to construct a right adjoint Lim to Δ. Thus every topos $\mathbf{E}$ has a colimit and limit for every diagram on a small category in $\mathbf{E}$. We say that $\mathbf{E}$ is *internally cocomplete* and *internally complete* (compare with Mac Lane (1971, Theorem 5.1) and Johnstone (1977, Ch. 2).

20.19 A *global section* of a diagram S on $\mathbf{A}$ is a global element of S in $\mathbf{E}^{\mathbf{A}}$, a natural transformation $b: \Delta 1 \longrightarrow S$; in other words, a $\mathbf{E}$ arrow $b: A_0 \longrightarrow S$ such that

$$\vdash s(b(o)) = o \qquad o.A_0$$

$$Df = o \vdash \mathrm{ac}_s(f, b(o)) = b(Cf) \qquad f.A_1$$

Show that if $v: S \longrightarrow W$ is a natural transformation and b is a global section of S, then $v \circ b$ is a global section of W.

20.20 Show that there is a functor $\Gamma: \mathbf{E}^{\mathbf{A}} \longrightarrow \mathbf{E}$ that takes each diagram to the object of its global sections, right adjoint to $\Delta: \mathbf{E} \longrightarrow \mathbf{E}^{\mathbf{A}}$, and so there is a geometric morphism $\Delta \dashv \Gamma: \mathbf{E}^{\mathbf{A}} \longrightarrow \mathbf{E}$.

I do not recommend formalizing this too strictly, but the reader may notice that the indexed product $\Pi_{A0} S$ is the object of selections of a value $b(o)$ in each fibre $S(o)$, so that the object of global sections of S is a sub-object of that $\Gamma S \rightarrowtail \Pi_{A0} S$. An $\mathbf{E}$ arrow $f: B \longrightarrow S$ gives an arrow

$$A_0 \times B \xrightarrow{\sim} B \times A_0 \xrightarrow{f \times A_0} (\Gamma S) \times A_0 \rightarrowtail (\Pi_{A0} S) \times A_0 \longrightarrow S$$

Reasoning internally to $\mathbf{E}$, this arrow takes any $\mathbf{A}$ object o and value y in B to $(f(y))(o)$; that is to the value in $S(o)$ of the global section $f(y)$. One can prove that arrow is a natural transformation from ΔB to S. Conversely, any natural transformation $v: \Delta B \longrightarrow S$ determines an arrow $f: B \longrightarrow \Gamma S$ which, put internally, takes each y in B to the section b defined by $b(o) = v(o, y)$.

20.21 This is an exercise on formalizing an abuse of notation in the internal language. Terms such as $f \circ g$, used in the axioms for a small category, are an abuse of notation, since the composition arrow of $\mathbf{A}$ is defined for terms of sort A_2, not $A_1 \times A_1$. Having $Df = Cg$ in the sequent does not eliminate the abuse, since that equation is a presupposition for definedness of $f \circ g$, and not just a condition for the truth of some claim about it.

Call a composition term such as $f \circ g$ a *pseudoterm*. If f and g are not bound variables then we can regard a sequent with $f \circ g$ in it as short for a sequent with a formula $(p_1 c = f \ \ \& \ \ p_2 c = g)$ added to the left-hand side, and every occurrence of $f \circ g$ replaced by $m(c)$, where c is a variable over A_2 not already occurring in the sequent. This must be done inductively, since a pseudoterm $s \circ t$ might have other pseudoterms embedded in s and t. Call the result the *expanded form* of the sequent with pseudoterms, and call a sequent with pseudoterms true iff its expanded form is. The expanded form of the associativity axiom is quite long.

Call $Df = Cg$ the *presupposition* of the pseudoterm $f \circ g$, and call a sequent with pseudoterms *proper* if its left-hand side includes the presupposition of each pseudoterm in the sequent.

Show that topos logic is sound for proper sequents; that is, applying any rule to true proper sequents in such a way as to yield a proper sequent always yields a true one.

Give analogous definitions for the pseudoterms $\mathrm{ac}(f, x)$ used with diagrams on $\mathbf{A}$, with presupposition $Df = o(x)$.

20.22 The cut rule is not sound for arbitrary sequents with pseudoterms because it can eliminate a pseudoterm the presupposition of which is required for the conclusion

For example, for any small category **A** the domain–codomain axioms imply that

$$i_o \circ i_{o'} = i_o \vdash o = o' \qquad o, o' . A_0$$

but the identity axiom implies that $i_o \circ i_{o'} = i_o$ so cut would give

$$\vdash o = o'$$

saying that all objects of **A** are equal. Show that adding the presuppositions turns this into a proof that $o = o'$ implies $o = o'$.

20.23 The reader familiar with Russell's theory of definite descriptions might enjoy proving that our expanded forms of sequents with pseudoterms '$f \circ g$' are equivalent to the result obtained by treating '$f \circ g$' as a definite description 'the h such that there is some c with $p_1 c = f$ and $p_2 c = g$ and $mc = h$', and spelling it out in Russell's way.

21

Topologies

In this chapter is described Lawvere and Tierney's elementary treatment of the topologies on toposes that Grothendieck first described. We use any base topos **E** and, given a topology j on **E** we construct a topos $\mathbf{E}_j$ which is '**E** modified by making j-true formulas actually true'. Grothendieck toposes are described.

21.1 Definition

A *topology* on **E** is an arrow $j:\Omega \longrightarrow \Omega$ such that:

$$j \circ t = t, \qquad j \circ j = j, \qquad j \circ \wedge = \wedge \circ (j \times j)$$

Let $J \rightarrowtail \Omega$ be the sub-object classified by j. Every sub-object $s: S \rightarrowtail A$ of any object A has a *j-closure*, written $\bar{s}: \bar{S} \rightarrowtail A$ and defined as the sub-object classified by $j \circ \chi_s$, where χ_s classifies s. Equivalently, $\bar{s}$ is the pullback of J along χ_s.

The j-closure operator has various nice properties:

THEOREM 21.1 For any topology j, sub-objects s and w of A, and arrow $f: B \longrightarrow A$:

(1) $s \subseteq \bar{s}$;
(2) $\bar{s} \equiv \bar{\bar{s}}$;
(3) $\overline{(s \cap w)} \equiv \bar{s} \cap \bar{w}$;
(4) if $s \subseteq w$ then $\bar{s} \subseteq \bar{w}$;
(5) the j-closure of $f^{-1}(s)$ is $f^{-1}(\bar{s})$.

We say that the j-closure operator is inflationary, idempotent, it preserves intersections, it preserves order, and it is stable under pullback.

PROOF Since $j \circ t = t$ we have $t \subseteq J$ as sub-objects of Ω, and (1) follows since pullback along χ_s preserves order. Clause (2) is immediate from $j \circ j = j$, and (3) from $j \circ \wedge = \wedge \circ (j \times j)$. Since $s \subseteq w$ means the same thing as $s \cap w \equiv s$, clause (3) implies (4). And (5) holds since both sub-objects are classified by $j \circ \chi_s \circ f$. □

A sub-object $s: S \rightarrowtail A$ is called *j-closed* if it is its own j-closure; that is, if $s \equiv \bar{s}$. It is called j-dense if its j-closure is all of A; that is, if $\bar{s} \equiv 1_A$.

Immediately from Theorem 21.1, if s is closed then so is any pullback $f^{-1}(s)$. If s is dense then so is any sub-object that contains s and any pullback $f^{-1}(s)$.

We need two technical lemmas:

LEMMA 21.2 Suppose that s is dense and w closed, and that the following square commutes:

$$\begin{array}{ccc} S & \xrightarrow{\ f\ } & W \\ {\scriptstyle s}\big\downarrow & & \big\downarrow{\scriptstyle w} \\ X & \xrightarrow[\ g\ ]{} & A \end{array}$$

Then there is a unique $u: X \longrightarrow W$ with $g = w \circ u$. It follows that $f = u \circ s$.

PROOF The pullback $g^{-1}(w)$ is closed, and also dense since it contains s; thus it is (up to equivalence) 1_X. So g factors through w, and uniquely since w is monic. To prove that $f = u \circ s$, compose both sides with the monic w. □

LEMMA 21.3 Any sub-object $w: W \rightarrowtail A$ is dense in its own closure.

PROOF The inclusion $W \rightarrowtail \bar{W}$ is the pullback of w along $\bar{w}$, so its closure is the pullback of $\bar{w}$ along $\bar{w}$. But that is $\bar{W}$. □

This use of terms from point-set topology has a history, but on our level it is more misleading than helpful. Lawvere suggests looking at a topology as a modal operator 'it is j-locally the case that'. We may shorten that to 'it is j-true that'. The arrow j appears internally as an operator applying to formulas and giving formulas, such that:

$$\vdash w \longrightarrow jw \qquad w, w' . \Omega$$

$$\vdash jjw \longleftrightarrow jw$$

$$\vdash j(w \,\&\, w') \longleftrightarrow (jw) \,\&\, (jw')$$

For example, take any $u: 1 \longrightarrow \Omega$. It is easy to verify that the arrow $(u \to \_): \Omega \longrightarrow \Omega$ is a topology. To be u-true means to have truth value at least u. This is called an *open topology*, since in the case where **E** is a topos of sheaves on a topological space, $\mathrm{Shv}(T)$ for some topological space T, u will correspond to an open subset U of T and the u-true formulas will be those true over at least U.

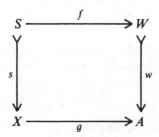

The arrow $(u \vee \_): \Omega \longrightarrow \Omega$ is the *closed topology* for u. A formula is true in this topology if its disjunction with u is true. If **E** is $\mathrm{Shv}(T)$ the formulas true for this topology are those true over at least the closed set the complement of U.

The double negation arrow $\sim\sim : \Omega \longrightarrow \Omega$ is a topology. In some toposes **E**, to be not-not-true means to be 'true so far as global elements in **E** are concerned'. If **E** is Shv(T) it means 'not false on any open set' or, in other words, 'true on some dense open subset of T'.

21.2 Sheaves

For any topos **E** and topology j on it, an **E** object **A** is a *sheaf* for j iff the following holds. For every object X and j-dense sub-object $s: S \rightarrowtail X$ and every $f: S \longrightarrow A$ there is a unique $g: X \longrightarrow A$ with $f = g \circ s$:

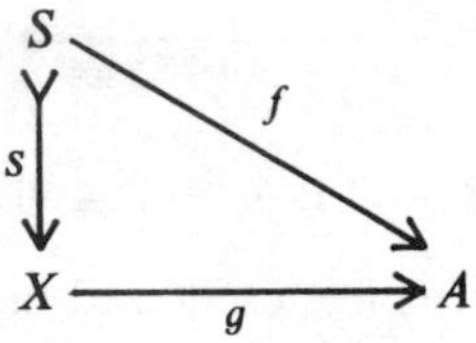

The idea is that, since s is j-dense, it is j-true that s is all of X. A j-sheaf A is an object that thinks s is all of X, so that an arrow from S to A fully determines an arrow from X to A.

Let $\mathbf{E}_j$ be the full subcategory of **E** the objects of which are the j-sheaves, and let $i: \mathbf{E}_j \longrightarrow \mathbf{E}$ be the inclusion functor.

THEOREM 21.4 $\mathbf{E}_j$ has all finite limits, and i preserves them.

PROOF Trivially, 1 is a j-sheaf and is terminal in $\mathbf{E}_j$. Given j-sheaves A and B, form their product $A \times B$ in **E**. For any j-dense $S \rightarrowtail X$, any arrow from S to $A \times B$ corresponds to arrows from S to A and to B, so $A \times B$ is a sheaf and is a product of A and B in $\mathbf{E}_j$. Similarly, the equalizer in **E** of any parallel pair of arrows from a sheaf A to a sheaf B is a sheaf (this uses the sheaf property of A and of B) and in fact is an equalizer in $\mathbf{E}_j$. □

COROLLARY An arrow f between sheaves is monic in $\mathbf{E}_j$ iff it is monic in **E**.

PROOF Clearly, if f is monic in **E** it is in $\mathbf{E}_j$. And since i preserves pullbacks it preserves monics (compare with Exercise 4.9). □

THEOREM 21.5 For any object A, if B is a j-sheaf then so is B^A. It follows that i preserves exponentials.

PROOF Take any j-dense $s: S \rightarrowtail X$ and arrow $f: S \longrightarrow B^A$. Then $s \times A: S \times A \rightarrowtail X \times A$ is j-dense, since it is the pullback of s along $p_1: X \times A \longrightarrow X$. The transpose $\bar{f}: S \times A \longrightarrow B$ determines a unique

$u: X \times A \longrightarrow B$, the transpose of which is the unique arrow showing that B^A is a sheaf. The second claim is left to the reader. □

Define $\Omega_j \rightarrowtail \Omega$ as the equalizer of j and 1_Ω. Since $j \circ t = t$, the arrow t factors through a unique $t: 1 \longrightarrow \Omega_j$. So Ω_j contains the truth values equivalent to their own j-truth, those w with $j(w) = w$. We will prove that Ω_j is a sub-object classifier for $\mathbf{E}_j$. First note that Ω_j classifies closed sub-objects in $\mathbf{E}$. For any monic $s: S \rightarrowtail A$, the classifying arrow χ_s factors through Ω_j iff $j \circ \chi_s = \chi_s$ and thus iff s is closed.

LEMMA 21.6 Let A be a sheaf and take any monic $w: W \rightarrowtail A$. Then W is a sheaf iff w is closed.

PROOF By Lemma 21.3, $i: W \rightarrowtail \bar{W}$ is dense. So if W is a sheaf there is a unique arrow $u: \bar{W} \longrightarrow W$ with $1_W = u \circ i$. But then $w \circ u$ and $\bar{w}$ both compose with i to give w, and since A is a sheaf this shows that $w \circ u = \bar{w}$. So w is equivalent to $\bar{w}$.

Then suppose that w is closed and take any dense $s: S \rightarrowtail X$ and arrow $f: S \longrightarrow W$. Apply Lemma 21.2 to s and w, with g the unique arrow to the sheaf A with $w \circ f = g \circ s$, to obtain some $u: X \longrightarrow W$ with $f = u \circ s$. For any $v: X \longrightarrow W$, if $f = v \circ s$ then g and $w \circ v$ have the same composite with s. So the sheaf property of A implies that $g = w \circ v$, and since w is monic $v = u$. □

THEOREM 21.7 For any monic $s: S \rightarrowtail A$ with S and A both sheaves, there is a unique $u: A \longrightarrow \Omega_j$ such that s is a pullback of t along u.

PROOF Since s is closed, its classifying arrow to Ω factors uniquely through Ω_j. □

Theorem 21.7 will imply that Ω_j is a sub-object classifier for $\mathbf{E}_j$, and so $\mathbf{E}_j$ is a topos, as soon as we show that Ω_j is itself a sheaf.

THEOREM 21.8 Ω_j is a sheaf.

PROOF For any dense $s: S \rightarrowtail X$ and any $f: S \longrightarrow \Omega_j$, let $w: W \rightarrowtail S$ be the sub-object classified by $S \longrightarrow \Omega_j \rightarrowtail \Omega$, so that w is closed. The pullback of $s \circ w$ along s is w, so the pullback of $\overline{s \circ w}$ along s is $\bar{w}$, and that is w. So $j \circ \chi_{(s \circ w)} \circ s$ is the classifying arrow of w. But $j \circ \chi_{(s \circ w)}$ factors through a unique $g: X \longrightarrow \Omega_j$, and this is the unique arrow with $f = g \circ s$. □

21.3 The sheaf reflection

Given a topology j we will construct a left adjoint L to the inclusion $i: \mathbf{E}_j \longrightarrow \mathbf{E}$ and prove that it preserves all finite limits. A left adjoint to the inclusion of a full subcategory is often called a *reflection* functor. It 'reflects' each object into the subcategory. We call L the sheaf reflection functor.

Separated objects have half of the sheaf property and make a good halfway point in constructing L. An object A is *separated* (for the topology j) iff: for every object X and j-dense sub-object $s: S \rightarrowtail X$, if arrows g and g' from X to A have $g \circ s = g' \circ s$ then $g = g'$. This is what you obtain by saying 'at most one arrow g' rather than 'a unique arrow g' in the definition of a sheaf. In particular, every sheaf for j is j-separated.

Let $\mathbf{E}_{\text{sep},\, j}$ be the full subcategory of $\mathbf{E}$ with the j separated objects. We often omit j in the subscript.

THEOREM 21.9 The following are equivalent:
(i) The object A is separated.
(ii) The equalizer of any parallel pair of arrows with codomain A is closed.
(iii) The diagonal $A \rightarrowtail A \times A$ is closed.
(iv) This sequent is true:

$$j(x = x') \vdash x = x' \qquad x, x' . A$$

PROOF Suppose that A is separated and take any $f, g: B \longrightarrow A$ and the equalizer $e: E \rightarrowtail B$, with its closure $\bar{e}: \bar{E} \rightarrowtail B$. The inclusion $s: E \rightarrowtail \bar{E}$ of E into its closure is dense, and we have

$$f \circ \bar{e} \circ s = f \circ e = g \circ e = g \circ \bar{e} \circ s$$

Since A is separated this implies that $f \circ \bar{e} = g \circ \bar{e}$. Thus $\bar{e}$ factors through e, proving that $e \equiv \bar{e}$. Clause (iii) follows from (ii), since the diagonal is the equalizer of the two projections to A. Clauses (iii) and (iv) are equivalent since the extension of $j(x = x')$ is the closure of the diagonal. Finally, assume that the diagonal is closed and consider any dense $s: S \rightarrowtail X$ and $f: S \longrightarrow A$. If $h \circ s = f$ and $k \circ s = f$, then $\langle h, k \rangle \circ s: S \longrightarrow A \times A$ factors through the diagonal, and so Lemma 21.2 shows that $\langle h, k \rangle$ already factors through it. But that means that $h = k$. So A is separated. □

In constructing the sheaf reflection functor we freely take the kernel pair of an arrow $f: A \longrightarrow B$ as a pair of arrows, h and k, to A or a sub-object $\langle h, k \rangle: P \rightarrowtail A \times A$, whichever is convenient. We slightly abuse our notation, using j for the image factorization

$$\Omega \xrightarrow{\;j\;} \Omega_j \rightarrowtail \Omega$$

of the topology $j: \Omega \longrightarrow \Omega$ (see Exercise 21.6). For each object A, our construction of $L(A)$ centres on the arrow

$$A \overset{\{\,\}}{\rightarrowtail} \Omega^A \xrightarrow{\;j^A\;} \Omega_j^A$$

The idea is that this composite coequalizes each pair of values in A that are j-equal, yielding a separated object MA embedded in the sheaf Ω_j^A. The closure of MA is the sheaf $L(A)$.

Lemma 21.10 The kernel pair of $j^A \circ \{\ \}$ is closure of the diagonal of $\mathbf{A}$.

Proof The kernel pair is the extension of

$$\{y.A \mid j(y = x)\} = \{y.A \mid j(y = x')\} \qquad x, x'.A$$

This formula is equivalent to $j(x = x')$, so its extension is the closure of the diagonal. □

Theorem 21.11 For any object A, let $q: A \twoheadrightarrow MA$ be the image of the arrow $j^A \circ \{\ \}$. Then MA is separated, and for every separated object C and arrow $g: A \longrightarrow C$ there is a unique $u: MA \longrightarrow C$ such that $g = u \circ q$.

Proof MA is separated since it is a sub-object of the sheaf Ω_j^A (see Theorems 21.5 and 21.8, and Exercise 21.5). Take any g from A to a separated object C and let $\langle h, k\rangle$ be its kernel pair. Every kernel pair includes the diagonal, and $\langle h, k\rangle$ is closed by Theorem 21.10 since it is the equalizer of $g \circ p_1: A \times A \longrightarrow C$ and $g \circ p_2$. So the closure of the diagonal is included in $\langle h, k\rangle$, and Exercise 21.7 completes the proof. □

By Theorem 10.3 there is a unique functor $M: \mathbf{E} \longrightarrow \mathbf{E}_{\text{sep}}$ left adjoint to the inclusion of $\mathbf{E}_{\text{sep}}$ into $\mathbf{E}$, with each arrow $q_A: A \longrightarrow MA$ universal to M. Each Mf is the unique arrow that makes the following square commute:

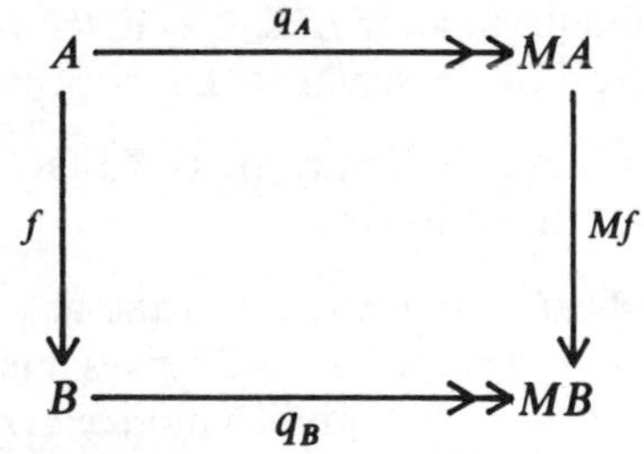

Theorem 21.12 M preserves monics. What is more, for any arrow $f: A \longrightarrow B$ we have Mf monic iff

$$fx = fx' \vdash j(x = x') \qquad x, x'.A$$

Proof The second claim implies the first, since the sequent holds for any monic f. So suppose the sequent holds. The following is sound in the internal language. Take any z and z' in MA such that $(Mf)(z) = (Mf)(z')$. Since q_A is epic there are some x and x' in A with $q_A(x) = z$ and $q_A(x') = z'$. And since $(Mf) \circ q_A = q_B \circ f$ it follows that $q_B(fx) = q_B(fx')$. But q_B is the quotient of the kernel pair of $j^B \circ \{\ \}$, so the last equation means that

$$\{w.B \mid j(w = fx)\} = \{w.B \mid j(w = fx')\}$$

and that is equivalent to

$$j(fx = fx')$$

By the assumption on f plus $jj = j$ this is equivalent to

$$j(x = x')$$

and so to

$$\{y.A \mid j(y = x)\} = \{y.A \mid j(y = x')\}$$

which proves that $z = q_A(x) = q_A(x') = z'$, so Mf is monic. The converse, if Mf is monic then the sequent in the statement is true, is left to the reader. □

THEOREM 21.13 For any object A, let LA be the closure of MA in Ω_j^A. So $s: MA \rightarrowtail LA$ is dense. Let $\eta: A \longrightarrow LA$ be $s \circ q_A$. Then LA is a sheaf and for every sheaf C and arrow $g: A \longrightarrow C$ there is a unique $v: LA \longrightarrow C$ with $g = v \circ \eta$.

PROOF LA is closed in the sheaf Ω_j^A and so, by Lemma 21.6, a sheaf. Consider any C and g as in the theorem. Since C is separated there is a unique $u: MA \longrightarrow C$ with $g = u \circ q_A$. Since s is dense and C is a sheaf there is a unique $v: LA \longrightarrow C$ with $u = v \circ s$. Thus $g = v \circ \eta$. Uniqueness is left to the reader. □

Therefore there is a unique functor $L: \mathbf{E} \longrightarrow \mathbf{E}_j$ left adjoint to the inclusion $i: \mathbf{E}_j \longrightarrow \mathbf{E}$ with the arrows η universal to L.

THEOREM 21.14 L preserves finite products. In fact, for any A and B the arrow $\langle \eta_A, \eta_B \rangle$ is universal to sheaves.

PROOF Since 1 is a sheaf, L preserves it. Take any sheaf C and arrow f: $A \times B \longrightarrow C$. Since C^B is a sheaf, $\bar{f}: A \longrightarrow C^B$ gives a unique $u: L(A) \longrightarrow C^B$. The double transpose $\bar{\bar{u}}: B \longrightarrow C^{L(A)}$ gives a unique $v: L(B) \longrightarrow C^{L(A)}$. Calculation shows that the twisted transpose of v, an arrow from $L(A) \times L(B)$ to C, is the unique arrow the composite of which with $\langle \eta_A, \eta_B \rangle$ is f. □

THEOREM 21.15 L preserves monics. In fact, Lf is monic iff Mf is.

PROOF The second claim implies the first, since M preserves monics. So suppose that Mf is monic. Lf is the unique arrow that makes the following diagram commute:

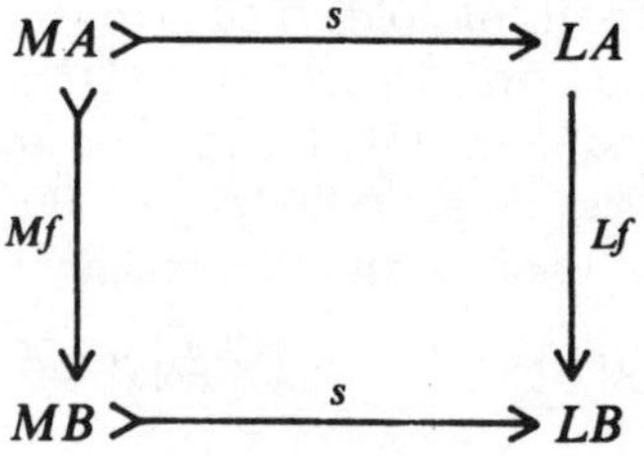

Consider any arrows $h, k: T \longrightarrow LA$ with $(Lf)\circ h = (Lf)\circ k$. The pullbacks of h and k along s are both dense sub-objects of T. So their intersection $i: I \rightarrowtail T$ is also dense, by Theorem 21.1. Let $h', k': I \longrightarrow MA$ be the restrictions to I of h and k. Chasing through the diagram shows that $s\circ(Mf)\circ h' = s\circ(Mf)\circ k'$ and since s and (Mf) are monic we have $h' = k'$. Thus

$$i\circ h = h'\circ s = k'\circ s = i\circ k$$

Since I is dense and LA is a sheaf, $h = k$. Conversely, if Lf is monic so is $(Lf)\circ s$, and thus so is Mf. □

THEOREM 21.16 A monic $s: S \rightarrowtail A$ is dense iff Ls is iso.

PROOF If s is dense then since LS is a sheaf there is an arrow $f: A \longrightarrow LS$ with $f\circ s = \eta_S$. Since η_A is universal to sheaves the corresponding $u: LA \longrightarrow LS$ is easily shown to be right inverse to the monic Ls. So Ls is iso.

Conversely, suppose that Ls is iso. Then η_S factors as

$$S \rightarrowtail A \longrightarrow LA \xrightarrow{\sim} LS$$

so for every sheaf B every arrow $f: S \longrightarrow B$ factors uniquely through A. Consider the following diagram:

$$S \rightarrowtail A \underset{\chi_s}{\overset{t}{\rightrightarrows}} \Omega \xrightarrow{j} \Omega_j$$

As S is the equalizer of the middle pair, so its closure is the equalizer of the two composites to Ω_j. But those each have the same composite with s and Ω_j is a sheaf. So $j\circ t = j\circ \chi_s$ and thus the equalizer is A. So s is dense. □

An arrow $f: A \longrightarrow B$ is called *almost monic* (for j) if

$$fx = fx' \quad \vdash j(x = x')$$

Theorems 21.12 and 21.15 show that f is almost monic iff Lf is monic. The arrow is called *almost epic* iff its image is j-dense in B or, equivalently, iff

$$\vdash j((\exists x\,.\,A)\, fx = y) \qquad y\,.\,B$$

And f is called *j-bidense* iff it is almost monic and almost epic.

THEOREM 21.17 For any arrow $f: A \longrightarrow B$, the arrow Lf is epic iff f is almost epic. So Lf is iso iff f is bidense.

PROOF Consider the image factorization $A \twoheadrightarrow I \rightarrowtail B$ of f in $\mathbf{E}$. L preserves monics, and any left adjoint preserves epics, so LI is an image factorization of Lf in $\mathbf{E}_j$. Thus Lf is epic iff $Li: LI \rightarrowtail LB$ is epic, and since Li is monic that means iff it is iso. Theorem 21.16 shows that this happens iff the image I is dense in B. Again, since monic–epics in a topos are iso, Lf is iso iff f is bidense. □

COROLLARY Each universal arrow η_A is bidense, since $LA \longrightarrow LLA$ is iso. □

Our proof that L preserves equalizers follows Veit (1981).

LEMMA 21.18 If B is a sheaf and $f: X \longrightarrow B$ is almost monic, then LX is iso to the closure of the image $I \rightarrowtail B$ of f.

PROOF If f is almost monic then (up to the iso $B \cong LB$) we have a monic $Lf: LX \rightarrowtail B$. Since $\bar{I}$ is a sheaf (being closed in the sheaf B) there is an arrow from LX to $\bar{I}$ and, in fact, a monic showing that $LX \subseteq \bar{I}$ as sub-objects of B. On the other hand, f factors through LX so $I \subseteq LX$, and since LX is closed in B this implies that $\bar{I} \subseteq LX$. Thus $\bar{I}$ and LX are equivalent sub-objects of B. □

THEOREM 21.19 L preserves equalizers.

PROOF Suppose that the top row here is an equalizer:

$$\begin{array}{ccccc} E & \overset{e}{\rightarrowtail} & A & \underset{g}{\overset{f}{\rightrightarrows}} & B \\ \downarrow & & \downarrow & & \downarrow \\ LE & \overset{Le}{\rightarrowtail} & LA & \underset{Lg}{\overset{Lf}{\rightrightarrows}} & LB \end{array}$$

Since $(Lf) \circ (Le) = (Lg) \circ (Le)$, the equalizer of Lf and Lg includes LE. Also note that $\eta_A \circ e$ is almost monic, so LE is equivalent as a sub-object of LA to the closure of the image of E in LA. Therefore we are done once we show that the equalizer of Lf and Lg is included in that image. Reasoning internally to $\mathbf{E}$, let x be any value in LA with $(Lf)(x) = (Lg)(x)$. As η_A is bidense we have

$$j((\exists y \,.\, A)\,\eta_A(y) = x)$$

But $\eta_A(y) = x$ implies that

$$(Lf)(\eta_A(y)) = (Lg)(\eta_A(y))$$

and then by naturality of η we obtain

$$\eta_B(f(y)) = \eta_B(g(y))$$

and since η_B is almost monic that gives

$$j(f(y) = g(y))$$

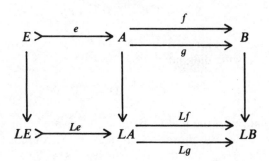

which implies that

$$j((\exists w\,.\,E)e(w) = y)$$

Summing up, we obtain

$$j((\exists w\,.\,E)\,\eta_A(e(w)) = x)$$

which says that x is in the closure of the image. (To be quite precise, summing up gives $j(\exists y\,.\,A)j(\exists w\,.\,E)(\eta_A(e(w) = x)$. But we drop the quantifier over y because y does not appear in the formula, and we use $jj = j$.) □

Altogether, $L \dashv i$ is a geometric morphism from $\mathbf{E}_j$ to $\mathbf{E}$.

21.4 Grothendieck toposes

The real strength of topologies appears in connection with diagram categories, as in Chapter 20. Algebraic geometers often work with diagrams on the dual to a given small category. A diagram on $\mathbf{A}^{\mathrm{op}}$ is called a *presheaf* on $\mathbf{A}$, and the category $\mathbf{E}^{\mathbf{A}^{\mathrm{op}}}$ is the presheaf category on $\mathbf{A}$. This dualizing is a minor detail, but coherence with the literature requires it.

Given a small category $\mathbf{A}$, a *sieve* on an $\mathbf{A}$ object o is a collection of $\mathbf{A}$ arrows with co-domain o, such that if g is in the collection and $Cf = Dg$ then $g \circ f$ is also in it. These are the duals to co-sieves, defined in Section 20.4. A *Grothendieck topology* on $\mathbf{A}$ assigns to each $\mathbf{A}$ object o a collection of sieves on it, called the *covering sieves* on o for that topology. The covering sieves must meet certain conditions, such as that the maximal sieve on an object is always covering, but we will not use these conditions. Makkai and Reyes (1977) work with them at length, and Artin *et al.* (1972) at greater length. Most books on toposes list them. Mumford and Tate (1978) discuss Grothendieck's use of the idea in algebraic geometry.

As a collection of sieves, a Grothendieck topology is contained in the subobject classifier $\Omega_{\mathbf{A}^{\mathrm{op}}}$ for the presheaf category $\mathbf{E}^{\mathbf{A}^{\mathrm{op}}}$. In fact, the conditions on a Grothendieck topolgy are equivalent to saying that it is a sub-presheaf $J \rightarrowtail \Omega_{\mathbf{A}^{\mathrm{op}}}$, the classifying arrow j of which is a topology on the topos $\mathbf{E}^{\mathbf{A}^{\mathrm{op}}}$ in the sense we have been using. We may call the Grothendieck topology J and use j for the corresponding topology on $\mathbf{E}^{\mathbf{A}^{\mathrm{op}}}$, but the two determine each other interchangeably.

A *site* in a topos $\mathbf{E}$ is a small category $\mathbf{A}$ in $\mathbf{E}$ plus a Grothendieck topology J on it. The *Grothendieck topos* over that site is the category of sheaves for j on $\mathbf{E}^{\mathbf{A}^{\mathrm{op}}}$. Call this topos $\tilde{\mathbf{A}}$. This is a very flexible device for constructing toposes over $\mathbf{E}$. We briefly discuss one kind of example in Chapter 24, namely models for synthetic differential geometry defined over a topos of sets.

The notion of covering in a Grothendieck topology is tied to the notion in Chapter 18 in that each J-covering of an object o in $\mathbf{A}$ gives an epic to $L(R_o)$, the 'sheafification' of the representable presheaf on o in $\tilde{\mathbf{A}}$ (that is, R_o is the representable functor on o in $\mathbf{A}^{op}$). All of the terms of external semantics have special forms for the case of Grothendieck toposes, and 'external semantics' is sometimes used just to mean this case (see Exercise 21.11).

Exercises

21.1 Show that for any $u: 1 \longrightarrow \Omega$, the arrow $(u \to \_): \Omega \longrightarrow \Omega$ is a topology by proving these sequents:

$$w \vdash u \longrightarrow w \qquad w, w'.\Omega$$

$$u \longrightarrow (u \longrightarrow w) \vdash u \longrightarrow w$$

$$u \longrightarrow (w \,\&\, w') \dashv\vdash (u \longrightarrow w) \,\&\, (u \to w')$$

Similarly, verify that $(u \vee \_)$ and $\sim\sim$ are topologies.

21.2 Verify the following. The identity arrow $\Omega \longrightarrow \Omega$ is a topology , and every $\mathbf{E}$ object is a sheaf for it. The constant arrow true, that is $t \circ !: \Omega \longrightarrow 1 \longrightarrow \Omega$, is a topology, called the *trivial topology*. Every sub-object is dense for the trivial topology, and only terminal objects are sheaves. The identity topology is the open topology for t, trivial is the closed topology for t.

21.3 Show that a sub-object is dense for the double negation topology iff its negation is empty. Show that $\binom{t}{fa}: 1 + 1 \rightarrowtail \Omega$ is dense for this topology. [Hint: prove that $\sim\sim(t \vee fa)$ in topos logic.] Conclude that $1 + 1$ is dense in $\Omega_{\sim\sim}$ and so $\mathbf{E}_{\sim\sim}$ is Boolean.

21.4 Take any $u: 1 \longrightarrow \Omega$ and the $U \rightarrowtail 1$ that it classifies. Let j_u^o be the open topology for u. Show the j_u^o-closure of any $s: S \rightarrowtail A$ is the material conditional $A \times U \Rightarrow S$ and so S is dense iff S contains $p_1: A \times U \rightarrowtail A$. So p_1 is the minimal j_u^o-cover of A. Show that an object B is a j_u^o-sheaf iff it is iso to B^U. [Hint: To say that B is a sheaf means that an arrow from the minimal cover $A \times U$ to B determines a unique arrow from A to B.]

Let $\mathbf{E}_u$ be the topos of sheaves. Show that $U^*: \mathbf{E} \longrightarrow \mathbf{E}/U$ restricted to $\mathbf{E}_u$ is full and faithful. Conclude that there is an adjoint equivalence $U^* \dashv \_^U: \mathbf{E}_u \longrightarrow \mathbf{E}/U$. One often says that (up to this adjoint equivalence) the slice $\mathbf{E}/U$ is the category of sheaves. Thus the sheaf reflection is a logical functor.

21.5 Show that every sub-object of a separated is separated. Show that $\mathbf{E}_{sep}$ has all finite limits, and that the inclusion into $\mathbf{E}$ preserves them. Also, if B is separated then for any object A the exponential B^A is separated. [Hint: follow the proofs of Theorems 21.4 and 21.5.]

21.6 In any category, an arrow $f: A \longrightarrow A$ with $f \circ f f$ is called *idempotent*. Show that if $e: \mathbf{E} \rightarrowtail A$ is an equalizer for an idempotent f and the identity 1_A, then e is split monic, so f has an image factorization $A \twoheadrightarrow E \rightarrowtail A$.

21.7 In any topos, suppose that $\langle h, k \rangle$ is the kernel pair of an epic $f: A \twoheadrightarrow B$, and that $\langle h', k' \rangle$ is the kernel pair of $g: A \longrightarrow C$. Show that $\langle h, k \rangle \subseteq \langle h', k' \rangle$ as sub-object of $A \times A$ iff g factors through f. [Hint: recall that in a topos every epic is coequalizer to its kernel pair.]

21.8 The sheaf reflection L for j is not logical unless j is an open topology. But $\mathbf{E}_j$ is always '$\mathbf{E}$ modified by making j-true formulas actually true' in the following two senses. The extension of a formula φ is taken by L to an identity iff $j\varphi$ is true in $\mathbf{E}$. Objects A and B of $\mathbf{E}$ become isomorphic in $\mathbf{E}_j$ iff it is j-true in $\mathbf{E}$ that they are iso. Expand on these.

21.9 A topology j is called *smaller* than j' iff

$$\vdash j(w) \longrightarrow j'(w) \qquad w.\Omega$$

Show that j is smaller than j' iff $J \subseteq J'$ as sub-objects of Ω, and so iff every j-cover is a j'-cover, and so iff every j'-sheaf is a j-sheaf.

21.10 Define the object of topologies $\mathrm{Top} \rightarrowtail \Omega^{\Omega}$ as the extension

$$[j.\Omega^{\Omega} \mid j(t) = t \quad \& \quad (\forall w.\Omega)(w \to j(w)) \quad \& \quad (\forall w, w'.\Omega)(j(w \,\&\, w') \leftrightarrow j(w) \,\&\, j(w'))]$$

Show that the transpose of any global element of Top is a topology.

Given any monic $F \rightarrowtail \Omega$, show that there is a smallest topology that takes everything in F to true. In fact, it is the intersection of all topologies with that property, so it is the classifier for the extension

$$[w.\Omega \mid (\forall j.\mathrm{Top})((\forall w'.\Omega)(w' \in F \longrightarrow j(w') = t) \longrightarrow j(w) = t]$$

Show that the classifier is indeed a topology.

For any arrow $f: A \longrightarrow B$, let F be the image of the classifying arrow for the extension $[y.B \mid (\exists x.A) fx = y] \rightarrowtail B$. Show that Lf is epic iff $F \subseteq J$. Conclude that there is a smallest topology that makes f epic.

For any arrow $f: A \longrightarrow B$, let $K \rightarrowtail A \times A$ be the kernel pair. Show that f is monic iff the inclusion of the diagonal $\Delta \rightarrowtail K$ is epic. Conclude that there is a smallest topology that makes f monic, and thus also a smallest one that makes f iso.

The smallest topology for a given condition is often called the *forcing* topology for the condition. (The term has its own plausibility and it is related to the set theorist's term, in that a classical set-theoretic forcing construction turns into a Boolean-valued model by passing to $\sim\sim$-sheaves.) But it could be trivial, and so collapse the topos to a trivial topos. The only topologies on a topos of sets are the identity and the trivial. So, forcing anything in **Set** that is not already true collapses **Set** to triviality.

21.11 Use the Yoneda lemma to prove that in a presheaf topos $\mathbf{E}^{\mathbf{A}^{\mathrm{op}}}$ the representables R_o form a class of generators. (If natural transformations have the same composites with all arrows from representables, they are equal.) Let $\varphi \longrightarrow \psi$ be any conditional formula in $\mathbf{E}^{\mathbf{A}^{\mathrm{op}}}$, the only free variable of which is x of sort S. Show that clause (iii) of Theorem 18.1 is equivalent to saying: a member c of $S(o)$ satisfies $\varphi \longrightarrow \psi$ iff, for every $f: o \longrightarrow o'$, if $Sf(c)$ satisfies φ then it satisfies ψ. [Hint: take representables as stages of definition, and remember to switch from $\mathbf{A}$ and $\mathbf{A}^{\mathrm{op}}$.] The reader familiar with Kripke models can compare this to them. External semantics for Grothendieck toposes is discussed in Kock (1981, Ch. II.8), for example.

Part IV

SOME TOPOSES

22

Sets

The distinguishing feature of sets is that 'a set is determined by its elements', which we express categorically by an axiom that says that every $f: A \longrightarrow B$ is fully determined by its effect on global elements of A. A topos satisfying this axiom is called well-pointed, and we will refer to its objects as sets and its arrows as functions.

Topos theorists sometimes use 'set' for an object in any topos (especially the effective topos). Lawvere calls the objects of any topos 'variable sets' and those of a well-pointed topos 'constant sets', the limiting case of variation. One way in which to see the sense of this is to look at sheaves on a topological space T as sets varying over T. If T is a one-point space the sheaves are constant sets. See Lawvere (1975, 1976) and elsewhere for more on this view. In these terms, in Section 22.1 constant sets are described, and in Section 22.2 some variable sets are constructed from them.

22.1 Axioms

We assume a topos **Set** satisfying the following axioms:

(SET_1) Well-pointedness, or 1 generates: for any parallel pair of arrows $f, g: A \longrightarrow B$ either $f = g$ or there is some $c: 1 \longrightarrow A$ with $f \circ c \neq g \circ c$.

(SET_2) Non-triviality: 1 is not iso to $\emptyset$.

(SET_3) Infinity: there is a natural number object N.

We show that **Set** has some familiar properties and that its internal logic matches its external description by global elements. No theorems in this chapter use the axiom of infinity, but applications here and later do.

THEOREM 22.1

1. Every set is either initial or has global elements.
2. The terminator 1 has (up to equivalence) only the two subsets $\emptyset \rightarrowtail 1$ and $1 \rightarrowtail 1$.
3. **Set** is Boolean.

PROOF Any set A has subsets $\emptyset$ and 1_A. Either the classifying functions are equal and $\emptyset \cong A$, or some $c: 1 \longrightarrow A$ distinguishes them. Clause (2) follows, since for any $u: U \rightarrowtail 1$ if there is any $c: 1 \longrightarrow U$ then $u = c^{-1}$. It follows that $t: 1 \longrightarrow \Omega$ and $fa: 1 \longrightarrow \Omega$ are the only global elements of Ω. In any topos $\sim\sim t = t$ and $\sim\sim fa = fa$, but 1 generates in **Set** so $\sim\sim$ is 1_Ω. Therefore **Set** is Boolean, by Theorem 17.8. □

Any topos in which 1 has (up to equivalence) only the subobjects $\emptyset$ and 1, or in other words the sub-object classifier has only t and fa as global elements, is called *two-valued*. The example of $\mathbf{Set}^\mathrm{M}$ in Section 22.2 shows that a two-valued topos need not be Boolean. In **Set** we write $1 + 1$ for the subset classifier. In the rest of this chapter we use 'elements' to mean 'global elements' and we write either $x \in A$ or $x: 1 \longrightarrow A$ to say that x is an element of A. We will not use other generalized elements in **Set**.

COROLLARY Elements of a set are equal or disjoint. That is, given $c: 1 \longrightarrow A$ and $d: 1 \longrightarrow A$, either $c = d$ or the intersection $c \cap d$ is $\emptyset$. □

COROLLARY If monics $s: S \rightarrowtail A$ and $w: W \rightarrowtail A$ have the same members they are equivalent subsets of A.

PROOF Suppose that every $c: 1 \longrightarrow A$ is in s iff it is in w. Then the classifying functions of s and w agree on all elements of A and thus are equal. □

THEOREM 22.2 A function $f: B \longrightarrow C$ is monic iff, for all elements b and b' of B, $fb = fb'$ implies $b = b'$. It is epic iff for every element c of C there is some element b of B with $fb = c$.

PROOF The first 'only if' holds in any category. Conversely, suppose that $fb = fb'$ implies $b = b'$ for all elements b and b' of B, and take any $h, k: D \longrightarrow B$ with $f \circ h = f \circ k$. For every element d of D we have $f \circ h \circ d = f \circ k \circ d$, which implies $h \circ d = k \circ d$, and so $h = k$, which proves that f is monic.

Suppose that f is epic. For any $c: 1 \longrightarrow C$, the pullback $S \longrightarrow 1$ of f along c is epic, so S is not initial and any element y of S has $f(s \circ y) = c$. Conversely, suppose that every element of C factors through f and suppose that $h, k: C \longrightarrow D$ have $h \circ f = k \circ f$. Then h and k agree on all elements of C, and so $h = k$. □

THEOREM 22.3 Let φ and ψ be closed formulas in **Set** and let θ be a formula with at most one free variable x, and that of sort A. Then we have:

$\vdash \sim\varphi$ iff not $\vdash \varphi$

$\vdash \varphi \,\&\, \psi$ iff both $\vdash \varphi$ and $\vdash \psi$

$\vdash \varphi \vee \psi$ iff either $\vdash \varphi$ or $\vdash \psi$

$\vdash \varphi \longrightarrow \psi$ iff either not $\vdash \varphi$ or $\vdash \psi$

$\vdash (\exists x\,.\,A)\theta$ iff there is some $c \in A$ such that $\vdash \theta(x/c)$

$\vdash (\forall x\,.\,A)\theta$ iff for every $c \in A \vdash \theta(x/c)$

PROOF The first four hold since a closed formula in any two-valued topos has extension $\emptyset \rightarrowtail 1$ or $1 \rightarrowtail 1$. The fifth follows since there is an epic from $[x\,.\,A|\theta]$ to the extension of $(\exists x\,.\,A)\theta$, and so either both are initial or both have elements. The sixth follows from the fifth since **Set** is Boolean and $(\forall x\,.\,A)\theta$ is equivalent to $\sim(\exists x\,.\,A) \sim \theta$. □

In short, there is no need to distinguish between internal and external descriptions in **Set**. Notably, the internal quantifiers over a set A can be read as ranging over global elements of A. Thus the classical methods of set theory apply directly (with bounded comprehension; see the comparison with membership-based set theory in Section 22.3 below). We have a natural number object, and since **Set** is Boolean the Dedekind reals will agree with the Cauchy reals, and all the usual results of analysis that do not use the axiom of choice will follow. Adding the axiom of choice as given in Chapter 17, saying that all epics split, gives all of classical analysis. We will not go into that: our example of working methods in **Set** will be to describe some of the toposes $\mathbf{Set}^{\mathbf{A}}$ for small categories **A** in **Set**. (Another good example is the proof of the independence of the continuum hypothesis from the axioms SET_{1-3} plus the axiom of choice in Tierney (1972).)

22.2 Diagram categories over Set

We summarize Chapter 20 as applied to **Set**. We can describe a small category **A** in **Set** by its objects o and arrows f, the (global) elements of A_0 and A_1. We write $h: o \longrightarrow o'$ to say that h is an **A** arrow with $Dh = o$ and $Ch = o'$. We call a functor F from **A** to **Set** a diagram on **A**. It assigns a set Fo to each **A** object o, and a function $Fh: Fo \longrightarrow Fo'$ to each **A** arrow h, preserving identities and composition. The category $\mathbf{Set}^{\mathbf{A}}$ of all diagrams on **A** and natural transformations between them is a topos.

A natural transformation $v: F \longrightarrow G$ has component functions $v_o: F(o) \longrightarrow G(o)$, and is monic iff every component is. The terminator of $\mathbf{Set}^{\mathbf{A}}$ is $\Delta 1$, the diagram assigning 1 to every object of **A** and consequently assigning the unique $1 \longrightarrow 1$ to every **A** arrow. Given diagrams F and G on **A**, the product $F \times G$ in $\mathbf{Set}^{\mathbf{A}}$ has $(F \times G)(o) = F(o) \times G(o)$ and $(F \times G)(h) = F(h) \times G(h)$ for every object and arrow of **A**. All limits and co-limits in $\mathbf{Set}^{\mathbf{A}}$ are calculated in this way.

Define the small category **2** as in Chapter 1, with two objects 0 and 1, and one arrow $\alpha: 0 \longrightarrow 1$ in addition to the identities. A diagram F on **2** amounts to two sets $F0$ and $F1$ and a function $F\alpha: F0 \longrightarrow F1$. The functions $F(id_0)$ and $F(id_1)$ have to be 1_{F0} and 1_{F1}. A global section of F, that is a natural transformation $\sigma: \Delta 1 \longrightarrow F$, is (uniquely determined by) any element x of $F0$ since the corresponding element of $F1$ has to be $F\alpha(x)$.

A subdiagram $S \rightarrowtail F$ consists of subsets $S0 \rightarrowtail F0$ and $S1 \rightarrowtail F1$ such that for any element x of $F0$, if x is a member of $S0$ then $F\alpha(x)$ is a member of $S1$. Of course, x might not be in $S0$ while $F\alpha(x)$ is in $S1$, or both might be out of S:

$$\begin{array}{ccc} S0 & \overset{v_0}{\rightarrowtail} & F0 \\ {\scriptstyle S\alpha}\big\downarrow & & \big\downarrow{\scriptstyle F\alpha} \\ S1 & \underset{v_1}{\rightarrowtail} & F1 \end{array}$$

The negation of S, the largest subdiagram disjoint from S, is a subdiagram $\sim s: \sim S \rightarrowtail F$ such that $\sim S0$ contains all elements x of $F0$ such that $F\alpha(x)$ is not in $S1$, and $\sim S1$ contains all the elements of $F1$ not in $S1$. Thus, if $F0$ has any elements x not in $S0$ but with $F\alpha(x)$ in $S1$, then they are omitted from $\sim S$ as well, so $\mathbf{Set}^2$ is not Boolean.

It is useful to think of a diagram F in $\mathbf{Set}^2$ as a 'set with one step of development'. Think of $F0$ as 'F now' and $F1$ as 'F later', so that any element x of F now becomes the element $F\alpha(x)$ of F later. In these terms two elements now may become one later (if $F\alpha$ is not monic), or an element may first arise later (if $F\alpha$ is not epic), but an element now cannot split into two later or disappear. If an element of F is in $S \rightarrowtail F$ now, it is also later, but not conversely.

For another example, consider a small category $\mathbf{M}$ with one object 0 and two arrows, the identity i and an arrow m with $m \circ m = m$. A diagram F on $\mathbf{M}$ is a set $F0$ and an idempotent function $Fm: F0 \longrightarrow F0$. The *fixed points* of F are those x in $F0$ with $Fm(x) = x$. Since Fm is idempotent every $Fm(x)$ is fixed. Since $\Delta 1$ in $\mathbf{Set}^{\mathbf{M}}$ just consists of one fixed point, a global element of F is (uniquely determined by) any fixed point of F. Thus supports split in $\mathbf{Set}^{\mathbf{M}}$.

One can think of a diagram on $\mathbf{M}$ as a 'set with one step of resolution'. Each x in $F0$ resolves itself into a fixed point $Fm(x)$. A subdiagram $S \rightarrowtail F$ consists of a subset $S0 \rightarrowtail F0$ such that every member of $S0$ resolves itself (under Fm) to a fixed point also in $S0$. The negation of S includes all of the elements of $F0$ that resolve to fixed points outside $S0$. An element x of $F0$ might be outside $S0$ but resolve into it. Then x is in neither S nor $\sim S$. So $\mathbf{Set}^{\mathbf{M}}$ is two-valued but not Boolean.

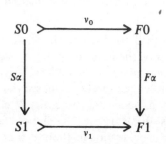

22.3 Membership-based set theory

The central difference between sets in a well-pointed topos and sets approached via membership as in Zermelo–Frankel (ZF) and other set theories is that membership is *local* in the topos, while it is *global* in membership-based theories. That is, in **Set** (or any topos) we can take a set (object) A and ask whether a given element of A is a member of a given subset of A, but it is pointless to ask whether an element of A is also an element of some other set B. An element of A is an arrow with codomain A, and it certainly does not also have another codomain B. In **Set** membership is a relation between elements and subsets of one given set.

In ZF and other membership-based set theories, elementhood and membership collapse into a single relation that holds between sets, and that is global. The elements of a set are sets themselves and it always makes sense to whether one set is a member of another. Every set is a member of a proper class of others.

The sets of a well-pointed topos are cardinals in Cantor's sense. For Cantor, a cardinal is 'a definite set composed out of mere ones (*lauter Einsen*)' (Cantor 1895, p. 482). Similarly, a set in **Set** has a definite multiplicity of (global) elements, distinct from one another but with no distinguishing features. Distinct elements of a ZF set, on the contrary, are distinguished in having different elements themselves.

The local character of membership in **Set** appears formally in the fact that, internally, **Set** has only bounded quantifiers. This entails some trivial differences from membership-based set theories. For example, the ZF axioms say that any sets x and y have a pair set $\{x, y\}$. The **Set** axioms only imply that any elements x and y of a given set A have a pair subset $\{x, y\}$ of A. Chapter 14 showed that the empty set, extensionality, pair set, union, and power set axioms of ZF are sound in bounded quantifier form in any topos, and in the case of **Set** we also have Boolean logic, and we can interpret the quantifiers as ranging over just the global elements. For these axioms the bounds on quantifiers make no serious difference in practice.

The bounded quantifiers do make a considerable difference between **Set** and ZF, which the cognoscenti will recognise in the finite axiom system for **Set**. These axioms do not have the strength of the separation axiom scheme, much less replacement.

The axiom scheme of separation can be stated for **Set** as follows. For every property P of functions with domain 1, and every set A, there exists a subset $S \rightarrowtail A$ the members of which are just those elements of A with property P. (Anyone familiar with axiomatic set theory will see how to formalize this, but note that it is *not* to be formalized in the internal language with its bounded quantifiers. That version is already a theorem. It is to be formalized in the first order categorical language used for the axioms SET_{1-3}.) The replacement scheme can be stated as follows. For every relation R between functions with domain 1 and sets, and every set A, if each element x of A has R to a unique set B_x then there is an A-indexed family of sets $C \longrightarrow A$ such that each fibre C_x is iso to the corresponding B_x. In short, if every element of A is R-related to a unique set, then there is a disjoint union C of all those sets. An easy exercise shows that replacement implies separation, since a subset $S \rightarrowtail A$ can be seen as an A-indexed family with the fibre $\emptyset$ over x if x is not in S and 1 if x is in S.

Since the finitely many category and topos axioms plus SET_{1-3} suffice for ordinary set-theoretic mathematics, category theorists have paid little attention to the axiom schemes.

Various extensions of SET_{1-2} have been shown to be equiconsistent with corresponding fragments and extensions of ZF (see Johnstone 1977; also Cole 1973, Mitchell 1972, Osius 1974). The earlier papers were testing grounds for the internal language of toposes, and for the logical character of toposes in general as opposed to their geometric character.

Exercises

22.1 In any topos **E** we say *supports split* if, for every object T, the image $i: T \twoheadrightarrow I$ of the arrow $!: T \longrightarrow 1$ is split epic. Show that **E** is well-pointed iff it is Boolean, two-valued, and supports split.

22.2 Show that every slice **Set**/A is Boolean. [Hint: show that Σ_A preserves complements.] It is not two-valued unless $A \cong 1$ in **Set**. Show that, for any $h, k: f \longrightarrow g$ in **Set**/A, either $h = k$ or there is some subterminator $i: I \rightarrowtail A$ and arrow $c: i \longrightarrow f$ with $h \circ c \neq k \circ c$.

Show that supports split in **Set**/$(1 + 1)$, and that the axiom of choice for **Set** is equivalent to saying that supports split in every slice **Set**/A.

22.3 Show that the only topologies on **Set** are the identity and the trivial.

22.4 Finite sets are co-products of 1 with itself. Suppose that a set A has exactly three elements, $x: 1 \longrightarrow A$, $y: 1 \longrightarrow A$, and $z: 1 \longrightarrow A$. Show that A is iso to $1 + 1 + 1$ with x, y, and z as the injections. [Hint: define a monic–epic function from the coproduct to A.] Thus we can define a function f with domain A by saying what each of $f(x), f(y)$, and $f(z)$ are.

Also conclude from stability of coproducts that for any $f: B \longrightarrow A$, the set B is iso to the coproducts of the fibres f_x, f_y, and f_z. Generalize to any finite number of elements.

22.5 Our first order theory of **Set** is not formally equipped to deal with infinite coproducts directly (except for internally indexed ones). But show that the axiom scheme of separation is equivalent to the following axiom scheme, which says—to the extent that our first order language can—that each set is the coproduct of its elements: for every relation R between functions with domain 1 and for any sets A and B, if every element of A is R-related to a unique element of B then there is a function $f: A \longrightarrow B$

such that every element x of A is R-related to fx. [Hint: in one direction construct the graph, in the other use $1 + 1$.]

22.6 Show that a weakened law of excluded middle in sound is $\mathbf{Set}^{\mathbf{2}}$ and $\mathbf{Set}^{\mathbf{M}}$. For any formula φ we have $\vdash \sim\sim\varphi \vee \sim\varphi$. This is sound in every topos in this chapter except one in Exercise 22.13.

22.7 Show that the representable diagram R_0 in $\mathbf{Set}^{\mathbf{2}}$ is the terminal diagram. Compare this, via the Yoneda lemma, with the remark that a global section of a diagram F in $\mathbf{Set}^{\mathbf{2}}$ amounts to an element of $F0$. Show that R_1 has $R_1(0) \cong \emptyset$ and $R_1(1) \cong 1$.

22.8 Let S and W be diagrams on $\mathbf{2}$. Define a diagram W^S by making $(W^S)0$ the set of natural transformations from S to W, $(W^S)1$ the set of functions from $S1$ to $W1$, and $(W^S)\alpha$ the function that takes a natural transformation v to its component v_1. Define $ev\colon W^S \times S \longrightarrow W$ by letting $ev_0(v, x)$ be $v_0(x)$ and letting $ev_1(f, y)$ be fy. Show that this is an exponential of W by S. (Compare with Chapter 20, but it is also worthwhile to verify it directly.)

22.9 Define a sub-object classifier Ω for $\mathbf{Set}^{\mathbf{2}}$. Make $\Omega 0$ a three-element set, $1 + 1 + 1$. Call the elements n, g, and a, to represent 'in now', 'only gets in later', and 'always out'. Let $\Omega 1$ be $1 + 1$ and call its elements i and o, for 'in' and 'out'. Define $F\alpha$ by

$$F\alpha(n) = F\alpha(g) = i \qquad \text{and} \qquad F\alpha(a) = o$$

Verify that the global section picking out n from $\Omega 0$ and i from $\Omega 1$ is a sub-object classifier for $\mathbf{Set}^{\mathbf{2}}$.

22.10 Define the poset $\mathbf{N}$ of natural numbers with set of objects N and set of arrows the extension of $\leq$. Generalize from Exercise 22.9 to describe the sub-object classifier for $\mathbf{Set}^{\mathbf{N}}$. (There is one truth value for each natural number.) Note that every representable in $\mathbf{Set}^{\mathbf{N}}$ is subterminal. Describe exponentials in $\mathbf{Set}^{\mathbf{N}}$. Generalize all this to any poset in place of $\mathbf{N}$.

22.11 Define a small category $\mathbf{Z}_2$ with one object 0 and two arrows, the identity i, and an arrow m with $m \circ m = i$. (This is the group $\mathbf{Z}_2$.) A diagram F on $\mathbf{Z}_2$ is a set $F0$ and $Fi\colon F0 \longrightarrow F0$, with Fi its own inverse. An arrow that is its own inverse is an *involution*. Show that the global sections of F are its fixed points, the x with $Fi(x) = x$. Show that the diagram category on $\mathbf{Z}_2$ is Boolean and two-valued.

Verify that the representable diagram R_0 on $\mathbf{Z}_2$ has two elements in its set, and that the involution reverses them. (In fact, it is the regular representation of $\mathbf{Z}_2$.) Conclude that supports do not split in the diagram category. In particular, we have

$$\vdash (\exists x \,.\, R_0)x = x$$

but R_0 has no global section.

Generalize to any group $\mathbf{G}$ in place of $\mathbf{Z}_2$. Show that $\mathbf{Set}^{\mathbf{G}}$ is Boolean and two-valued, and that supports do not split unless $\mathbf{G}$ is the singleton group. Define the exponential W^S in $\mathbf{Set}^{\mathbf{G}}$ by letting the set $W^S(0)$ contain all functions f from the set $S0$ to $W0$, with arrows g in $\mathbf{G}$ acting so that $W^S g(f)$ is the function that takes each x in $S0$ to $Wg^{-1}(f(Wg(x)))$.

22.12 One might generalize Exercise 22.11 and $\mathbf{Set^M}$ to any monoid $\mathbf{H}$ in place of $\mathbf{Z}_2$ or $\mathbf{M}$. Take a single object, and the elements of $\mathbf{H}$ as arrows, with multiplication xy as composition $x \circ y$. $\mathbf{Set^H}$ is Boolean iff $\mathbf{H}$ is a group. This and other examples are discussed in the introduction to Lawvere (1972).

22.13 Define a small category $\mathbf{S}$ with three objects 0, 0′, and 1, and two arrows, $f: 0 \longrightarrow 1$ and $f': 0' \longrightarrow 1$, in addition to the identities. Describe the representable diagrams R_0 and $R_{0'}$. Show that their coproduct has an epic natural transformation to the terminator, but no global section. Roughly, this and Exercise 22.11 give the two reasons why internal existence in a Grothendieck topos need not be global. The element may exist internally but not as a fixed point, or it may exist on overlapping parts of a cover of 1 but not continuously over all of 1 (think of 1 as a space, with regions as parts).

Calculate R_0 and $R_{0'}$ in $\mathbf{Set}^{\mathbf{S}^{op}}$, the presheaf category in $\mathbf{S}$. Show that each is the negation of the other as sub-objects of $\Delta 1$, so that the weak law of excluded middle in Exercise 22.6 fails in $\mathbf{Set}^{\mathbf{S}^{op}}$.

22.14 Define the global section functor from $\mathbf{Set^2}$ to $\mathbf{Set}$, taking each diagram F to $F0$. Verify directly that the constant diagram functor is left adjoint to it.

22.15 Show that every diagram category over $\mathbf{Set}$ has as natural number object its constant diagram ΔN. In fact, show that for $\mathbf{E}$ any topos with natural number object every Grothendieck topos over $\mathbf{E}$ has a natural number object.

23

Synthetic differential geometry

The synthetic differential geometry axioms (SDG) describe a topos in which each object has a differentiable structure, each arrow has a derivative, and the basic rules of calculus are simple calculations with infinitesimals. For this chapter we assume that we have a particular topos **Spaces** that satisfies the axioms. We refer to objects of **Spaces** as spaces, to its arrows as maps, and to global elements $p: 1 \longrightarrow M$ as points of the space M.

The axioms posit a space R, with addition and multiplication making a kind of number line. But R is unlike the standard reals (i.e. the Cauchy or Dedekind reals in any topos) in that the subspace $D \rightarrowtail R$ of x in R with $x^2 = 0$ is not just $\{0\}$. We call D the space of infinitesimals of square zero, and the axiom SDG_2 says that every map from D to R has a well-defined slope. We describe some non-classical spaces in addition to D, and some models for SDG built in **Set**.

23.1 A ring of line type

We assume axioms for a non-trivial topos, a topos with $\emptyset$ not iso to 1. The axiom SDG_1 says:

(SDG_1) The space R has selected points $0: 1 \longrightarrow R$ and $1: 1 \longrightarrow R$ (do not confuse the point 1 on R with the terminator 1) and maps $-: R \longrightarrow R$, $+: R \times R \longrightarrow R$, and $.: R \times R \longrightarrow R$ that make R a non-trivial ring. For variables x, y, and z over R:

$$0 + x = x \quad x + (-x) = 0 \quad x + y = y + x \qquad 1.x = x \quad x.y = y.x$$

$$(x + y) + z = x + (y + z) \qquad (x.y).z = x.(y.z)$$

$$x.(y + z) = (x.y) + (x.z)$$

And non-triviality means the internal negation

$$\vdash \sim (0 = 1)$$

Let $D \rightarrowtail R$ be the equalizer $[x.R | x^2 = 0]$. Clearly $0 \in D$. We call D the space of *infinitesimals of square zero*. Since D is a subspace of R we freely mix variables over R and D in expressions such as '$x + d$' or '$d.y$'. (Formally, we should write '$x + i(d)$' or '$i(d).y$' to show the monic $i: D \rightarrowtail R$.)

Define a map $\bar{\alpha}: R \times R \times D \longrightarrow R$ by $\bar{\alpha}(a, b, d) = a + d.b$. The transpose $\alpha: R \times R \longrightarrow R^D$ is described internally by

$$\alpha(a, b) = (\lambda d.D)(a + d.b)$$

The crucial axiom for SDG is the *axiom of line type*:

$$(\mathrm{SDG}_2) \quad \alpha: R \times R \longrightarrow R^D \text{ is an isomorphism.}$$

THEOREM 23.1 The axiom SDG_2 is equivalent to any of the following:

$$\vdash (\forall f.R^D)(\exists! a, b.R) f = \alpha(a, b)$$

$$\vdash (\exists! a, b.R)(\forall d.D) f(d) = a + d.b \qquad f.R^D$$

$$\vdash (\exists! b.R)(\forall d.D) f(d) = f(0) + d.b$$

PROOF The existence of parameters $\langle a, b \rangle$ for each f in R^D says that α is epic, and uniqueness says that it is monic. The second formula is equivalent by extensionality. And if $f(d) = (a + d.b)$ then $f(0) = a$. □

The rest of this section describes D in more detail, and is not actually used in the next section. Pictorially, SDG_2 says that D is too small to bend but larger than the point 0, so that its images in R have well-defined slopes. It does not follow that anything in D is affirmatively unequal to 0, since D is not decidable. Thus **Spaces** is not Boolean.

THEOREM 23.2 D is not $\{0\}$. In fact, internally:

$$\vdash \sim (\forall d.D) d = 0$$

PROOF If we put 1_R for f in SDG_2 the assumption $(\forall d.D)d = 0$ would imply $(\forall d.D)(d = d.0 \,\&\, d = d.1)$ and so $0 = 1$ (see Exercise 23.4). This contradicts SDG_1. □

THEOREM 23.3 D is not decidable. What is more, the internal negation is true, where $d \neq 0$ abbreviates $\sim (d = 0)$,

$$\vdash \sim (\forall d.D)(d = 0 \vee d \neq 0)$$

PROOF If we assume that

$$(\forall d.D)(d = 0 \vee d \neq 0)$$

then a functional relation defines a map $f: D \longrightarrow R$ such that

$$f(0) = 0 \quad \& \quad (d \neq 0 \longrightarrow f(d) = 1)$$

It follows from SDG_2 by topos logic that there is some b in R with

$$d \neq 0 \longrightarrow f(d) = 1 = d.b$$

But, by SDG_1, if $1 = d.b$ then $1 = (d.b)^2 = 0$, a contradiction. So we conclude that $\sim\sim(d=0)$. Our assumption gives $d=0$ and then $(\forall d\,.\,D)d=0$, contradicting Theorem 23.2. □

Nor is R decidable, since it includes D. Axiom SDG_4 implies that everything in D is not-not-equal to 0 (see Exercise 23.1).

23.2 Calculus

For any map $f: R \longrightarrow R$ and any x in R, there is a map $g: D \longrightarrow R$ defined by $g(d) = f(x+d)$ and so a unique b in R with

$$f(x+d) = g(d) = g(0) + d\,.\,b = f(x) + d\,.\,b \qquad d\,.\,D$$

Since each x has a unique b, there is a unique map $f': R \longrightarrow R$ such that

$$f(x+d) = f(x) + d\,.\,f'(x) \qquad x\,.\,R\,d\,.\,D$$

Call f' the *derivative* of f (Exercise 23.2). Then f' also has a derivative. In fact, every function from R to R is *smooth*, in the sense of having derivatives of all (finite) orders.

The familiar rules of calculus follow quickly. Suppose that f is constant, that is, for some c in R we have $f(x) = c$ for all x. Then

$$f(x+d) = c = f(x) + d\,.\,0$$

and so $f'(x) = 0$, for all x. The identity function 1_R has

$$1_R(x+d) = x + d\,.\,1$$

so the derivative of 1_R is the constant 1.

Given functions f and g from R to R and a value b in R, define $(f+g)$ by the rule $(f+g)(x) = (f(x) + g(x))$. Define $f.g$ by $(f.g)(x) = (f(x).g(x))$, and $b.f$ by $(b.f)(x) = b.(f(x))$. Then:

$$\begin{aligned}(f+g)(x+d) &= f(x) + d\,.\,f'(x) + g(x) + d\,.\,g'(x)\\ &= (f+g)(x) + d\,.\,(f'+g')(x)\\ (f.g)(x+d) &= (f(x) + d\,.\,f'(x))\cdot(g(x) + d\,.\,g'(x))\\ &= (f.g)(x) + d\,.\,(f.g' + f'.g)(x)\\ (b.f)(x+d) &= (b.f)(x) + d\,.\,(b.f')(x)\end{aligned}$$

The product rule uses the step $d^2.g(x).f(x) = 0$. We conclude that

$$\begin{aligned}(f+g)' &= f' + g'\\ (f.g)' &= f.g' + f'.g\\ (b.f)' &= b.f'\end{aligned}$$

The chain rule is also a simple calculation:

$$(f \circ g)(x+d) = f(g(x) + d.g'(x))$$
$$= (f \circ g)(x) + d.g'(x).f'(g(x))$$

The second step uses $(d.g'(x))^2 = 0$, so $d.g'(x)$ is in D. Thus

$$(f \circ g)' = (f' \circ g).g'$$

which is the classical chain rule.

We give R an order relation compatible with the ring structure:

(SDG$_3$) There is a relation $<$ on R such that for all x, y, and z in R:

$$(x < y \,\&\, y < z) \longrightarrow x < z$$

$$\sim (x < x) \qquad 0 < 1$$

$$x < y \longrightarrow (x+z) < (y+z) \quad \text{and} \quad (x < y \,\&\, 0 < z) \longrightarrow (x.z) < (y.z)$$

THEOREM 23.4 For any x in R and d in D we have

$$0 < x \longrightarrow 0 \neq x \quad \text{and} \quad 0 < x \longrightarrow -x < 0 \quad \text{and} \quad \sim (d < 0 \vee 0 < d)$$

PROOF The first follows from $\sim (x < x)$, and the second by adding $-x$ to both sides of $0 < x$. For the third, using the second claim plus $(-1)^2 = 1$, either $d < 0$ or $0 < d$ implies that $0 < d^2$, contradicting $0 = d^2$. □

There cannot be a trichotomy law, saying that for any x and y either $x < y$ or $x = y$ or $y < x$, since this would imply that, for every d in D, $d = 0$. (See the discussion of smoothness in parameters at the end of Section 23.3.) The second clause in the next axiom partly replaces trichotomy, while the first says that R is a field:

$$(\text{SDG}_4) \vdash x \neq 0 \longrightarrow (\exists y.R)x.y = 1 \qquad x.R$$
$$\vdash 0 < x \vee x < 1$$

Given any k in R with $k \neq 0$, we write $1/k$ for the inverse of k. For any h and k in R we write (h, k) for $\{x.R | h < x < k\}$, the open interval from h to k. Now an axiom can relate derivatives to local behaviour:

(SDG$_5$) Internally, for any f in R^R and x in R, if $0 < f'x$ there is some $\varepsilon > 0$ such that for any y and w such that

$$(x - \varepsilon) < y < w < (x + \varepsilon)$$

we have $fy < fw$.

In words, if f' is greater than 0 at x then f is increasing on an interval around x. This implies a mean value theorem, and that every map from R to R is continuous. We state both internally.

THEOREM 23.5 For any f in R^R and any x in R, if $h < f'x < k$ there is some $\varepsilon > 0$ such that for any y and z with

$$x - \varepsilon < y < z < x + \varepsilon$$

we have

$$h.(z - y) < fz - fy < k.(z - y)$$

PROOF Apply SDG_5 to the maps $fx - (k.x)$ and $(h.x) - fx$. □

So upper and lower bounds on the derivative at x become upper and lower bounds on the total rate of change on some interval around x.

THEOREM 23.6 Every f in R^R is continuous. That is, for any x in R and any h, k in R, if $h < fx < k$ there is some $\varepsilon > 0$ with $h < fy < k$ for all y in $(x - \varepsilon, x + \varepsilon)$.

PROOF By algebra we can reduce this to the following. For every g, if $0 < gx$ there is some $m > 0$ with $0 < gy$ for every y in $(x - m, x + m)$. So suppose that $0 < gx$. Take any r in R with $0 < r$ and $-r < g'x < r$ (see Exercise 23.3) and any $\varepsilon > 0$ such that the inequality

$$-r.(z - y) < gz - gy < r.(z - y)$$

holds for all $y < z$ in the interval $(x - \varepsilon, x + \varepsilon)$. Take any m with

$$0 < m < \frac{gx}{4.k} \qquad \text{and } m < \varepsilon$$

Put x for y, and $x + m$ for z, in the inequality, to show that $g(x + m)$ is greater than $(3.gx)/4$. Then since $-\varepsilon < -m$ it follows for every z in the interval $(x - m, x + m)$ that

$$g(x + m) - gz < r.((x + m) - z)$$

but $2.r.m$ is less than $(gx)/2$, and so we obtain $0 < (gx)/4 < gz$ for all z in that interval. □

From R we construct other spaces. There is the plane $R \times R$ and, more generally, R^n for any natural number n. The 2-sphere, S^2, is defined as a subspace of R^3 by the equalizer

$$[\langle x, y, z\rangle . R^3 | (x^2 + y^2 + z^2) = 1]$$

There are quotients such as the projective plane P^2, the quotient of S^2 by the relation identifying each point $\langle x, y, z\rangle$ with $\langle -x, -y, -z\rangle$. There are function spaces such as R^R. The exponential adjunction forces the smooth curves $c: R \longrightarrow R^R$ to be the transposes of smooth maps $\bar{c}: R \times R \longrightarrow R$. (This characterizes the Fréchet topology among the many different topologies that functional analysts have for the space of maps from R to R.) The axioms SDG_{1-5} give these spaces many of their classical properties, but not all.

Different further axioms suit different purposes, and no one axiom set for SDG has become standard. Some further axioms deal with non-classical spaces. For example, there is a space $[x.R|x^3 = 0]$, called D_2, or the space of infinitesimals with cube zero. In general, D_n is $[x.R|x^{n+1} = 0]$. A higher axiom of line type says that every map $f: D_2 \longrightarrow R$ is quadratic with unique coefficients: there are unique a, b, c in R such that for any d in D_2, $f(d) = a + d.b + d^2.c$. Therefore D_2 is just large enough for a map defined on it to have a slope and a rate of change for the slope, but no more. Similar axioms say that any D_n is just large enough to support n derivatives. This generalizes further to more variables and mixed orders of derivatives (see Kock 1981). There are also richer non-classical spaces such as $[x.R| \sim \sim (x = 0)]$, sometimes called the germ neighbourhood of 0 (see Penon (1981) or Bunge and Dubuc (1987) and references therein).

Theorem 23.6 shows that every map $f: R \longrightarrow R^n$ is continuous (if fx lies in any open rectangle of R^n then for some interval around x all values of f lie in that rectangle), since its composite with any projection $p_i: R^n \longrightarrow R$ is continuous. But it does not show that maps from R^m to R^n are continuous. That claim follows if we also assume that the sphere S^{m-1} is compact. We could also take an axiom that says that every differential equation on a vector space R^n has a local solution at every point (i.e. for every initial condition). For any $f: R \longrightarrow R$ an integral of f is a map $g: R \longrightarrow R$ with $g' = f$. As this is a differential equation on g it follows that f has a definite integral over any interval on R. Using compactness, the classical reduction of partial differential equations to ordinary ones goes through in SDG, and the inverse and implicit function theorems follow, since they are themselves partial differential equations. These axioms are pursued in McLarty (1983).

23.3 Models over Set

To say that a topos **G** models SDG means that all of the SDG axioms are true of some ring R in **G**. There are Grothendieck toposes constructed in **Set** (or using ZF or any reasonable set theory), each satisfying all the axioms mentioned in this chapter, proving the axioms consistent if the **Set** axioms are (or whatever theory is in use). More than that, we briefly describe a class of models that relate SDG closely to analytic differential geometry.

Take the category **Man**, the objects of which are manifolds and the arrows of which are the infinitely differentiable maps (see, for example, the first three chapters of Spivak 1970). There is the standard real line, the set of real numbers $\mathbb{R}$. Every manifold $\mathbb{M}$ has a tangent bundle $T\mathbb{M}$. To say that **G** is *fully well-adapted* means that it is a model with a full and faithful functor

$\mathbf{I}:\mathbf{Man}\longrightarrow\mathbf{G}$ with various properties, including:
(i) $\mathbf{I}\mathbb{R}\cong R$;
(ii) for every $f:\mathbb{R}\longrightarrow\mathbb{R}$ in **Man** with derivative f', we have $(\mathbf{I}f)'=\mathbf{I}(f')$;
(iii) for every manifold $\mathbb{M}$, $\mathbf{I}(T\mathbb{M})\cong(\mathbf{I}(\mathbb{M}))^D$.
That is, **I** takes the line as a manifold $\mathbb{R}$ to the ring of line type R, it takes the derivative of f calculated in **Man** to the derivative of $\mathbf{I}f$ calculated in **G**, and it takes the tangent bundle $\mathbb{M}$ of M in **Man** to the tangent bundle $(\mathbf{I}(\mathbb{M}))^D$ of $\mathbf{I}(\mathbb{M})$ in **G**. In short, the differentiable structure in **G** of the images under **I** of manifolds agrees with their differentiable structure in **Man**. Such models exist to satisfy any of an array of extensions of the SDG axioms (see Moerdijk and Reyes, 1991).

Construction of such a model begins with the co-ordinate ring $\mathbf{C}(\mathbb{M})$ of each manifold $\mathbb{M}$; that is, the ring of smooth maps from $\mathbb{M}$ to $\mathbb{R}$ with pointwise addition and multiplication. There is no manifold like D, but we do know what the co-ordinate ring of D should be. A smooth map from D to $\mathbb{R}$ should be given by a pair $\langle a,b\rangle$, thought of as a base point and slope, or as the constant and linear coefficients of a Taylor series. They should add and multiply as those coefficients do; that is,

$$\langle a_1,b_1\rangle+\langle a_2,b_2\rangle=\langle a_1+a_2,b_1+b_2\rangle$$
$$\langle a_1,b_1\rangle.\langle a_2,b_2\rangle=\langle a_1.a_2,(a_1.b_2+b_1.a_2)\rangle$$

A smooth map $f:\mathbb{M}\longrightarrow\mathbb{M}'$ gives a ring homomorphism $\mathbf{C}(f):\mathbf{C}(\mathbb{M}')\longrightarrow\mathbf{C}(\mathbb{M})$ that takes each g in $\mathbf{C}(\mathbb{M}')$ to $g\circ f$, so **C** is a contravariant functor from **Man** to the category **Ring** of rings. Therefore we take a non-full subcategory **A** of **Ring**, including all co-ordinate rings of manifolds, and every ring we would like to be the co-ordinate ring of an infinitesimal space such as D. We include only those homomorphisms which ought to correspond to smooth maps. Think of the dual category $\mathbf{A}^{\mathrm{op}}$ as a category of spaces including the manifolds and the new infinitesimal spaces. This duality between rings and spaces is the basic reason why algebraic geometers so often work with both a category and its dual.

We define a suitable Grothendieck topology on $\mathbf{A}^{\mathrm{op}}$ and define **G** as the category of sheaves for this topology. This gives a topos the objects of which are 'locally' like the spaces of $\mathbf{A}^{\mathrm{op}}$. The topos has a richer structure than $\mathbf{A}^{\mathrm{op}}$, such as Cartesian closedness, and the effect is that **G** has all the properties we have described for **Spaces**.

At a first glance, classical manifolds behave differently in **Spaces**, and thus in **G**, than they do in **Man**. For example, classically, for any x on the line $\mathbb{R}$, either $x=0$ or $x\neq 0$, and it is not so in **G**. But this difference is only apparent. Variables over R in **G** can be replaced by arbitrary generalized elements of R, that is by arbitrary maps to R from other spaces while, classically, x is to be a point. An internal satement in **G** with a variable over R

is true iff the corresponding statement in **Man** is true *smoothly in parameters*, meaning that it not only holds for all points of $\mathbb{R}$ but is locally true for all maps to $\mathbb{R}$. A classical map $f: \mathbb{R} \longrightarrow \mathbb{R}$ may have $f(x^\circ) = 0$ for some point x° and while not being constantly 0 on any neighbourhood of x°. Then $f = 0$ is not locally true at x°—nor is $f \neq 0$— and so '$f = 0 \vee f \neq 0$' is not locally true. This is also why we have no trichotomy law on R. For maps f and g there may be points x° such that no neighbourhood of x° verifies any single one of $f < g$ or $f = g$ or $g < f$. But every point x° either has a neighbourhood with $0 = f$ or one with $f < 1$, since f is continuous.

A statement that is true for all points of $\mathbb{R}$ in **Man** is true for all global elements of R in the usual models. This is related to the fact that those models are two-valued, and so a closed law of excluded middle is sound in them:

$$\text{For any closed formula } \varphi \text{ we have } \vdash \varphi \vee \sim \varphi$$

Note that φ itself may have quantifiers, but we cannot assume excluded middle *inside* the scope of any. The closed law of excluded middle implies, for any (global) point $p^\circ: 1 \longrightarrow R$,

$$\vdash p^\circ = 0 \vee p^\circ \neq 0$$

It implies that any double-negation true formula with free variables over R is actually true for any points substituted for the variables. This idea is applied in Bunge and Heggie (1984) and McLarty (1987).

Exercises

23.1 Using $\mathrm{SDG}_{1,2,4}$, prove that $(\forall x . R)(x^2 = 0 \longrightarrow \sim\sim (x = 0))$. Show that if we assume two-valuedness, then 0 is the only point of D.

23.2 Section 23.2 begins by reasoning internally. Formalize the first paragraph as proving that

$$\vdash (\forall f . R^R)(\exists! f' . R^R)(\forall x . R)(\forall d . D) f(x + d) = f(x) + d . f'(x)$$

Use global elements of R^R to show that for any actual map $f: R \longrightarrow R$ in **Spaces** there is an actual map $f': R \longrightarrow R$ with the relevant property.

23.3 There is no absolute value function in SDG, since it would not have a derivative at 0. We replace it by an positive upper bound function. Define $S(x)$ to be $2 + x^2$. Use SDG_{1-4} to show that, for all x in R, we have $0 < S(x)$ and also $-S(x) < x < S(x)$. (The second clause of SDG_4 is crucial.)

23.4 Certainly if $d^2 = 0$ then $d.b = d.c$ does not imply that $b = c$. (Let $b = d$ and $c = 0$.) But use SDG_2 to show that $(\forall d . D)(d.b = d.c)$ does imply that $b = c$. Conclude that, internally, $\sim (\forall d . D)(\forall e . D)(d.e = 0)$. [Hint: show that $(\forall d . D)(\forall e . D)(d.e = 0)$ implies that $(\forall d . D)(d = 0)$.]

23.5 Show that for every f in R^R, x in R, and d and e in D:

$$f(x + d + e) = fx + (d + e).f'x + (1/2)(d + e)^2.f''x$$

Generalize to higher order terms of the Taylor series by using the sum of more terms in D.

23.6 Consider the exponential map $+^R: R^R \times R^R \longrightarrow R^R$. Show that it defines addition of maps pointwise; that is:

$$\vdash (f +^R g)(x) = f(x) + g(x) \qquad x.Rf, g.R^R$$

Show that R^R is a ring with points 0^R and $1^R: 1 \longrightarrow R^R$ and operations $+^R$, $.^R$, and $-^R$. (The proof is nearly just that exponentiation preserves products.)

23.7 Use SDG_3 to prove that $0 < 1 + 1$ and so $0 \neq 1 + 1$.

23.8 For each natural number n, define a function $\_^n: R \longrightarrow R$ by

$$x^0 = 1 \qquad \text{and} \qquad x^n = x.x^{(n-1)}$$

For each $n > 0$ prove that the derivative of $\_^n$ is $n.(\_^{(n-1)})$. Conclude with the general rule for derivatives of polynomials. This reasoning can be given internally if we assume a natural number object N and recursively define a map $N \longrightarrow R^R$ that takes each n to $\_^n$. Calculate the derivative of, say, $(x^3)^2$ by the chain rule.

23.9 Assuming a natural number object N, there is a map $i: N \longrightarrow R$ defined by $i(0) = 0$ and $i(n + 1) = i(n) + 1$, where 0 and $+$ on the left are in N, while those on the right are in R. Prove by induction that i preserves addition and multiplication; that is, $i(n + m) = i(n) + i(m)$ and $i(n.m) = i(n).i(m)$. Show that SDG_{1-3} imply that i is monic.

23.10 Still assuming a natural number object, define the nth power map $N \longrightarrow R^R$ as in Exercise 23.8. Show, for any natural number n, that the function x^n is not equal to $p(x)$ for p any polynomial function of degree less than n. [Hint: consider the nth derivative.]

For any n, define $P: R^n \times R \longrightarrow R$ by

$$P(\langle x_1, \ldots, x_n \rangle, y) = x_1 + (x_2.y) + \cdots + (x_n.y^{n-1})$$

Show that the transpose of P is monic $R^n \rightarrowtail R^R$. Show that it is linear (for the operations defined in Exercise 23.6). Conclude that SDG_{1-3} plus the axiom of infinity imply that R^R is infinite-dimensional.

23.11 For any h and k with $h < k$, for any x in R, either $h < x$ or $x < k$. The interval (h, k) is empty iff $\sim(h < k)$. For the first, apply SDG_4 to $(x - h)/(k - h)$. For the second, one can show that $h < k$ holds iff the mean $(h + k)/2$ is $> h$ and $< k$.

23.12 Define a *tangent vector* on any space M to be a map $v: D \longrightarrow M$, and the *base point* of v to be $v(0)$. Intuitively, v is an infinitesimal straight stroke on M based at $v(0)$. Define the *tangent space* on M to be M^D. Show that $M^D \longrightarrow M$, the exponential of M by $0: 1 \longrightarrow D$, takes each tangent vector to its base point. Show that the tangent space on R is $R \times R$, and that for any R^n it is R^{2n} since exponentiation preserves products.

23.13 Take the projections $p_i: R^3 \longrightarrow R$ for $i = 1, 2, 3$, and for any tangent vector $v: D \longrightarrow R^3$ define the components of v as the composites $v_i = p_i \circ v: D \longrightarrow R$. By SDG_2 we have some b_1 and b_2 and b_3 such that

$$v_i(d) = v_i(0) + d.b_i$$

Show that v factors through the equalizer defining S^2 iff

$$v_1(0)^2 + v_2(0)^2 + v_3(0)^2 = 1$$

and

$$(v_1(0).b_1) + (v_2(0).b_2) + (v_3(0).b_3) = 0$$

In other words, v is a tangent vector to S^2 iff its base point is in S^2 and $\langle b_1, b_2, b_3 \rangle$ has inner product 0 with the radius vector. [Hint: use the fact, which follows from SDG_{1-4}, that $2.x = 0$ implies $x = 0$.] Compare with Exercise 21 in Chapter 3 of Volume 1 of Spivak (1970).

23.14 For those familiar with the terms, generalize Exercise 23.13 to any subspace of R^n defined by an equation $f(x_1, \ldots, x_n) = 0$. This gives the classical tangent plane at points at which f is non-singular (i.e. at which f gives a classical manifold), and all tangent vectors to R^n at any point at which f is singular.

24

The effective topos

In the effective topos, **Eff**, described by Hyland using suggestions from Dana Scott, Church's thesis is true in the strong sense that every function from the naturals to themselves in recursive. In fact, all functions between objects built from the natural numbers, including the rationals and the real numbers, are recursive. Thus those objects are naturally suited to modeling data types in programming languages, and **Eff** is a natural setting for higher order recursion theory.

We construct **Eff** out of sets by a method due to Freyd, and compare this to Hyland's construction. To date there are no such decisive categorical axioms for **Eff** as for **Set** and **Spaces**. We discuss this briefly at the end of the chapter.

We assume basic recursive function theory: Kleene (1952, Chs. IX and XI) or Mendelson (1987, Ch. 5) provide more than enough. We take a fixed enumeration of the unary partial recursive functions and write $n(m)$ for the value at m of the nth function, if that value is defined. We also use a recursive pairing function $(\_,\_)\colon N \times N \longrightarrow N$ and recursive functions $\ell\colon N \longrightarrow N$ and $\imath\colon N \longrightarrow N$ called *left* and *right*, such that for any natural numbers n and m we have $\ell(n, m) = n$ and $\imath(n, m) = m$ and $n = (\ell n, \imath n)$. So (n, m) is a natural number which codes the ordered pair $\langle n, m \rangle$, and ℓ and $\imath$ serve as projections. Note that every number n codes a pair (see Exercise 24.1).

24.1 Constructing the topos

An *assembly* consists of a set A called the *ambient* set and a sequence of subsets of it, A_n for each natural number n, called the *caucuses*. The union of all the caucuses is the *carrier* of the assembly. The caucuses need not be disjoint, and the carrier need not be all of A. We give assemblies names such as (A), and write A_n for the nth caucus of (A) and $|A|$ for its carrier. The ambient is just a technical convenience. We often define an assembly by giving its caucuses, and then the ambient may be their union, the carrier. An arrow between assemblies $f\colon (A) \longrightarrow (B)$ is a function $f\colon |A| \longrightarrow |B|$ between their carriers which has at least one *modulus*, a partial recursive function φ such that for every $x \in |A|$ if $x \in A_n$ then φn is defined and $fx \in B_{\varphi n}$. One arrow may have many different moduli. Arrows compose by composing the functions, and the composite of moduli is necessarily a modulus for the composite arrow. Call the category of assemblies and their arrows **A**. There is obviously an underlying set functor from **A** to **Set** that takes each assembly to its carrier and each arrow to itself as function.

The *restriction* of (A) to a subset $S \rightarrowtail |A|$ of the carrier is the sub-assembly $s:(S) \rightarrowtail (A)$, where each caucus S_n is the intersection of A_n with S. Then s has the identity function as a modulus.

THEOREM 24.1 **A** has all finite limits, and the underlying set functor preserves them.

PROOF For the assembly (1) let each caucus be the singleton 1. Given assemblies (A) and (B), define $(A \times B)$, where each caucus $(A \times B)_n$ is the product $A_{\ell n} \times B_{\imath n}$. The projection p_1 has modulus ℓ, and p_2 has $\imath$. For the equalizer (E) of $f, g:(A) \longrightarrow (B)$, restrict (A) to the set equalizer of f and g as functions on $|A|$. For any $h:(T) \longrightarrow (A)$ with $f \circ h = g \circ h$, the obvious $u:(T) \longrightarrow (E)$ has at least the same moduli as h. □

Any product (C) for (A) and (B) is isomorphic to $(A \times B)$ in **A** but it could have its caucuses arranged differently and not have ℓ and $\imath$ as moduli for the projections. We take the given construction of $(A \times B)$ as canonical, and similarly for equalizers and other constructions to follow.

THEOREM 24.2 **A** has a natural number object (N), where each caucus N_n is $\{n\}$, so the carrier is the set of natural numbers N.

PROOF Let $0:(1) \longrightarrow (N)$ be the function $0:1 \longrightarrow N$ in **Set**. The constant function 0 is the modulus. Let $s:(N) \longrightarrow (N)$ be the successor function in **Set** with itself as modulus. Take any assembly (A) and arrow $x:(1) \longrightarrow (A)$ and $f:(A) \longrightarrow (A)$ with moduli φ and ψ respectively. The functions x and f induce a unique $u: N \longrightarrow |A|$ in **Set** and recursion data $\theta 0 = \varphi 0$ and $\theta(n+1) = \psi \theta n$ give a recursive θ, which is a modulus for u. □

THEOREM 24.3 The arrows $f:(N) \longrightarrow (N)$ are the recursive functions $f: N \longrightarrow N$ of **Set**, since each arrow from (N) to (N) must be its own modulus. □

The category **A** is Cartesian closed (see Exercise 24.5) but not a topos. It has no sub-object classifier but, more to the point, it does not have quotients for all equivalence relations. We define **Eff** by adding quotients to **A** by the method of Chapter 25. That is, we define **Eff** to be what Chapter 25 calls Map(**A**) and we freely use results from that chapter here. In short, there is a full monic functor $\mathbf{A} \rightarrowtail \mathbf{Eff}$ such that every equivalence relation $(R) \rightarrowtail (A \times A)$ in **A** has a quotient (A/R) in **Eff** and every **Eff** object is the quotient of some equivalence relation in **A**.

The inclusion $\mathbf{A} \rightarrowtail \mathbf{Eff}$ preserves all limits, so it preserves monics. On the other hand, every **Eff** sub-object of an assembly is (iso to) an assembly. Therefore the intersections, unions, material conditionals, and complements in **Eff** of sub-assemblies of an assembly (A) already exist in **A**. Some are calculated directly in Exercises 24.7 and 24.8.

The quotients which do exist in **A**, namely the quotients of double-negation closed equivalence relations according to Exercise 24.9, remain quotients in **Eff**. Once we prove that **Eff** is a topos it will follow that an arrow between assemblies is epic in **Eff** iff it is surjective in **A**. Hyland (1982, Prop. 3.3) shows that the inclusion preserves the natural number object (N), so in **Eff** too the arrows from the natural number object to itself are the recursive functions of **Set**.

The next theorem shows that the results of Chapter 25 do apply to **A**:

THEOREM 24.4 **A** has stable surjective images and thus is regular. In fact, every arrow $f:(A)\longrightarrow(B)$ has an image factorization $f = i \circ g$, where g is a quotient for the kernel pair of f.

PROOF Given f construct (I), where each I_n contains all those $y \in |B|$ such that some $x \in A_n$ has $fx = y$. Define $g:(A)\longrightarrow(I)$ by $|g| = |f|$. Then g has the identity as a modulus and the inclusion $i:(I)\rightarrowtail(B)$ has the same moduli as f. For any monic $j:(J)\rightarrowtail(B)$, if f factors through j then $i \subseteq j$, with the same moduli as the factorization of f through j. The reader can show that g is a co-equalizer, and thus is surjective, and that every pullback of g is surjective. □

Call this factorization the canonical image of f.

COROLLARY The arrow f is surjective iff its canonical image is iso to (B), and iff f is a quotient for its kernel pair.

PROOF Lemmas 25.1 and 25.2 prove the first claim. The second follows directly. □

Note that a relation of assemblies $(F)\rightarrowtail(A \times B)$ is a series of **Set** relations. For any modulus φ of the monic, each caucus F_n is a relation between the sets $A_{\ell\varphi n}$ and $B_{\imath\varphi n}$. We write $xF_n y$ to say that $\langle x, y\rangle$ is in F_n.

THEOREM 24.5 **Eff**, or Map(**A**) in the notation of Chapter 25, is a topos.

PROOF By Theorem 25.25 it suffices to show that **Eff** has **A**-relation classifiers, and this is done in Exercises 24.14–24.15. □

The next two theorems help to draw conclusions about **Eff** by working with **A**, and are heavily used in Hyland (1982), for example:

THEOREM 24.6 The full monic $\mathbf{A}\rightarrowtail\mathbf{Eff}$ is (up to equivalence) the insertion of separated objects for the double-negation topology. In other words, the separated objects are precisely those iso to assemblies.

PROOF See Theorem 25.26 and Exercise 24.9. □

COROLLARY The inclusion $\mathbf{A} \rightarrowtail \mathbf{Eff}$ preserves exponentials, as does the inclusion of separateds for any topology on any topos. □

Exercises 24.11 and 24.16 define a full monic functor $\nabla: \mathbf{Set} \rightarrowtail \mathbf{Eff}$ right adjoint to the global section functor $\Gamma: \mathbf{Eff} \longrightarrow \mathbf{Set}$, so that there is a geometric morphism $\Gamma \dashv \nabla: \mathbf{Set} \longrightarrow \mathbf{Eff}$ which is in fact the insertion of double-negation sheaves:

THEOREM 24.7 The full monic $\mathbf{Set} \rightarrowtail \mathbf{Eff}$ is (up to equivalence) the double-negation sheaf insertion.

PROOF Exercise 24.18 shows that sheaves are the objects iso to assemblies which look like sheaves to assemblies. The reader can show that codiscretes look like sheaves to assemblies. Every assembly (A) has a double-negation dense monic $i:(A) \rightarrowtail \nabla|A|$ to the codiscrete on its carrier. If (A) looks like a sheaf to $\nabla|A|$ then i is iso. So sheaves are codiscrete. □

24.2 Realizability

Hyland (1982) describes **Eff** in terms of realizability, which nicely motivates many constructions in **Eff**. We compare it briefly with our approach. Hyland defines an **Eff** object as a set X plus a suitable assignment of a set of natural numbers to each pair $\langle x, x' \rangle$ in $X \times X$. Write $[x =_X x']$ for the set of numbers assigned to $\langle x, x' \rangle$. We say that n realizes $x =_X x'$ if n is in $[x =_X x']$. The conditions on $=_X$ are as follows:

(Symmetry) For some natural number m, whenever n realizes $x =_X x'$ the value $m(n)$ is defined and realizes $x' =_X x$.

(Transitivity) For some natural number m, whenever ℓn realizes $x =_X x'$ and $\imath n$ realizes $x' =_X x''$, the value $m(n)$ is defined and realizes $x =_X x''$.

An **Eff** object $(X, =_X)$ in Hyland's sense gives an assembly (A) with X as ambient, where each caucus A_n contains all x such that n realizes $x =_X x$. It is easy to see that $(=_X) \rightarrowtail (A \times A)$ is an equivalence relation on the assembly (A) when we define the caucuses by

$$x(=_X)_n x' \qquad \text{iff } n \text{ realizes } x =_X x'$$

Then the symmetry and transitivity clauses above just say that there are moduli for inclusions

$$(=_X) \subseteq (=_X{}^o) \qquad \text{and} \qquad (=_X \circ =_X) \subseteq (=_X)$$

proving symmetry and transitivity on (A). Reflexivity follows from the construction of (A). So we have an **Eff** object $(A/=_X)$ in our sense. Conversely, a quotient (A/R) gives an **Eff** object $(|A|, =_A)$ in Hyland's sense, where n realizes $x =_A x'$ iff $xR_n x'$. Hyland's definition of a functional relation from an **Eff** object $(X, =_X)$ in his sense to another $(Y, =_Y)$ is ours, only expressed in terms of realizability.

The assembly (N) corresponds to an **Eff** object $(N, =_N)$ in Hyland's sense, where $=_N$ is defined by caucuses

$$m(=_N)_n p \qquad \text{iff}\, n = m = p$$

This is Kleene's definition of 'n realizes $m = p$'. Anyone familiar with realizability will easily read the symmetry and transitivity clauses as saying that the formulas

$$(\forall x, x')(x = x' \longrightarrow x' = x)$$

$$(\forall x, x', x'')[(x = x' \,\&\, x' = x'') \longrightarrow x = x'']$$

are realized. Hyland (1982) gives more details, and remarks that a closed formula of first order arithmetic is true in **Eff** iff it is realized in Kleene's sense.

24.3 Features of Eff

The Kleene T-predicate and Kleene's output function U are both primitive recursive, so they exist in any topos with natural number object. In effect, $T(e, y, z)$ says that the eth partial recursive function applied to input y gives a completed calculation coded by z, and $U(z)$ gives the output value for z (as it might be, the number in the last line of z).

In **Eff**, Church's thesis holds in the most naive form:

$$\vdash (\exists e\,.(N))(\forall n\,.(N))(\exists y\,.(N))(T(e, n, y) \,\&\, Uy = fn) \qquad f.(N^N)$$

That is, internally every f in (N^N) is the eth partial recursive function for some e. (Of course, every f in (N^N) is in fact total.) See Hyland 1982.

This strong Church's thesis is inconsistent with Boolean logic. For example, reasoning internally with Boolean logic we could define an $f\colon(N) \longrightarrow (N)$ by saying: if

$$(\exists z\,.N)(T(n, n, z) \,\&\, Uz = m)$$

then $fn = m + 1$ and otherwise $fn = 0$. In words, if the nth partial recursive function is defined for input n and its output value for n is m, $fn = m + 1$, otherwise $fn = 0$. Then f would be defined for all N, but it could not have any code e since that would imply

$$fe = (fe) + 1$$

In **Set**, with Boolean logic, this proves that not all functions from N to N are recursive. In any topos where the strong internal Church's thesis is true it shows that Boolean logic cannot hold.

Call an **Eff** object *modest* if it is iso to an assembly in which each caucus is at most a singleton. These are often called *modest sets*, and are *effective* objects in Hyland (1982).

THEOREM 24.8 An **Eff** object is modest iff it is a quotient of a sub-object of (N) by a double-negation closed equivalence relation, and iff it is separated and every **Eff** arrow to it from $\nabla 2$ is constant, where 2 is a two-element set.

PROOF If (M) has each caucus at most a singleton then let $(S) \rightarrowtail (N)$ be the obvious sub-assembly with the same caucuses non-empty as (M), and let $q:(S) \longrightarrow (M)$ be the surjection with the identity modulus. Theorem 25.19 shows that (M) is the quotient in **Eff** of the kernel pair of q, which is closed by Exercise 24.10. For the second condition: separateds are iso to assemblies, and it is easy to see that all arrows from $\nabla 2$ to an assembly are constant iff all caucuses of the assembly are at most singleton. □

Carboni *et al.* (1988) and Hyland *et al.* (1990) show that modestness in **Eff** is also equivalent to having all arrows from the sub-object classifier constant. This may be the right definition for generalizing to other toposes, but we do not need it here.

Let $\mathbf{Mod} \rightarrowtail \mathbf{A}$ be the full subcategory of modest objects, so that it is also a full subcategory of **Eff**. Each arrow $g:(M) \longrightarrow (M')$ of **Mod** is partial recursive in the sense that it is fully determined by any one of its moduli. Hyland says that it is 'recursive in the indices', meaning the indices of the caucuses in our terms. Not every partial recursive function will be the modulus of an arrow from (M) to (M'), and one arrow may have many different moduli. In short, an arrow from (M) to (M') corresponds to a suitable set of partial recursive functions—all the moduli for the arrow. Hyland (1982, p. 198) shows how to express this internally as a generalized strong Church's thesis for modest objects.

THEOREM 24.9 **Mod** is Cartesian closed and has quotients for double-negation closed equivalence relations, and $\mathbf{Mod} \rightarrowtail \mathbf{A}$ preserves those structures; i.e. finite limits, exponentials, and closed quotients are calculated the same way whether in **Mod** or **A**, or **Eff**.

PROOF Use the characterization by at most singleton caucuses. □

In fact, **Mod** is closed under all products indexed in **Eff**, not only finite ones. If all arrows from $\nabla 2$ to each of a family of objects M_i is constant, then every arrow to the product of the family is also constant. (This is intuitive internally in **Eff**: see also Exercise 24.19.) Since **Eff** has limits for all internal diagrams, and the limits are equalizers of products, every functor from a small category in **Eff** to **Mod** has a modest limit: that is, **Mod** is internally complete in **Eff** and the inclusion into **Eff** preserves all small limits.

The crucial example in **Eff** is the exponential (N^N). In many toposes with natural number object N, such as **Set**, N^N is much larger than N. But in **Eff** the exponential is the object of recursive functions, and so it is a quotient of a sub-object of (N). The theorems prove this, but see also Exercise 24.21. The rational numbers (Q) in **Eff** are a closed quotient of $(N \times N)$, so they are modest (cf. the argument in Chapter 19 showing that the rationals are decidable). Hyland (1982) shows that the Cauchy reals (R_c) are modest in **Eff** and that they coincide with the Dedekind reals (R_d). So all of arithmetized analysis in **Eff** takes place in **Mod**, where all arrows are recursive. Hyland notes that analysis in **Eff** essentially 'is constructive real analysis in the sense of Markov' (p. 166).

Modest objects are plausible models of programming data types. We can look at a computer as basically working with natural numbers, coded in binary or whatever. Other data types allow certain codes as values, and may treat some pairs of different codes as representing the same value. From this point of view, data types are quotients of subsets of the natural numbers, and of course the computable functions between them are the ones recursive in the codes. Few, if any, programming languages have all of the power of modest sets, but the sense of the model is obvious.

Given any $(C) \rightarrowtail (N)$ and equivalence relation (R) on (C), the quotient (C/R) is, viewed internally to **Eff**, a collection of disjoint sub-objects of (C); namely, the equivalence classes, and thus a collection of disjoint sub-objects of (N). In fact, every collection of disjoint sub-objects of (N) arises in this way: take their union as (C) and define the equivalence 'xRy' to mean 'x and y are together in some one of the sub-objects'. Note that this is done in the internal language of **Eff**, not to be confused with the construction out of assemblies.

Therefore a modest object is internally a sub-object $Q \rightarrowtail P(N)$ of the power object satisfying formulas for non-empty disjointness, and closedness of the resulting relation:

$$(\forall S, S'.P(N))[(S \in Q \,\&\, S' \in Q) \longrightarrow (S = S' \longleftrightarrow (\exists x.N)(x \in S \,\&\, x \in S')]$$

$$(\forall S, S'.P(N))[(S \in Q \,\&\, S' \in Q) \longrightarrow (\sim\sim S = S' \longrightarrow S = S')]$$

We can use these formulas to define a sub-object $IM \rightarrowtail PP(N)$ of the double power object, the sub-object of all *internal modests*. Each global element of IM gives a modest object, and every modest is iso to one of these.

Given two modests, represented as sub-objects $Q \rightarrowtail P(N)$ and $Q' \rightarrowtail P(N)$, an arrow between the quotients appears as a relation from (N) to (N); namely, a relation such that for any sub-object S in Q there is a unique sub-object S' in Q' such that any x in S is related to y iff y is in S'. This amounts to the same thing as a functional relation between the quotients.

There is a sub-object of $PP(N \times N)$ that contains just these relations from (N) to (N).

Altogether then, we have an **Eff** object of representatives (up to isomorphism) of all modest objects, and another for arrows between them. Domain, codomain, and composition functions are easily defined internally. So we have a small category **IM** in **Eff** of all internal modests and arrows between them.

Intuitively, **IM** has a functor equivalence to **Mod**, the functor that takes each internal modest to the actual modest object. The functor is full and faithful, and every actual modest is iso to one of the values of the functor. Furthermore, **IM** has a striking property for a small category. Freyd proved that any small category in **Set** with a limit for every functor from a small category is a preorder. Any object in it has at most one arrow to any other (see Mac Lane 1971, p. 110). Preorders are not very interesting categories. This result extends to small categories in any Grothendieck topos over **Set**, but not to **Eff**:

THEOREM 24.10 **IM** is complete in **Eff**. Every functor from a small category **D** in **Eff** to **IM** has a limit.

PROOF A functor from **D** to **IM** gives one from **D** to **Mod**, which we have already seen has a limit, and that is iso to some internal modest in **IM**. □

Therefore **IM** has limits over diagrams as large as itself and even larger, and this makes some constructions in **IM** very easy. For example, it implies that every functor $\mathbf{T}: \mathbf{IM} \longrightarrow \mathbf{IM}$ in **Eff** has a fixed point, a modest object (M) such that $\mathbf{T}(M)$ is iso to (M) (see the conclusion of Hyland 1988). This in turn makes recursive definitions of modest objects very easy (see Freyd 1990). For example, it quickly gives a non-singleton modest object (D) with (D) iso to (D^D), so that (D) makes a model of untyped lambda calculus—a non-trivial 'set' iso to the 'set' of all functions from itself to itself.

Much work with modests uses the relations that they are quotients of. These are called *partial equivalence relations*, or Pers, on (N), and they are often described independently, outside the context of the effective topos. The very clear discussion in Robinson (1989) takes the approach that 'from our normal point of view' we deal with partial equivalence relations but 'inside this nonstandard model of set theory' which is **Eff** we work with their quotients. Other descriptions of applications are in Carboni *et al.* (1988), Longo and Moggi (1991), and numerous papers on polymorphism, Pers, and recursive type definition in IEEE (1990).

The proof of Theorem 24.10 is taken from Carboni *et al.* (1988). It has a robust working quality, but passes over genuine problems relating the internal category **IM** to the external category **Mod**. We have really only discussed global elements of the sub-object $IM \rightarrowtail PP(N)$, and thus global objects of the small category **IM**: we have ignored the recursive structure. The details in Hyland (1988) and Hyland *et al.* (1990) are complicated, and the subject has probably not reached its final form.

24.4 Further features

The topos **Eff** is a natural context for higher order recursion theory. Rosolini (1986) axiomatizes a fair part of this theory and derives various classical theorems. Using this approach applied to **Eff** we have genuine exponentials such as (N^N), and so we can work directly with functions rather than with their codes. Recursive sets of numbers or functions simply become the complemented sub-objects of (N) or (N^N).

Hyland (1982) mentions various applications to recursion theory. For one, he follows Powell and Pitts to show that the Turing degrees appear not only as sub-objects of the sub-object classifier in **Eff** (compare with Exercise 24.15) but as topologies on **Eff**. For any set A of natural numbers there is a sheaf topos over **Eff** where: the natural number object is the sheafification $L(N)$, its complemented sub-objects are just the sets of natural numbers Turing reducible to A, and the arrows from it to itself are the functions recursive relative to A.

Rosolini's axioms for higher order recursion far from uniquely characterize **Eff**. That is nice for the generality of his axioms, but not for our understanding of the topos. It has been suggested that excessive attention to the subjective aspect of recursiveness as computability has been one obstacle to axiomatizing **Eff**. Compare the way in which so much of Kleene realizability, invented to formalize the idea that every fact should come with effective instructions for its proof, falls out of the definition of assembly plus the considerations in Chapter 25. The latter considerations are in some sense more 'objective', and were discovered with no view towards recursiveness or provability at all. Perhaps less 'computational' aspects of **Eff**, such as that every object is a quotient of a separated, will yield useful axioms.

We end this chapter by quoting the conclusion of Hyland (1982). There has been great progress in understanding and applying **Eff** however, still 'what we lack, above all, in our treatment of the effective topos, is any real information about axiomatization'.

Exercises

24.1 Show that the function $(((n + m)^2 + n + m)/2) + m$ is an isomorphism from $N \times N$ to N. Describe ℓ and $\imath$ in the inverse $\langle \ell, \imath \rangle$ from N to $N \times N$. [Hint: recall that $(k^2 + k)/2$ is the sum of the first k natural numbers.]

24.2 Show that any assembly (A) with $|A|$ a singleton is iso to (1), no matter what the ambient or how the caucuses are arranged.

24.3 Show that (1) is a generator in **A**: in fact every assembly (A) and $x \in |A|$ there is an arrow $x:(1) \longrightarrow (A)$ that picks out x. Conclude that an **A** arrow $f:(A) \longrightarrow B$ is monic iff the function f is monic in **Set**.

24.4 Construct a canonical pullback of $f:(A) \longrightarrow (C)$ and $g:(B) \longrightarrow (C)$ using the canonical product and equalizer from Theorem 24.1. Show that each caucus $P_{(h,k)}$ is

$$\{\langle x, y\rangle | x \in A_h \quad \text{and } y \in B_k \quad \text{and } fx = gy\}$$

and that ℓ and $\imath$ are moduli for the projections.

24.5 Given (A) and (B), define (B^A) by letting the nth caucus B^A_n contain those arrows from (A) to (B) for which the nth partial recursive function is a modulus. Define the evaluation arrow and transposes of arrows $f:(C \times A) \longrightarrow (B)$. Show that the carrier of (N^N) is the set of recursive functions.

24.6 Show that any assemblies (A) and (B) have a coproduct $(A + B)$ with caucuses $(A + B)_{(0,n)} = A_n$ and $(A + B)_{(1,n)} = B_n$, and other caucuses empty. Show that no assembly has more than countably many arrows to the coproduct $(1 + 1)$.

24.7 Let $(S) \longrightarrow (A)$ and $(T) \longrightarrow (A)$ be any sub-assemblies of (A). Show that their intersection, calculated by the canonical pullback, has for its nth caucus the intersection $S_{\ell n} \cap T_{\imath n}$. Show they have a union, a smallest sub-assembly containing both, namely the image of their coproduct. (The canonical construction, following Exercise 24.6, gives caucuses $(S \cup T)_{(0,n)} = S_n$ and $(S \cup T)_{(1,n)} = T_n$ and the rest empty.) Anyone familiar with realizability can compare this to realizability for conjunctions and disjunctions.

24.8 Show that every monic $(S) \rightarrowtail (A)$ of assemblies has a negation $(\sim S) \rightarrowtail (A)$; that is, a largest sub-assembly disjoint from it. For a canonical complement, let each caucus $\sim S_n$ contain all those $x \in A_n$ which are not in $|S|$. (For comparison with realizability, $x \in \sim S_n$ iff $x \in A_n$ and for no m is $x \in S_m$.)

Show that a sub-assembly (S) has $(S) \equiv (\sim \sim S)$ iff it is (equivalent to) a restriction of (A). We call (S) *double-negation closed* if $(S) \equiv (\sim \sim S)$. Therefore the double-negation closed sub-assemblies of (A) correspond to subsets of the carrier.

24.9 Suppose that $(R) \rightarrowtail (A \times A)$ is double-negation closed. Show that (R) is an equivalence relation in **A** iff the underlying relation $|R| \rightarrowtail |A \times A|$ on the carrier is an equivalence relation in **Set**. Show that every double-negation closed equivalence relation in **A** has a quotient in **A**. [Hint: assume that (R) is a restriction of $(A \times A)$ and put the natural caucus structure on the quotient set of $|R| \rightarrowtail |A \times A|$.]

24.10 Show that the double-negation closed equivalence relations of **A** are precisely the kernel pairs.

24.11 For every set S, let ∇S be the assembly every caucus of which is S. Call this *the codiscrete assembly* on S. Define a full monic functor ∇: **Set** $\longrightarrow$ **A** that takes each set to its codiscrete assembly.

Generally, call an assembly (A) *codiscrete* if it is iso to ∇S for some set S. Show that an assembly is codiscrete iff at least one caucus contains the whole carrier.

24.12 Let $q: P \twoheadrightarrow Q$ be an epic in any topos. For any sub-objects i and j of Q, show that $q^{-1}(i) \equiv q^{-1}(j)$ iff $i \equiv j$. (What are the classifying arrows?) Conclude that

$q^{-1}(i) \subseteq q^{-1}(j)$ iff $i \subseteq j$. Conclude, for any given topology, that i is closed iff $q^{-1}(i)$ is, and dense iff $q^{-1}(i)$ is. Compare with Exercise 25.3.

24.13 Given relations $(F) \rightarrowtail (A \times B)$ and $(G) \rightarrowtail (B \times C)$, show that the composite $(G, F) \rightarrowtail (A \times C)$ defined in Theorem 25.4, if constructed using canonical pullbacks and images in **A**, has caucuses

$$x(G \circ F)_n z \qquad \text{iff for some } y \quad xF_{\ell n}y \quad \text{and} \quad yG_{\imath n}z$$

(Note that this chapter uses '$\circ$' and reverses the order of F and G compared with the notation in Chapter 25.)

24.14 Let (W) be the codiscrete assembly on the power set of N. That is, each caucus W_n contains all subsets of N. Define an assembly (M) with M_n containing all those subsets S of N with $n \in S$, and let $(M) \rightarrowtail (W)$ be the inclusion. Show that for any monic $(C) \rightarrowtail (B)$ there is at least one arrow $(B) \longrightarrow (W)$ making (C) a pullback of (M). [Hint: for each $x \in |B|$ consider the set S of all n with $x \in C_n$.]

For any assembly (A) let $(V) \rightarrowtail (A \times W^A)$ be the pullback of (M) along $ev \circ tw: (A \times W^A) \longrightarrow (W)$. Conclude directly from the result above that for any relation $(C) \rightarrowtail (A \times B)$ there is at least one arrow $c: (B) \longrightarrow (W^A)$ such that (C) is a pullback of (V) along $1_{(A)} \times c$.

24.15 For (W) as defined in Exercise 24.14, define a relation $(h, k): (E) \rightarrowtail (W \times W)$ by caucuses, by letting SE_nS' iff: for every m in S the value $\ell n(m)$ is defined and is in S', and for every m in S' the value $\imath n(m)$ is defined and is in S. Show, for any assembly (B) and arrows f and g both from (B) to (W), that (M) has the same pullbacks along f and g iff there is some $r: (B) \longrightarrow (E)$ with $h \circ r = f$ and $k \circ r = g$. Conclude that (E) is an equivalence relation on (W).

For any assembly (A) consider the exponential $(E^A) \rightarrowtail (W^A \times W^A)$. Conclude from general properties of exponentials that, for any assembly (B), any arrows f and g both from (B) to (W^A) correspond to the same subobject of $(B \times A)$ iff there is some $r: (B) \longrightarrow (E^A)$ with $h^A \circ r = f$ and $k^A \circ r = g$.

Use Lemma 25.24 to define an $\in^A$ such that the following is a pullback in **Eff**:

$$\begin{array}{ccc}
(M) & \longrightarrow & \in^A \\
\downarrow & & \downarrow \\
(A) \times [A] & \longrightarrow & (A) \times [A]/E
\end{array}$$

Since the quotient (W^A/E^A) has kernel pair h^A, k^A, conclude that (W^A/E^A) is an A-relation classifier for A in **Eff**. Thus, by Theorem 25.23 the quotient (W^A/E^A) is a power object for A in **Eff**, and (W/E) is a subobject classifier.

24.16 We also use ∇ to name the composite of ∇ in Exercise 24.11, with the inclusion $\mathbf{A} \rightarrowtail \mathbf{Eff}$. Thus $\nabla: \mathbf{Set} \rightarrowtail \mathbf{Eff}$ is monic and full.

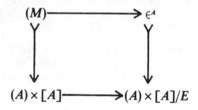

Show that a global element $s: (1) \longrightarrow (A/R)$ of an **Eff** object (A/R) corresponds to a subset S of the carrier $|A|$ such that: for some $x \in S$, and all $y \in |A|$, we have $y \in S$ iff there is some n such that $xR_n y$. (In Hyland's terms, for some x in $|A|$, S contains all the y such that $x =_A y$ is realized.)

Let Γ: **Eff** $\longrightarrow$ **Set** be the global section functor, so that each $\Gamma(A/R)$ is the set of global sections, and for each $f:(A/R) \longrightarrow (B/R')$ the function Γf takes each global section s to $f \circ s$. Show that there is an adjunction $\Gamma \dashv \nabla$: **Set** $\longrightarrow$ **Eff**, and in fact a geometric morphism, since any global section functor preserves limits.

24.17 Conclude from Exercise 24.6 that no **Eff** object has more than countably many arrows to $(1 + 1)$, and therefore that **Eff** contains no infinite co-products of non-empty assemblies. Thus the global section functor has no left adjoint and **Eff** is not a Grothendieck topos over **Set**.

24.18 Let **E** be any topos and j any topology on it, with separated reflection functor m: **E** $\longrightarrow$ $\mathbf{E}_{\text{sep},j}$. For any j-dense $s: S \rightarrowtail X$ show that $Ms: MS \rightarrowtail MX$ is also dense. [Hints: $S \subseteq q^{-1}(MS)$ for $q: X \twoheadrightarrow MX$ the universal arrow. Use Exercise 24.12.] Show that an **E** object B is a j-sheaf iff it is j-separated and 'looks like a sheaf to separateds' in the following sense: for any separated object X and dense monic $s: S \rightarrowtail X$ and arrow $f: S \longrightarrow B$, there is a unique arrow $g: X \longrightarrow B$ with $f = s \circ g$. [Hint: f factors through the universal $q: S \longrightarrow MS$.]

24.19 For any topos **E** and object A of it, show that the functors A^* and Π_A take $\sim\sim$-separated objects to $\sim\sim$-separated objects. (Hint: Recall that A^* is a logical functor. For Π_A use the definition of separatedness, noting that A^* takes dense monics to dense monics.) For any **E** objects B and C, if the projection $C \times B \longrightarrow C$ has an isomorphism as transpose $C \xrightarrow{\sim} C^B$ then we say that internally every arrow from B to C is constant. In the internal language this says 'for every g in C^B there is a unique y in C such that g is the constant function with value y'. For any objects B of **E** and f of **E**$/A$, show that if internally all arrows from A^*B to f are constant then so are all from B to $\Pi_A f$. (Hint: Argue internally, or else show that $(\Pi_A f)^B$ is naturally isomorphic to $\Pi_A(f^{A^*B})$ by showing that as functors of f both have the same left adjoint.)

Show that an object of **Eff** is modest iff it is separated and internally every arrow from $\nabla 2$ to it is constant. By an (A/R)-indexed *family of modest objects* we mean any $\sim\sim$-separated object f of **Eff**$/(A/R)$ such that internally every arrow from $(A/R)^* \nabla 2$ to f is constant. Show this implies for every $c: 1 \longrightarrow (A/R)$ the fiber f_c is modest in **Eff**. Show it implies that $\Pi_{(A/R)} f$ is modest in **Eff**.

24.20 **Mod** is not complete for small categories in **Set**, and does not even have products over all sets. Show that neither **Eff** nor **Mod** has a product of countably many copies of $1 + 1$. [Hint: such a product would have to be an assembly.]

24.21 Let $(C) \rightarrowtail (N)$ be the sub-assembly of codes of (total) recursive functions and $(R) \rightarrowtail (C \times C)$ the relation which identifies m and n iff they code the same recursive function. Show that the exponential (N^N) constructed as in Exercise 24.5 is also the quotient of (C) by (R).

Show that (N^N) is not iso to (N). In fact, there is no surjection from (N) to (C), nor from (N) to (N^N). [Hint: neither sort of surjection could have a modulus.]

25

Relations in regular categories

Regular categories, defined below, support a nice theory of relations. We follow Freyd and Scedrov (1990) in extending any regular category **C** to a regular category Map(**C**) with quotients for all equivalence relations. Then we give a necessary and sufficient condition for Map(**C**) to be a topos. We begin by assuming that **C** has all finite limits, and that it is regular, once we have defined that.

25.1 Categories of relations

An arrow $q: A \longrightarrow Q$ is *surjective*, a *surjection*, if the smallest sub-object of Q it factors through is 1_Q.

LEMMA 25.1 Any surjection is epic. Any surjective monic is iso.

PROOF Let $q: A \longrightarrow Q$ be surjective. If $h \circ q = k \circ q$ then the equalizer of h and k is (up to equivalence) 1_Q, so $h = k$. If q is monic itself then $q \equiv 1_Q$. □

LEMMA 25.2 If $q: A \longrightarrow Q$ is surjective then an arrow $p: Q \longrightarrow P$ is surjective iff $p \circ q$ is.

PROOF $p \circ q$ factors through a sub-object $M \rightarrowtail P$ iff q factors through $p^{-1}(M)$. □

A *surjective image* of an arrow $f: A \longrightarrow B$ is an image factorization $i \circ q: A \twoheadrightarrow Q \rightarrowtail B$ with q surjective.

LEMMA 25.3 If f has surjective image i, q then i is the smallest sub-object that f factors through. So surjective images are unique up to equivalence.

PROOF If q factors through $i \cap i'$ then $i \cap i' \equiv i$. □

A surjection $q: A \twoheadrightarrow Q$ is *stable* if for every $s: Q' \longrightarrow Q$ the pullback of q is surjective to Q'. A category is *regular* if it has all finite limits, all surjections are stable, and every arrow has a surjective image. Hereafter we assume that **C** is regular.

The next theorem defines composition of relations in **C**. For any relation $R \rightarrowtail A \times B$, and T-elements x of A and y of B, we write xRy to say that $\langle x, y \rangle \in R$.

THEOREM 25.4 Given $F \rightarrowtail A \times B$ and $G \rightarrowtail B \times C$, with common object B, there is a unique (up to equivalence) $FG \rightarrowtail A \times C$ such that, for any T and T-elements x of A and z of C, we have $xFGz$ iff there is some surjective $s: S \twoheadrightarrow T$ and some $y: S \longrightarrow B$ such that $(x \circ s)Fy$ and $yG(z \circ s)$.

PROOF Given F and G and having formed the pullback

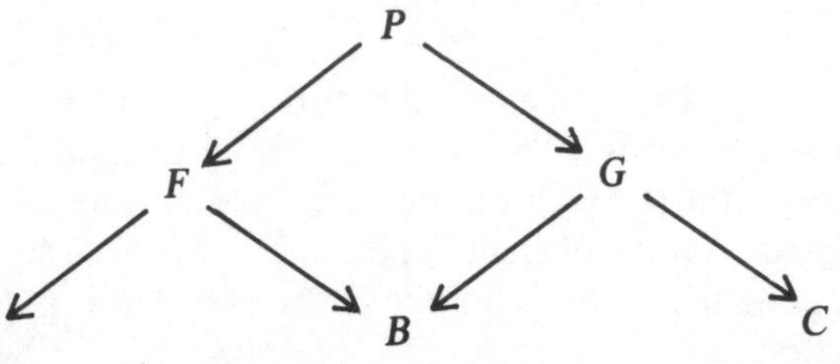

let $FG \rightarrowtail A \times C$ be the surjective image of $P \longrightarrow A \times C$. Suppose that T-elements x and z have $xFGz$, so that they factor through some $u: T \longrightarrow FG$. Let $s: S \twoheadrightarrow T$ be the pullback of $P \twoheadrightarrow FG$ along u, and let $y: S \longrightarrow P \longrightarrow B$ be the obvious composite. Then s and y are as in the theorem. Conversely, suppose that s is surjective and $(x \circ s)Fy$ and $yG(z \circ s)$. Then $\langle x, z \rangle \circ s$ factors through P and thus through FG. By Exercise 25.2, the pair $\langle x, z \rangle$ factors through FG. Any relation is determined up to equivalence by its members. □

In short, $xFGz$ means that 'there exists a y with xFy and yGz' in the sense of existence on a surjective cover s (cf. existential quantifiers in Chapter 18). In fact, to prove general inclusions or equivalences of relations one can ignore covers and stages of definition altogether (see Exercise 25.5).

COROLLARY For any composable F, G, and H, we have $F(GH) \equiv (FG)H$. Diagonals $A \rightarrowtail A \times A$ compose as identities: for any $F \rightarrowtail C \times A$ and $G \rightarrowtail A \times B$ we have $FA \equiv F$ and $G \equiv AG$. □

So there is a category Rel(**C**) with the same objects as **C** and with **C** relations as arrows. A relation $F \rightarrowtail A \times B$ has domain A and co-domain B. We say that F and $F' \rightarrowtail A \times B$ give the same Rel(**C**) arrow iff $F \equiv F'$ over $A \times B$. If F and G are composable then define the composite $G \circ F$ to be FG.

Unfortunately, we reverse the order of F and G in forming the composite, but it is a mere matter of notation. In our notation since Chapter 1, we have written $g \circ f$ for the composite of $f: A \longrightarrow B$ and $g: B \longrightarrow C$. Using the opposite order, writing the composite as fg, would spare us the present embarrassment but would fly in the face of common notation in other branches of mathematics. Writing $G \circ F$ for FG would consort awkwardly with our notation xFy for relations. Try the experiment. In fact, part of the category literature uses one order for composites, and part uses the other, and there is no option but to become familiar with both. In this chapter we generally write composites FG as in Theorem 25.4, and only use $G \circ F$ as a formality to define categories of relations. Chapter 24 uses the dot notation and the order $G \circ F$.

LEMMA 25.5 For any **C** arrow $f: A \longrightarrow B$ with graph Γ_f, and T-elements x of A and y of B, we have $x\Gamma_f y$ iff $y = fx$. □

It is easy to see that **C** is a subcategory of Rel(**C**), where each **C** arrow $f: A \longrightarrow B$ appears as its graph $\Gamma_f \rightarrowtail A \times B$. The composite of graphs $\Gamma_f \Gamma_g$ in Rel(**C**) is the graph of the composite $g \circ f$ in **C**, which follows from Lemma 25.5 or from the following pullback:

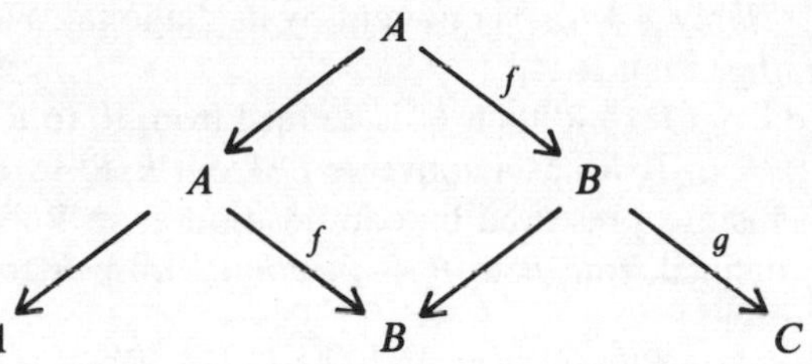

Every **C** relation $F \rightarrowtail A \times B$ has a converse F^o and **C** relations are ordered by inclusion. If $F \subseteq F'$ then $F^o \subseteq F'^o$. Composition preserves converses and inclusion:

THEOREM 25.6 Assuming that the relations are composable: $(FG)^o$ is equivalent to $(G^o F^o)$. If $F \subseteq F'$ then $FG \subseteq F'G$, and similarly if $G \subseteq G'$. □

THEOREM 25.7 $R \rightarrowtail A \times A$ is an equivalence relation iff

$$A \subseteq R \qquad \text{and} \qquad R \subseteq R^o \qquad \text{and} \qquad R \circ R \subseteq R$$

where A stands for the diagonal $A \rightarrowtail A \times A$. The first two clauses express reflexivity and symmetry as in Chapter 5. The third expresses transitivity. For one direction, if R is transitive then $xRRz$ implies that $(x \circ s)R(z \circ s)$, and by Exercise 25.2 it implies that xRz. □

Note that symmetry implies that $R \equiv R^o$. Reflexivity and transitivity imply that $(R) \equiv (R \circ R)$.

Given **C** equivalence relations $R \rightarrowtail A \times A$ and $R' \rightarrowtail B \times B$, a relation $F \rightarrowtail A \times B$ is *defined from* R to R' iff:

$$RF \equiv F \equiv FR'$$

Put roughly: if xRy and yFz then already xFz, and similarly for R'.

THEOREM 25.8 If F is defined from R to R' and G from R' to R'', then FG is defined from R to R''. Any equivalence relation $R \rightarrowtail A \times A$ is defined from itself to itself and composes as an identity with any relation defined to or from R. □

Therefore we can extend Rel (**C**) to a new category Eq (**C**). Each **C** equivalence relation $R \rightarrowtail A \times A$ is an object of Eq (**C**). When we think of R as an Eq (**C**) object we may write it as A/R to suggest the quotient of A by R. An Eq (**C**) arrow from A/R to B/R' is a **C** relation F defined from R to R', and we say that two relations give the same arrow iff they are equivalent sub-objects of $A \times B$. Again, to suit Chapter 1, given F defined from R to R' and G from R' to R'', the composite $G \circ F$ is FG as defined in Theorem 25.4.

Every **C** relation $R \rightarrowtail A \times B$ is defined from the diagonal on A to that on B, since $AF \equiv F \equiv FB$. So Rel (**C**) is a full subcategory of Eq (**C**) when we identify each assembly A with its quotient by its diagonal. We write A for the Eq (**C**) object rather than A/A.

If F is defined from R to R' then F^o is defined from R' to R, so every Eq (**C**) arrow F from A/R to B/R has a converse F^o from B/R' to A/R. And Eq (**C**) arrows have inclusions preserved by composition as in Rel (**C**).

A relation F defined from R to R' is *functional* from R to R' iff:

$$R \subseteq FF^o \qquad \text{and} \qquad F^oF \subseteq R'$$

Briefly, F is totally defined consistently with R-relatedness and single-valued up to R'-relatedness.

THEOREM 25.9 If F is functional from R to R', and G from R' to R'', then FG is functional from R to R''. Any equivalence relation is functional from itself to itself.

PROOF Both claims are calculations. For example,

$$R \subseteq FF^o \equiv FR'F^o \subseteq FGG^oF^o \equiv (FG)(FG)^o$$

establishes half of the first. □

Therefore functional relations form a subcategory of Eq (**C**) which we call Map (**C**). In fact, **C** is full in Map (**C**):

THEOREM 25.10 A relation $F \rightarrowtail A \times B$ is functional from the diagonal on A to that on B iff it is (equivalent to) the graph of a (necessarily unique) arrow $f: A \longrightarrow B$.

PROOF By Theorem 25.4, graphs are functional. Conversely, let F have projection arrows $h: F \longrightarrow A$ and $k: F \longrightarrow B$. Assuming that $A \subseteq FF^o$, apply Theorem 25.4 to 1_A as x to see that h is surjective. For any arrows v and w, if $h \circ v = h \circ w$ then $(k \circ v)F^oF(k \circ w)$. If we also assume that $F^oF \subseteq B$ then $k \circ v = k \circ w$, and h, k are jointly monic so $v = w$. So h is monic and thus iso, and F is the graph of $k \circ h^{-1}$. □

25.2 Map(C)

In short, we have four categories with inclusions as shown:

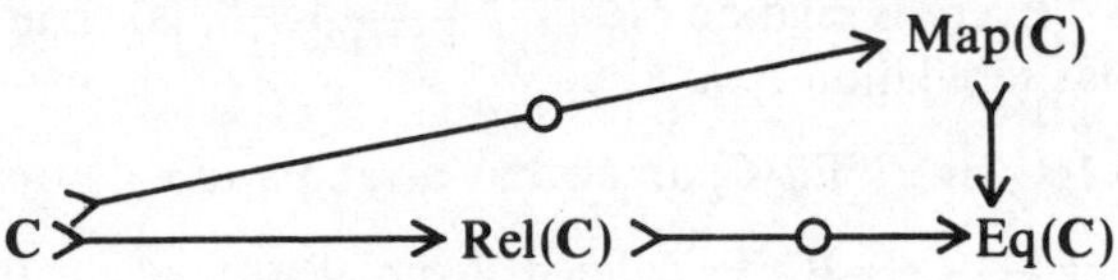

Those marked with circles are full. Each **C** arrow appears in the other categories as identified with its graph, and each **C** object A appears in Eq(**C**) and Map(**C**) as identified with its quotient by its diagonal. Given a **C** arrow $f: A \longrightarrow B$, we may write $f: A \longrightarrow B$ for its image in the other three categories.

We now prove that Map(**C**) is regular and has quotients for all equivalence relations, and that every Map(**C**) object A/R is the quotient of A by the equivalence relation R.

LEMMA 25.11 Take any functional relations $F: A/R \longrightarrow B/R'$ and $G: A/R \longrightarrow B/R'$. If $F \subseteq G$ then $F \equiv G$. And if F^o is functional it is inverse to F.

PROOF If $F \subseteq G$ then $F^o \subseteq G^o$ and so

$$G \subseteq FF^oG \subseteq FG^oG \subseteq F$$

If F^o is functional so are FF^o and F^oF, so that $R \equiv FF^o$ and $F^oF \equiv R'$. □

LEMMA 25.12 If $F: A/R \longrightarrow B/R'$ is functional and G and H are both defined from B/R' to some C/R'', then $F(G \cap H) \equiv FG \cap FH$.

PROOF For any F we have $F(G \cap H) \subseteq (FG \cap FH)$. Use Theorem 25.4 or Exercise 25.5 to verify that $(FG \cap FH) \subseteq F(G \cap H)$ given that $F^oF \subseteq R'$. □

LEMMA 25.13 For any F and G, both defined from A/R to B/R',

$$R \cap [(F \cap G)(F \cap G)^o] \equiv R \cap [FG^o]$$

as verified by Theorem 25.4 or Exercise 25.5, using $RG \subseteq G$ and $R^oF \subseteq F$. □

LEMMA 25.14 If $F: A/R \longrightarrow B/R'$ and $G: A/R \longrightarrow C/R''$ are functional and $(FF^o \cap GG^o) \equiv R$ then F, G are jointly monic in Map(**C**).

PROOF Subject to those conditions, let H and K be functional, with $HF \equiv KF$ and $HG \equiv KG$. From $H \equiv HR$ and Lemma 25.12 deduce that $H \equiv K$. □

A *tabulation* of an Eq(**C**) arrow $F: A/R \longrightarrow B/R'$ is a Eq(**C**) object P/V and arrows $S: P/V \longrightarrow A/R$ and $T: P/V \longrightarrow B/R'$ such that $S^o T \equiv F$ and $SS^o \cap TT^o \equiv V$.

LEMMA 25.15 Every **C** relation $\langle h, k \rangle: F \rightarrowtail A \times B$, taken as Eq(**C**) arrow $F: A \longrightarrow B$, has tabulation F, h, k. □

THEOREM 25.16 Every Eq(**C**) arrow has a tabulation.

PROOF Let $F \rightarrowtail A \times B$ be defined from R to R', with projections $h: F \longrightarrow A$ and $k: F \longrightarrow B$. Theorems 25.6 and 25.7 show that hRh^o and $kR'k^o$ are equivalence relations on F as **C** object. Thus so is their intersection, which we will call V, while hR is functional from V to R, and kR' from V to R', and these arrows tabulate F. □

THEOREM 25.17 Map(**C**) has all finite limits. Specifically, given Map(**C**) arrows $F: A/R \longrightarrow C/R''$ and $G: B/R' \longrightarrow C/R''$, a tabulation for FG^o is a pullback for F and G.

PROOF Clearly, 1 is terminal in Map(**C**). Let P/V, S, T tabulate FG^o as above. Prove that $SF \equiv TG$ from $TG \subseteq SS^o TG \equiv SFG^o G \subseteq SF$, since both sides are functional. For any Map(**C**) object D/W and arrows $H: D/W \longrightarrow A/R$ and $K: D/W \longrightarrow B/R'$, the induced Map(**C**) arrow is $(HS^o \cap KT^o)$. Abbreviate this as $I: D/W \longrightarrow P/V$. To see that $W \subseteq II^o$, use Lemma 25.13 to verify that

$$W \cap II^o \equiv W \cap (HS^o TK^o) \equiv W \cap (KGG^o K^o) \equiv W$$

And because composition preserves order we have

$$I^o I \subseteq (SH^o HS^o) \cap (TG^o GT^o) \subseteq (SS^o TT^o) \equiv V$$

Trivially, $IS \subseteq H$ and $IT \subseteq K$, and so $IS \equiv H$ and $IT \equiv K$. Finally, I is unique, as tabulations are jointly monic. □

COROLLARY A Map(**C**) arrow $F: A/R \longrightarrow B/R'$ is monic iff $FF^o \equiv R$.

PROOF By Exercise 4.9, F is monic iff its kernel pair is R, R, two copies of the identity on A/R. But that is iff $FF^o \equiv R$. □

THEOREM 25.18 Every Map(**C**) arrow $F: A/R \longrightarrow B/R'$ has a surjective image

$$A/R \xrightarrow{FS^o} P/V \overset{S}{\rightarrowtail} B/R'$$

where P/V, S, S tabulates $F^o F$.

PROOF Since $F^o F \subseteq R'$ the tabulation is two copies of a single Map(**C**) arrow $S: P/V \rightarrowtail B/R'$, monic since tabulations are jointly monic. Then $FS^o S \equiv FF^o F \equiv F$. Functionality of FS^o follows easily. Take any Map(**C**) arrow $G: A/R \longrightarrow C/R''$ and monic $H: C/R'' \rightarrowtail B/R'$, with $F \equiv GH$. Then $SS^o \equiv V$ and $HH^o \equiv R''$ show that SH^o is functional. Trivially, $SH^oH \subseteq S$ and so they are equal as Map(**C**) arrows, showing that $P/V \rightarrowtail C/R''$ in Map(**C**). □

COROLLARY F is surjective in Map(**C**) iff $F^oF \equiv R'$. □

COROLLARY Surjections are stable in Map(**C**).

PROOF Use the notation of Theorem 25.17 and suppose that G is surjective. From $S^oT \equiv FG^o$ conclude that $R \subseteq FF^o \equiv FG^oGF^o \equiv S^oTT^oS$. By Theorem 25.4 and joint monicness of S and T, conclude that $R \subseteq S^oS$, and thus $S^oS \equiv R$. □

THEOREM 25.19 A Map(**C**) arrow $F: A/R \longrightarrow B/R'$ is surjective iff it is a quotient of its kernel pair.

PROOF Every quotient is surjective. Conversely, let F be surjective with kernel pair S, T. Let $K: A/R \longrightarrow C/R''$ have $SK \equiv TK$. Calculation shows that F^oK is functional from R' to R''. By Theorem 25.9 we have $K \equiv FF^oK$. Since F is surjective, F^oK is unique. □

COROLLARY Take any **C** equivalence relation $\langle h, k\rangle: R \rightarrowtail A \times A$. Then $R: A \longrightarrow A/R$ is a quotient for h and k in Map(**C**).

PROOF Since h and k tabulate R, which is equivalent to RR^o, they are a kernel pair. And R is subjective since $R^oR \equiv R$. □

THEOREM 25.20 Every equivalence relation in Map(**C**) has a quotient.

PROOF Verify that a jointly monic pair in Map(**C**)

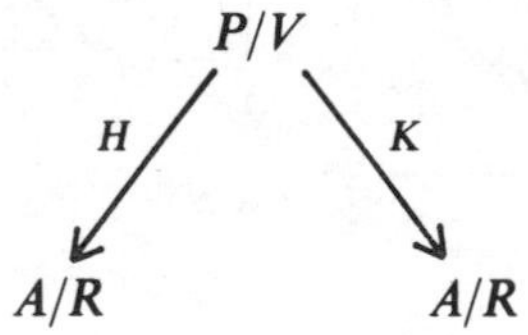

is an equivalence relation in Map(**C**) iff H^oK is an equivalence relation on A. In that case, A/H^oK is a coequalizer for H and K, with them as its kernel pair. □

The inclusion $\mathbf{C} \rightarrowtail$ Map(**C**) adds no sub-objects of **C** objects:

LEMMA 25.21 Every Map(**C**) monic $H: B/R \rightarrowtail A$ to a **C** object is equivalent to a **C** monic.

PROOF Given $H \rightarrowtail B \times A$ in **C**, monic from R to A in Map(**C**), the pullback H^oH is a **C** object and $H^oH \subseteq A$ makes it a sub-object of A.

Abbreviate the sub-object as $S \rightarrowtail A$. Then H factors through $B \times S \rightarrowtail B \times A$ and for the relation $H \rightarrowtail B \times S$ we have $HH^o \equiv R$ and $H^oH \equiv S$. Thus $H: B/R \xrightarrow{\sim} S$ is an equivalence of sub-objects of A. The lemma is proved, but we also note that $S \rightarrowtail A$ is the image of $H \longrightarrow A$. □

Corollary A **C** arrow $f: A \longrightarrow B$ is surjective in **C** iff it is surjective in Map(**C**), since B has the same sub-objects in each. □

Theorem 25.22 The full monic $\mathbf{C} \rightarrowtail \mathrm{Map}(\mathbf{C})$ preserves finite limits, images, and whatever quotients exist in **C**.

Proof Limits are left to the reader. The last corollary proves it for images. Then, by Theorem 25.19, it preserves quotients. □

25.3 When Map (C) is a topos

For any **C** object A, we say that a Map(**C**) object PA and monic $\in^A \rightarrowtail A \times PA$ classifies **C**-relations with A in Map(**C**) iff for every **C** object B and **C** relation $C \rightarrowtail A \times B$ there is a unique Map(**C**) arrow $U: B \longrightarrow PA$, with C a pullback of $\in^A$ along $1_A \times U$.

Lemma 25.23 If PA, $\in^A$ is a **C** relation classifier for A then it is a power object for A in Map(**C**).

Proof For any Map(**C**) object B/R and monic $C/R' \rightarrowtail A \times B/R$, take pullbacks along the product of the coequalizer presenting B/R with A.

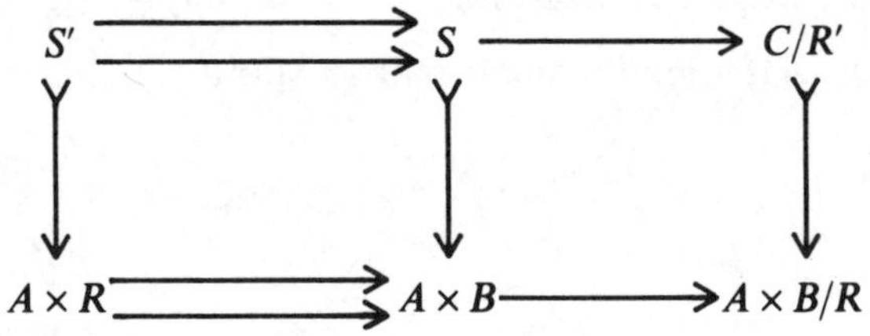

By Lemma 25.21, we can assume that S and S' are in **C**. The classifier $f: B \longrightarrow PA$ of S coequalizes the two projections $R \longrightarrow B$, since its composite with either one classifies S'. Thus it induces an arrow $u: B/R \longrightarrow PA$, which classifies some relation, and Exercise 25.3 shows that the relation is C/R'. □

Lemma 25.24 This holds in any regular category where all equivalence relations have quotients (and thus in Map(**C**)). Let q be a surjection with kernel pair h, k and let i be a monic as shown. Suppose that i has the same pullback along h as along k, so that there is a monic m and there are arrows h' and k' that make corresponding squares pullbacks, as follows:

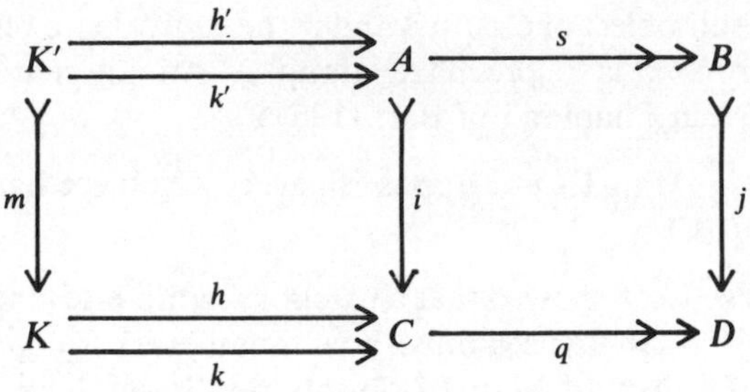

Then i, s, q, j is a pullback, where s, j is the image of $q \circ i$.

PROOF First we show, on the assumptions given, that h', k' is a kernel pair for s. From

$$j \circ s \circ h' = q \circ i \circ h' = q \circ h \circ m = q \circ k \circ m$$

and monicness of j, deduce that $s \circ h' = s \circ k'$. Then take any T and T-elements x and y of A such that $s(x) = s(y)$. Since $q \circ i(x) = q \circ i(y)$ and K is a kernel pair, there is some T-element z of K with $h(z) = i(x)$ and $k(z) = i(y)$. Since K' is a pullback, there is some T-element w of K' with $m(w) = z$ and $h'(w) = x$. But also $i \circ k'(w) = i(y)$ and so $k'(w) = y$. Uniqueness is left to the reader, and completes the proof that h', k' is a kernel pair. Since s is surjective, it is a quotient of h', k'.

For the next step let x and y be T-elements of B and C such that $j(x) = q(y)$. Let

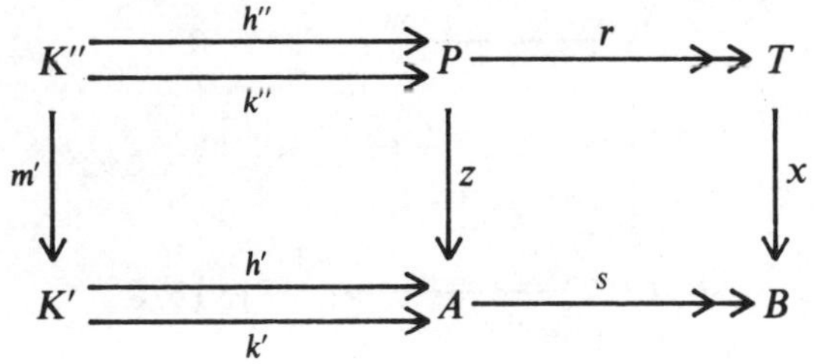

be the pullback of s and its kernel pair along x. So h'', k'' is also kernel pair to its quotient r. (Think of r as pulling x and y back to a stage P covering T.) Since $q \circ i(z) = q(y \circ r)$ there is some P-element v of K with $h(v) = i(z)$ and $k(v) = y \circ r$, and so some P-element b of K' with $m(b) = v$ and $h'(b) = z$. Simple chasing shows that $s \circ k' \circ b = x \circ r$ and $i \circ k' \circ b = y \circ r$, and since i is monic $k' \circ b$ is unique with those composites. Further chasing shows that $k' \circ b \circ h'' = k' \circ b \circ k''$, so that $k' \circ b$ factors through a unique $u: T \longrightarrow A$, and since r is epic we have $s \circ u = x$ and $i \circ u = y$. This proves the pullback property. □

In short, any sub-object of C closed under the equivalence relation given by the kernel pair of q is a preimage along q. An alternative proof uses Metatheorem 6.8 in Chapter 3 of Barr (1971).

THEOREM 25.25 Map(**C**) is a topos iff every **C** object has a **C** relation classifier in Map(**C**).

PROOF Exercise 13.16 showed that a category with finite limits is a topos iff it has power objects. So we need only show that if every **C** object has a power object in Map(**C**) then all Map(**C**) objects do. So let R be $\langle h, k\rangle: R \rightarrowtail A \times A$. Let $T_h \rightarrowtail R \times PA$ be the pullback of $\in^A$ along $h \times 1_{PA}$ and let $Ph: PA \longrightarrow PR$ classify T_h. Similarly, define Pk classifying T_k.

Define $e: P(A/R) \rightarrowtail PA$ as the equalizer of Ph and Pk. Define $\in^{A/R} \rightarrowtail (A/R) \times P(A/R)$ as the image of the following pullback:

$$\begin{array}{ccccc}
K & \twoheadrightarrow & \in^{A/R} & \rightarrowtail & (A/R) \times P(A/R) \\
\downarrow & & & & \downarrow {\scriptstyle (A/R)\times e} \\
\in^{A} & \rightarrowtail & A \times PA & \longrightarrow & (A/R) \times PA
\end{array}$$

Given a Map(**C**) monic $C/R'' \rightarrowtail (A/R) \times (B/R')$, take the pullback

$$\begin{array}{ccc}
S & \longrightarrow & C/R'' \\
\downarrow & & \downarrow \\
A \times (B/R') & \longrightarrow & (A/R) \times (B/R')
\end{array}$$

and let $u: (B/R') \longrightarrow PA$ classify S. But $Ph \circ u$ classifies the pullback of T_h along $R \times u$, and that is the pullback of S along $h \times 1_{C/R'}$. Similarly, $Pk \circ u$ classifies the pullback of S along $k \times 1_{C/R'}$. And S is already the pullback of C/R'' along an arrow that coequalizes $h \times 1_{C/R'}$ and $k \times 1_{C/R''}$, so $Ph \circ u$ and $Pk \circ u$ classify the same relation and are equal. Therefore there is a unique $w: (B/R') \longrightarrow PA$ with $u = e \circ w$.

By Exercise 25.3 we can prove that w classifies C/R'' by proving that S is the pullback of $\in^{A/R}$ along

$$A \times (B/R') \longrightarrow (A/R) \times (B/R') \xrightarrow{(A/R)\times w} (A/R) \times P(A/R)$$

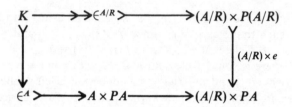

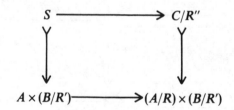

So it suffices to show that the outer square here is a pullback:

$$\begin{array}{ccccc}
S & \rightarrowtail & K & \longrightarrow & \in^{A/R} \\
\downarrow & & \downarrow & & \downarrow \\
A \times (B/R') & \xrightarrow[A \times w]{} & A \times P(A/R) & \longrightarrow & (A/R) \times P(A/R)
\end{array}$$

for K the pullback of $\in^A$ defined above. The left-hand side is a pullback because $u = e \circ w$, while the right is by Lemma 25.24. Indeed, K has the same pullback along both of $h \times P(A/R)$ and $k \times P(A/R)$, since T_h and T_k have the same pullback along $R \times e$, since T_h and T_k are pullbacks of $\in^R$ along two arrows equalized by $R \times e$. □

THEOREM 25.26 Suppose that Map (**C**) is a topos and let j be a topology on it such that an equivalence relation on a **C** object has a quotient in **C** iff it is j-closed. Then a Map (**C**) object is j-separated iff it is iso to a **C** object.

PROOF An object is j-separated iff its diagonal is j-closed. By Exercises 25.3 and 25.6 then, a Map (**C**) object A/R is separated iff R is j-closed in $A \times A$. But then, by supposition and by Theorem 25.22, the quotient of R is a **C** object. □

Exercises

25.1 Show that epics in a topos are surjective, so that toposes are regular. Use the category **2** to show that epics in a regular category need not be surjective. Show that pullback preserves surjective images in any regular category.

25.2 Give $i: I \rightarrowtail A$ and $x: T \longrightarrow A$, and a surjection $s: S \twoheadrightarrow T$, show that if $x \circ s \in I$ then $x \in I$.

25.3 Let $f: A \longrightarrow B$ be surjective in a regular category. Given sub-objects $i: I \rightarrowtail B$ and $j: J \rightarrowtail B$, show that $f^{-1}(i) \subseteq f^{-1}(j)$ iff $i \subseteq j$. [Hint: if $f^{-1}(i) \subseteq f^{-1}(j)$ then i is the surjective image of $f^{-1}(I) \longrightarrow J$.]

25.4 Show that every coequalizer in a regular category is surjective.

25.5 Let M and N be terms formed of converses, intersections, and composites of relations of **C**, and suppose that from vMw you can deduce vNw by these formal rules:

$xF^o y$ iff yFx
$x(F \cap G)z$ iff both xFz and xGz
$xFGz$ iff for some y both xFy and yGz

Conclude that $M \subseteq N$. In short, if you can prove that $M \subseteq N$ (or $M \equiv N$) by these rules, ignoring stages of definition, then suitable covering stages can always be filled in. (Use Exercise 25.2.)

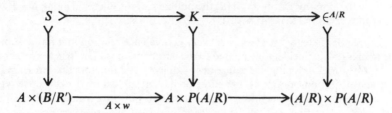

25.6 In any category with all finite limits, show that the pullback of a diagonal $B \rightarrowtail B \times B$ along a product arrow $f \times f: A \times A \longrightarrow B \times B$ is the kernel pair of f. Conclude that for any **Eff** object (A/R) the pullback of the diagonal along $R \times R$ is the equivalence relation $(R) \rightarrowtail (A \times A)$.

25.7 A regular category is *effective* if every equivalence relation in it has a quotient. (The term is only coincidentally related to the effective topos.) Let **Reg**, the category of regular categories, have regular categories as objects, and functors that preserve finite limits and surjections as arrows. Let **EReg** be the full sub-category of effective regular categories. Show that the inclusion $\mathbf{C} \rightarrowtail \mathrm{Map}(\mathbf{C})$ is universal from **C** to the inclusion functor $\mathbf{EReg} \rightarrowtail \mathbf{Reg}$, so that the inclusion has a left adjoint $\mathrm{Map}: \mathbf{Reg} \longrightarrow \mathbf{EReg}$. Map(**C**) is sometimes called the effective reflection of **C**.

Further reading

The standard text on category theory is Mac Lane (1971). The original Eilenberg and Mac Lane (1945), Freyd (1964), Herrlich and Strecker (1973), and Barr and Wells (1990), which aims at computer science, must also be mentioned. Our approach in Part II is based on Lawvere (1963, 1966). Johnstone (1977) is the standard reference on topos theory, with an exhaustive bibliography up to its publication date. For history see Mac Lane (1988) and McLarty (1990*b*).

Toposes in general

Freyd (1972) works with toposes by the elegant use of set-valued functors, including a representation theorem embedding toposes into products of copies of **Set**. Barr and Wells (1985) use an algebraic approach for quick proofs of the central theorems. Bell (1988*b*) works almost exclusively with the internal language. Schlomiuk (1977) (see also Schlomiuk 1974) introduces the internal language piece by piece as needed for each result. Goldblatt (1979) emphasizes the topos **Set**. Freyd and Scedrov (1990) is a highly personalized run through numerous topics, including toposes, with numerous applications to logic. Mac Lane and Moerdijk (1992) introduces toposes by means of Grothendieck topos constructions.

Two works on particular topics stand out for introducing a good part of the general theory; Kock (1981) on synthetic differential geometry, and Rosolini (1986) on the effective topos and another called the recursive topos.

Makkai and Reyes (1977) gives the basic results on Grothendieck toposes over **Set**. The original Artin, Grothendieck, and Verdier (1972), is rich and demanding. The case of sheaves on a complete Heyting algebra with the canonical topology (but the base topos is not assumed to be **Set**) is covered in Fourman and Scott (1979).

For the internal language see among other sources Boileau and Joyal (1981); and Osius (1975). Lambek and Scott (1986) covers an array of topics in the higher order logic of Cartesian closed categories and of toposes, and has a long bibliography. Scott (1979) axiomatizes a form of topos logic using existence predicates.

Topics

On geometric morphisms, see Johnstone (1977) and Mac Lane and Moerdijk (1992), both of which include the powerful theorem originally in Diaconescu (1975*b*). Wraith (1975) gives a survey without proofs. For recent advances, see Joyal and Tierney (1984) and Moerdijk (1988) and references therein. Joyal and Moerdijk (1990) describe toposes and geometric morphisms in a very geometrically motivated way.

Various proofs that the axiom of choice implies excluded middle agree at a certain level but differ in style. The one in Johnstone (1977) is similar to that in Diaconescu (1975*a*), as is ours. Scott (1979) reduces the theorem to the case of sub-objects of 1 (in any topos). Barr and Wells (1985) use a representation theorem to reduce to sub-objects of 1 in **Set**. Both Lambek and Scott (1986) and Bell (1988*b*) work with a choice principle in the internal language. Goodman and Myhill (1978) adapts the result to intuitionistic set theory.

On natural numbers

Johnstone (1977) covers many topics, including inductive definition of infinite structures (such as free groups) in toposes. Lambek and Scott (1986) describes computability in Cartesian closed categories and in toposes. Coste-Roy *et al.* (1980) study natural number objects and aspects of intuitionistic analysis. Rousseau (1978) proves theorems of classical analysis by interpreting them as simpler theorems in the internal languages of toposes of sheaves on suitable spaces.

On small categories

Almost the whole theory of small categories in **Set** is already sound in any topos with natural number object, including all of our Part II. (We avoided non-constructive theorems on functor equivalences, which would be valid in **Set** but not, say, in **Eff**.) Discussions of small categories in toposes tend to stick close to categorical primitives, since that is all that is new, and risk making the subject seem less familiar than it is. Our Chapter 20 does this, and Johnstone (1977) does it to a much greater extent. Wraith (1975) discusses the topic without proofs, and Diaconescu (1975*b*) deals with it. Moerdijk (1985) works freely with categories in an arbitrary topos.

Non-small categories

Over arbitrary toposes, these present greater problems. For example, given a non-trivial topos **E**, the category of groups in **E** is not small. There is no one **E** object of all **E** groups. Tools to work with indexed families of groups or homomorphisms, for example, have not found their final form: see the appendix in Johnstone (1977), Johnstone and Paré (1978), Bénabou (1985), and references therein.

On topologies

Johnstone (1977) gives more examples and theorems, and a description of sheaves for a given topology as a *category of fractions.* Tierney (1973) also describes sheaves in this way. For topologies on posets, see Fourman and Scott (1979).

On **Set**

See Lawvere (1964), and references in Chapter 22. Mac Lane (1986) proposes this as a foundation for mathematics. Little is written on that, since for working purposes **Set** is not very different from other set theories, being just a little closer to naive practice.

On **Spaces**

Most publications on SDG presume knowledge of basic differential geometry, as in Spivak (1970). Lawvere's original talks on the subject eventually appeared as Lawvere (1979): see also Lawvere (1980). Kock (1981) introduces the axiomatic theory and models. Lavendhomme (1987) discusses the axiomatic theory, with an extensively annotated bibliography. The first paper on fully well-adapted models, Dubuc (1979), is also nice reading. Moerdijk and Reyes (1991) go deeply into a wide array of models, including some with both the infinitesimals of SDG and those of non-standard analysis.

Bunge (1983), Bunge and Dubuc (1987), and Bunge and Gago (1988) use the germ neighbourhood of 0 for classical theorems in the synthetic setting. The collections of Kock (1979, 1983) cover a great deal. Bell (1988*a*) puts ideas from SDG into a general discussion of infinitesimals.

On **Eff**

Hyland (1982) uses the version of topos logic with existence predicates, developed in Scott (1979). In this version any formula of first order arithmetic is true in **Eff** iff it is realized. To relate this to our version, see the remarks on existence predicates in Boileau and Joyal (1981). Hyland *et al.* (1980) gives a general setting for constructing **Eff**, such as we obtain from regular categories.

Categorical logic

Categorical logic treats theories T as categories. An interpretation of T in another theory T' becomes a functor $T \longrightarrow T'$, a model in sets becomes a functor $T \longrightarrow$ **Set**, and so on. This began with the work of Lawvere (1963) on algebraic theories; that is, theories such as the theory of groups, axiomatized by equations between operators. Lawvere showed, among other things, when a category is (equivalent to) the category of models of an algebraic theory and how to recover the theory in that case. Barr and Wells (1985) treat algebraic theories and other weak fragments of first order logic. Makkai and Reyes (1977) gives definability and completeness results for finitary and infinitary first order classical and intuitionistic logic, with applications to Grothendieck toposes. See also Kock and Reyes (1977) and Pitts (1989).

Any first order theory T gives a category Mod(T), the objects of which are models of T in **Set**, and the arrows of which are elementary embeddings. There is also a notion of ultraproducts of models. Makkai (1987) shows just what information is needed to recover a first order theory from its models. Given any category with a specified 'ultraproduct' structure, he shows how to tell if it is (equivalent to) the category of models of a theory and how to recover the theory if it is. See also Makkai and Paré (1989).

For categorical treatments of various logics applied to semantics for programming languages, see papers in Gray and Scedrov (1989). G. Reyes (1991) presents a modal logic and a theory of reference in a topos context, and M. Reyes (forthcoming) applies similar ideas to metaphor and literary reference.

Bibliography

Artin, M., Grothendieck, A., and Verdier, J. L. (eds) (1972). *Théorie des topos et cohomologie etale des schémas SGA 4*. Lecture Notes in Mathematics No. 269. Springer-Verlag, Berlin.

Barr, M. (1971). Exact categories. In *Exact categories and categories of sheaves* (ed. M. Barr *et al.*) Lecture Notes in Mathematics No. 236. Springer-Verlag, Berlin.

Barr, M. and Wells, C. (1985). *Toposes, triples, and theories*. Springer-Verlag, Berlin.

Barr, M. and Wells, C. (1990). *Category theory for computer scientists*. Prentice-Hall, Englewood Cliffs, New Jersey.

Bell, J. L. (1982). Categories, toposes and sets. *Synthese*, **51**, 293–337.

Bell, J. L. (1986). From absolute to local mathematics. *Synthese*, **69**, 409–26.

Bell, J. L. (1988*a*). Infinitesimals. *Synthese*, **75**, 285–316.

Bell, J. L. (1988*b*). *Toposes and local set theories*. Oxford University Press.

Bénabou, J. (1985). Fibered categories and the foundations of naive category theory. *Journal of Symbolic Logic*, **50**, 10–37.

Beth, E. W. (1968). *The foundations of mathematics*. North-Holland, Amsterdam.

Blanc, G. and Donnadieu, M. R. (1976). Axiomatisation de la catégorie des catégories. *Cahiers de Topologie et Géométrie Différentielle*, **XVII-2**, 1–35.

Boileau, A. and Joyal, A. (1981). La logique des topos. *Journal of Symbolic Logic*, **46**, 6–16.

Bunge, M. (1983). Synthetic aspects of C^∞-mappings. *Journal of Pure and Applied Algebra*, **28**, 41–63.

Bunge, M. and Dubuc, E. J. (1987). Local concepts in synthetic differential geometry and germ representability. In *Mathematical logic and theoretical computer science* (ed. D. Kueker *et al.*), pp. 39–158. Marcel Dekker, New York.

Bunge, M. and Gago, F. (1988). Synthetic aspects of C^∞-mappings II: Mather's theorem. *Journal of Pure and Applied Algebra*, **55**, 213–50.

Bunge, M. and Heggie, M. (1984). Synthetic calculus of variations. In *Mathematical applications of category theory*. Contemporary Mathematics Series, Vol. 30. American Mathematical Society.

Cantor, G. (1895). Beiträge zur Begründung der transfiniten Mengenlehre. *Mathematische Annalen*, **46**, 481–512. Reprinted in: Zermelo, E. (ed.) (1966). *Georg Cantor, Gesammelte Abhandlungen*. Georg Olms.

Carboni, A., Freyd, P., and Scedrov, A. (1988). A categorical approach to realizability and polymorphic types. In *Proceedings of the 3rd ACM Workshop on Mathematical Foundations of Programming Language Semantics* (ed. M. Morin *et al.*). Lecture Notes in Computer Science No. 298. Springer-Verlag, Berlin.

Cole, J. C. (1973). Categories of sets and models of set theory. In *Proceedings of the Bertrand Russell Memorial Logic Conference, Uldum 1971*. (ed. J. Bell and A. Slomson) N.P., Leeds.

Coste-Roy, M.-F., Coste, M., and Mahé, L. (1980). Contribution to the study of the natural number object in elementary topoi. *Journal of Pure and Applied Algebra*, **17**, 35–68.

Diaconescu, R. (1975*a*). Axiom of choice and complementation. *Proceedings of the American Mathematical Society*, **51**, 176–8.

Diaconescu, R. (1975*b*). Change of base for toposes with generators. *Journal of Pure and Applied Algebra*, **6**, 191–218.

Dubuc, E. J. (1979). Sur les modèles de la géométrie différentielle synthétique. *Cahiers de Topologie et Géométrie Différentielle*, **XX-3**, 231–79.

Ehresmann, C. (1965). *Catégories et structures.* Dunod, Paris.

Eilenberg, S. and Mac Lane, S. (1945). The general theory of natural equivalences. *Transactions of the American Mathematical Society*, **58**, 231–94. Reprinted in: Eilenberg, S. and Mac Lane, S. (1986). *Eilenberg–Mac Lane: collected works.* Academic Press, New York.

Fourman, M. P. and Scott, D. S. (1979). Sheaves and logic. In *Applications of sheaves, proceedings of the L.M.S. Durham Symposium 1977* (ed. M. P. Fourman, C. Mulvey, and D. S. Scott), pp. 302–402. Lecture Notes in Mathematics No. 753, Springer-Verlag, Berlin.

Fourman, M. P., Mulvey, C., and Scott, D. S. (eds) (1979). *Applications of sheaves, proceedings of the L.M.S. Durham Symposium, 1977.* Lecture Notes in Mathematics No. 753. Springer-Verlag, Berlin.

Freyd, P. (1964). *Abelian categories.* Harper and Row, New York.

Freyd, P. (1972). Aspects of topoi. *Bulletin of the Australian Mathematics Society*, **7**, 1–76 and 467–80.

Freyd, P. (1990). Recursive types reduced to inductive types. In *Fifth Annual IEEE Symposium on Logic in Computer Science*, pp. 498–507. IEEE Computer Society Press.

Freyd, P. and Scedrov, A. (1990). Categories and allegories. North-Holland, Amsterdam.

Goldblatt, R. (1979). *Topoi: the categorical analysis of logic.* North-Holland, Amsterdam.

Goodman, N. and Myhill J. (1978). Choice implies excluded middle. *Zeitschrift für mathematische Logik und Grundlagen der Mathematik*, **24**, 461.

Gray, J. W. and Scedrov, A. (eds) (1989). *Categories in computer science and logic.* Contemporary Mathematics Series No. 92. American Mathematical Society.

Grothendieck, A. and Verdier, J. L. (1972). Topos. In *Théorie des topos et cohomologie etale des schémas SGA 4* (ed. M. Artin, A. Grothendieck, and J. L. Verdier), pp. 299–519. Lecture Notes in Mathematics No. 269. Springer-Verlag, Berlin.

Hatcher, W. S. (1982). *The logical foundations of mathematics.* Pergamon, Oxford.

Herrlich, H. and Strecker, G. E. (1973). *Category theory.* Allyn and Bacon, Boston.

Hyland, J. M. E. (1982). The effective topos. In *The L. E. J. Brouwer Centenary Symposium* (ed. A. S. Troelstra, and D. van Dalen), pp. 165–217. North-Holland, Amsterdam.

Hyland, J. M. E. (1988). A small complete category. *Annals of Pure and Applied Logic*, **40**(3), 135–65.

Hyland, J. M. E., Johnstone, P. T., and Pitts, A. (1980). Tripos theory. *Mathematical Proceedings of the Cambridge Philosophical Society*, **88**, 205–32.

Hyland, J. M. E., Robinson, E. P., and Rosolini, G. (1990). The discrete objects in the effective topos. *Proceedings of the London Mathematical Society*, **60**, 1–36.

IEEE (1990). *Fifth Annual IEEE Symposium on Logic in Computer Science*. IEEE Computer Society Press.

Johnstone, P. T. (1977). *Topos theory*. Academic Press, London.

Johnstone, P. T. (1979). Conditions relating to De Morgan's law. In *Applications of sheaves, proceedings of the L.M.S. Durham Symposium 1977* (ed. M. P. Fourman, C. Mulvey, and D. S. Scott), pp. 479–91. Lecture Notes in Mathematics No. 753. Springer-Verlag, Berlin.

Johnstone, P. T. and Paré, P. (1978). *Indexed categories*. Lecture Notes in Mathematics No. 661. Springer-Verlag, Berlin.

Joyal, A. and Moerdijk, I. (1990). Toposes as homotopy groupoids. *Advances in Mathematics*, **80**, 22–38.

Joyal, A. and Tierney, M. (1984). *An extension of the Galois theory of Grothendieck*, Memoirs of the American Mathematical Society, Vol. 51, No. 309, September.

Kant, I. (1781). *Kritik der reinen Vernunft*. Hartnoch. (Trans. N. Kemp Smith (1929), as *Critique of pure reason*. Macmillan.)

Kleene, S. C. (1952). *Introduction to metamathematics*. Van Nostrand, New York.

Kock, A. (1981). *Synthetic differential geometry*. London Mathematical Society. Lecture Note Series, Vol. 51. Cambridge University Press.

Kock, A. (ed.) (1979). *Topos theoretic methods in geometry*. Aarhus University Matematisk Institut various publications series no. 30. Aarhus, Denmark.

Kock, A. (ed.) (1983). *Category theoretic methods in geometry*. Aarhus University Matematisk Institut various publications series no. 35. Aarhus, Denmark.

Kock, A. and Reyes G. (1977). Doctrines in categorical logic. In *Handbook of mathematical logic* (ed. J. Barwise), pp. 284–313. North-Holland, Amsterdam.

Lambek, J. and Scott, P. J. (1986). *Introduction to higher order categorical logic*. Cambridge University Press.

Lavendhomme, R. (1987). *Leçons de géométrie différentielle synthétique naïve*, Institut de Mathematique, Louvain-La-Neuve, Belgium.

Lawvere, F. W. (1963). Functorial semantics of algebraic theories. Unpublished dissertation, Columbia University.

Lawvere, F. W. (1964). An elementary theory of the category of sets. *Proceedings of the National Academy of Sciences, U.S.A.*, **52**, 1506–11.

Lawvere, F. W. (1966). The category of categories as a foundation for mathematics. In *Proceedings of the Conference on Categorical Algebra in La Jolla, 1965* (ed. S. Eilenberg *et al.*), pp. 1–21.

Lawvere, F. W. (1969). Adjointness in foundations. *Dialectica*, **23**, 281–96.

Lawvere, F. W. (1970). Equality in hyperdoctrines and the comprehension schema as an adjoint functor. In *Proceedings of the New York Symposium on Applications of Categorical Algebra, (ed. A. Heller), pp.* 1–14. American Mathematical Society.

Lawvere, F. W. (ed.) (1972). *Toposes, algebraic geometry and logic*. Lecture Notes in Mathematics No. 274. Springer-Verlag, Berlin.

Lawvere, F. W. (1975). Continuously variable sets: algebraic geometry = geometric logic. In *Proceedings of the ASL Logic Colloquium, Bristol 1973* (ed. H. E. Rose and J. C. Shepherdson), pp. 135–56. North-Holland, Amsterdam.

Lawvere, F. W. (1976). Variable quantities and variable structures in topoi. In *Algebra, topology and category theory: a collection of papers in honor of Samuel Eilenberg* (ed. A. Heller and M. Tierney), pp. 101–31. Academic Press, New York.

Lawvere, F. W. (1979). Categorical dynamics. In *Topos theoretic methods in geometry* (ed. A. Kock *et al.*), pp. 1–28. Aarhus University Matematisk Institut various publications series no. 30. Aarhus, Denmark.

Lawvere, F. W. (1980). Towards the description in a smooth topos of the dynamically possible motions and deformations of a continuous body. *Cahiers de Topologie et Géométrie Différentielle*, **XXI-4**, 377–92.

Lawvere, F. W. and Schanuel, S. (eds) (1986). *Categories in continuum physics.* Lecture Notes in Mathematics No. 1174. Springer-Verlag, Berlin.

Lawvere, F. W. *et al.* (eds) (1975). *Model theory and topoi.* Lecture Notes in Mathematics No. 445. Springer-Verlag, Berlin.

Longo, G. and Moggi, E. (1991). Constructive natural deduction and its 'modest' interpretation. *Mathematical Structures in Computer Science*, **1**, 215–54.

Mac Lane, S. (ed.) (1969). *Reports of the Midwest Category Seminar III.* Lecture Notes in Mathematics No. 106. Springer-Verlag, Berlin.

Mac Lane, S. (1971). *Categories for the working mathematician.* Springer-Verlag, Berlin.

Mac Lane, S. (1986). *Mathematics: form and function.* Springer-Verlag, Berlin.

Mac Lane, S. (1988). Concepts and categories in perspective. In ed. P. Duren *et al. A century of mathematics, part I*, pp. 323–65. American Mathematical Society.

Mac Lane, S. and Moerdijk, I. (1992). *Sheaves in geometry and logic: a first introduction to topos theory.* Springer-Verlag, Berlin.

Makkai, M. (1987). Stone duality for first order logic. *Advances in Mathematics*, **65**, 97–170.

Makkai, M. and Paré, R. (1989). *Accessible categories.* Contemporary Mathematics No. 104. American Mathematical Society.

Makkai, M. and Reyes, G. (1977). *First order categorical logic.* Lecture Notes in Mathematics No. 611. Springer-Verlag, Berlin.

McLarty, C. (1983). Local, and some global, results in synthetic differential geometry. In *Category theoretic methods in geometry* (ed. A. Kock *et al.*), pp. 226–56. Aarhus University Matematisk Institut various publications series no. 35, Aarhus, Denmark.

McLarty, C. (1987). Elementary axioms for canonical points of toposes. *Journal of Symbolic Logic*, **52**, 202–4.

McLarty, C. (1988). Defining sets as sets of points of spaces. *Journal of Philosophical Logic*, **17**, 75–90.

McLarty, C. (1990*a*). Review of Bell (1988). In *Notre Dame Journal of Formal Logic*, **31**(1), 150–61.

McLarty, C. (1990*b*). The uses and abuses of the history of topos theory. *British Journal of the Philosophy of Science*, **41**, 351–75.

McLarty, C. (1991). Axiomatizing a category of categories. *Journal of Symbolic Logic*, **56**, 1243–60.

Mendelson, E. (1987). *Introduction to mathematical logic*, 3rd Edn. Wadsworth and Brooks/Cole, California.

Mitchell, W. (1972). Boolean topoi and the theory of sets. *Journal of Pure and Applied Algebra*, **2**, 261–74.

Moerdijk, I. (1985). An elementary proof of the descent theorem for Grothendieck toposes. *Journal of Pure and Applied Algebra*, **37**, 185–91.

Moerdijk, I. (1988). Toposes and groupoids. In *Categorical algebra and its applications* (ed. F. Borceux). Lecture Notes in Mathematics No. 1348. Springer-Verlag, Berlin.

Moerdijk, I. and Reyes, G. (1991). *Models for smooth infinitesimal analysis.* Springer-Verlag, Berlin.

Mumford, D. and Tate, J. (1978). Fields medals. *Science*, **202**, 737–9.

Osius, G. (1974). Categorical set theory. *Journal of Pure and Applied Algebra*, **4**, 79–119.

Osius, G. (1975). Logical and set theoretic tools in elementary topoi. In *Model theory and topoi* (ed. F. W. Lawvere *et al.*), pp. 297–346. Lecture Notes in Mathematics No. 445. Springer-Verlag, Berlin.

Penon, J. (1981). Infinitesimaux et intuitionnisme. *Cahiers de Topologie et Géométrie Différentielle*, **XXII**, 67–72.

Pitts, A. M. (1989). Conceptual completeness for intuitionistic logic. *Annals of Pure and Applied Logic*, **41**(1), 31–81.

Rasiowa, H. and Sikorski, R. (1963). *The mathematics of metamathematics*. Polish Scientific Publishers.

Reyes, G. (1991). A topos theoretic approach to reference and modality. *Notre Dame Journal for Formal Logic*, **32**, 359–91.

Reyes, M. (forthcoming). Referential structure of fictional texts. In *Logic and cognition* (ed. J. MacNamara and G. Reyes). Oxford University Press.

Robinson, A. (1974). *Non-standard analysis*. North-Holland, Amsterdam.

Robinson, E. (1989). How complete is PER? In IEEE. *Fourth Annual IEEE Symposium on Logic in Computer Science*. IEEE Computer Society Press.

Rosolini, G. (1986). Continuity and effectiveness in topoi. Doctoral dissertation, Oxford. Also preprint, Department of Computer Science, Carnegie–Mellon University.

Rousseau, Christiane (1978). Topos theory and complex analysis. *Journal of Pure and Applied Algebra*, **10**(3), 299–313.

Scedrov, A. (1988). A guide to polymorphic types. In *Logic and Computer Science* (ed. S. Homer *et al.*). Lecture Notes in Mathematics No. 1429. Springer-Verlag, Berlin.

Schlomiuk, D. I. (1974). Topos di Grothendieck, topos di Lawvere e Tierney. *Rendiconti di Matematica di Roma, Seria 6*, **VII**(2), 1–41.

Schlomiuk, D. I. (1977). *Logique des topos*. Presses de l'Université de Montréal.

Scott, D. S. (1979). Identity and existence in intuitionistic logic. In *Applications of sheaves, proceedings of the L.M.S. Durham Symposium 1977* (ed. M. P. Fourman, C. Mulvey, and D. S. Scott), pp. 660–97. Lecture Notes in Mathematics No. 753. Springer-Verlag, Berlin.

Spivak, M. (1970). *Differential geometry*, Vol. 1. Publish or Perish Inc., Boston.

Tennison, B. R. (1975). *Sheaf theory*. London Mathematical Society Lecture Notes Series No. 20. Cambridge University Press.

Tierney, M. (1972). Sheaf theory and the continuum hypothesis. In *Toposes, algebraic geometry and logic* (ed. F. W. Lawvere), pp. 13–42. Lecture Notes in Mathematics No. 274. Springer-Verlag, Berlin.

Tierney, M. (1973). Axiomatic sheaf theory. In *Proceedings of the CIME Conference on Categories and Commutative Algebra, Varenna, 1971*, pp. 249–36. Edizione Cremonese. (Reviewed in *Mathematical Reviews*, **50**, (1975) 7277.)

van der Hoeven, G. F. and Moerdijk, I. (1984). Sheaf models for choice sequences. *Annals of Pure and Applied Logic*, **27**, 63–107.

Veit, B. (1981). A proof of the associated sheaf theorem by means of categorical logic. *Journal of Symbolic Logic*, **46**, 45–55.

Wraith, G. C. (1975). Lectures on elementary topoi. In *Model theory and topoi* (ed. F. W. Lawvere *et al.*), pp. 114–206. Lecture Notes in Mathematics No. 445. Springer-Verlag, Berlin.

Symbol index

Subject index